T0234334

Universitext

Mario Lefebvre

Applied Stochastic Processes

 Springer

Mario Lefebvre
Département de mathématiques
  et de génie industriel
École Polytechnique de Montréal, Québec
C.P. 6079, succ. Centre-ville
Montréal H3C 3A7
Canada
mlefebvre@polymtl.ca

Mathematics Subject Classification (2000): 60-01, 60Gxx

Library of Congress Control Number: 2006935530

ISBN 978-0-387-34171-2   ISBN 978-0-387-48976-6 (eBook)
DOI 10.1007/978-0-387-48976-6
Printed on acid-free paper.

9 8 7 6 5 4 3 2 1

springer.com   (TXQ)

*Life's most important questions are, for the most part,
nothing but probability problems.*

Pierre Simon de Laplace

Preface

This book is based on the lecture notes that I have been using since 1988 for the course entitled *Processus stochastiques* at the École Polytechnique de Montréal. This course is mostly taken by students in electrical engineering and applied mathematics, notably in operations research, who are generally at the master's degree level. Therefore, we take for granted that the reader is familiar with elementary probability theory. However, in order to write a self-contained book, the first chapter of the text presents the basic results in probability.

This book aims at providing the readers with a reference that covers the most important subjects in the field of stochastic processes and that is accessible to students who don't necessarily have a sound theoretical knowledge of mathematics. Indeed, we don't insist very much in this volume on rigorous proofs of the theoretical results; rather, we spend much more time on applications of these results.

After the review of elementary probability theory in Chapter 1, the remainder of this chapter is devoted to random variables and vectors. In particular, we cover the notion of conditional expectation, which is very useful in the sequel.

The main characteristics of stochastic processes are given in Chapter 2. Important properties, such as the concept of independent and stationary increments, are defined in Section 2.1. Next, Sections 2.2 and 2.3 deal with ergodicity and stationarity, respectively. The chapter ends with a section on Gaussian and Markovian processes.

Chapter 3 is the longest in this book. It covers the cases of both discrete-time and continuous-time Markov chains. We treat the problem of calculating the limiting probabilities of the chains in detail. Branching processes and birth and death processes are two of the particular cases considered. The chapter contains nearly 100 exercises at its end.

The Wiener process is the main subject of Chapter 4. Various processes based on the Wiener process are presented as well. In particular, there are subsections on models such as the geometric Brownian motion, which is very important in financial mathematics, and the Ornstein–Uhlenbeck process. White noise is defined, and first-passage problems are discussed in the last section of the chapter.

In Chapter 5, the Poisson process, which is probably the most important stochastic process for students in telecommunications, is studied in detail. Several generalizations of this process, including nonhomogeneous Poisson processes and renewal processes, can be found in this chapter.

Finally, Chapter 6 is concerned with the theory of queues. The models with a single server and those with at least two servers are treated separately. In general, we limit ourselves to the case of exponential models, in which both the times between successive customers and the service times are exponential random variables. This chapter then becomes an application of Chapter 3 (and 5).

In addition to the examples presented in the theory, the book contains approximately 350 exercises, many of which are multiple-part problems. These exercises are all problems given in exams or homework and were mostly created for these exams or homework. The answers to the even-numbered problems are given in Appendix B.

Finally, it is my pleasure to thank Vaishali Damle, Julie Park, and Elizabeth Loew from Springer for their work on this book.

Mario Lefebvre
Montréal, November 2005

Contents

List of Tables

List of Figures

1

Review of Probability Theory

1.1 Elementary probability

Definition 1.1.1. *A* **random experiment** *is an experiment that can be repeated under the same conditions and whose result cannot be predicted with certainty.*

Example 1.1.1. We consider the following three classical experiments:
E_1: a coin is tossed three times and the number of "tails" obtained is recorded;
E_2: a die is thrown until a "6" appears, and the number of throws made is counted;
E_3: a number is taken at random in the interval $(0, 1)$.

Remark. A closed interval will be denoted by $[a, b]$, whereas we write (a, b) in the case of an open interval, rather than $]a, b[$, as some authors write.

Definition 1.1.2. *The* **sample space** S *of a random experiment is the set of all possible outcomes of this experiment.*

Example 1.1.2. The sample spaces that correspond to the random experiments in the example above are the following:
$S_1 = \{0, 1, 2, 3\}$;
$S_2 = \{1, 2, \ldots\}$;
$S_3 = (0, 1)$.

Definition 1.1.3. *An* **event** *is a subset of the sample space* S. *In particular, each possible outcome of a random experiment is called an* **elementary event**.

The number of elementary events in a sample space may be finite (S_1), countably infinite (S_2), or uncountably infinite (S_3).

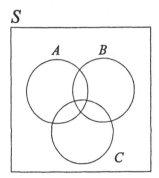

Fig. 1.1. Venn diagram.

We often use Venn[1] diagrams in elementary probability: the sample space S is represented by a rectangle and the events A, B, C, etc. by circles that overlap inside the rectangle (see Fig. 1.1).

Example 1.1.3. We can define, in particular, the following events with respect to the sample spaces associated with the random experiments of Example 1.1.1:
$A_1 = $ "tails" is obtained only once, that is, $A_1 = \{1\}$;
$A_2 = $ six or seven throws are made to obtain the first "6," that is, $A_2 = \{6, 7\}$;
$A_3 = $ the number taken at random is smaller than $1/2$, that is, $A_3 = [0, 1/2)$.

Notations

Union: $A \cup B$ (corresponds to the case when we seek the probability that one event *or* another one occurred, or that both events occurred).

Intersection: $A \cap B$ or AB (when we seek the probability that an event *and* another one occurred). If two events are *incompatible* (or *mutually exclusive*), then we write that $A \cap B = \emptyset$ (the empty set).

Complement: A^c (the set of elementary events that do *not* belong to A).

Inclusion: $A \subset B$ (when all the elementary events that belong to A also belong to B).

Definition 1.1.4. *A* **probability measure** *is a function P of the subsets of a sample space S, associated with a random experiment E, that possesses the following properties:*

Axiom I*: $P[A] \geq 0$ $\forall$ $A \subset S$;*

Axiom II*: $P[S] = 1$;*

---

[1] John Venn, 1834–1923, was born and died in England. He was a mathematician and priest. He taught at the University of Cambridge and worked in both mathematical logic and probability theory.

Axiom III: *If $A_1, A_2, \ldots$ is an infinite sequence of events that are all incompatible when taken two at a time, then*

$$P\left[\bigcup_{k=1}^{\infty} A_k\right] = \sum_{k=1}^{\infty} P[A_k] \tag{1.1}$$

In particular,

$$P[A \cup B] = P[A] + P[B] \quad if \ A \cap B = \emptyset \tag{1.2}$$

If the number n of elementary events is *finite* and if these events are *equiprobable* (or *equally likely*), then we may write that

$$P[A] = \frac{n(A)}{n} \tag{1.3}$$

where $n(A)$ is the number of elementary events in A. However, in general, the elementary events are *not* equiprobable. For instance, the four elementary events of the sample space S_1 in Example 1.1.2 have the following probabilities: $P[\{0\}] = P[\{3\}] = 1/8$ and $P[\{1\}] = P[\{2\}] = 3/8$ (if we assume that the coin is *fair*), and not $P[\{k\}] = 1/4$, for $k = 0, 1, 2, 3$.

Remark. It is said that the French mathematician d'Alembert[2] believed that when a fair coin is tossed twice, then the probability of getting one "tail" and one "head" is equal to 1/3. His reasoning was based on the fact that there are three possible outcomes in this random experiment: getting two "tails"; two "heads"; or one "tail" and one "head." It is easy to determine that the probability of obtaining one "tail" and one "head" is actually 1/2, because here there are four equiprobable elementary events: H_1H_2 (that is, "heads" on the first and on the second toss); H_1T_2; T_1H_2; and T_1T_2. Finally, the event A: getting one "tail" and one "head" corresponds to two elementary events: H_1T_2 and T_1H_2.

Proposition 1.1.1. *We have*

1) $P[A^c] = 1 - P[A] \quad \forall \ A \subset S.$

2) For all events A and B,

$$P[A \cup B] = P[A] + P[B] - P[AB] \tag{1.4}$$

3) For any three events A, B, and C,

$$P[A \cup B \cup C] = P[A] + P[B] + P[C] - P[AB] - P[AC] - P[BC] + P[ABC] \tag{1.5}$$

---

[2] Jean Le Rond d'Alembert, 1717–1783, was born and died in France. He was a prolific mathematician and writer. He published books, in particular, on dynamics and on the equilibrium and motion of fluids.

Counting

Proposition 1.1.2. *If k objects are taken at random among n distinct objects and if the order in which the objects were drawn does matter, then the number of different* **permutations** *that can be obtained is given by*

$$n \times n \times \ldots \times n = n^k \tag{1.6}$$

if the objects are taken **with** *replacement, and by*

$$n \times (n-1) \times \ldots \times [n-(k-1)] = \frac{n!}{(n-k)!} := P_k^n \quad \text{for } k = 0, 1, \ldots, n \tag{1.7}$$

when the objects are taken **without** *replacement.*

Remark. If, among the n objects, there are n_i of type i, where $i = 1, 2, \ldots, j$, then the number of different permutations of the entire set of objects is given by (the *multinomial coefficients*)

$$\frac{n!}{n_1! n_2! \cdots n_j!} \tag{1.8}$$

Example 1.1.4. The combination of a certain padlock is made up of three digits. Therefore, *theoretically* there are $10^3 = 1000$ possible combinations. However, in practice, if we impose the following constraint: the combination cannot be made up of two identical consecutive digits, then the number of possible combinations is given by $10 \times 9 \times 9 = 810$. This result can also be obtained by subtracting from 1000 the number of combinations with at least two identical consecutive digits, namely, 10 (with three identical digits) + $10 \times 1 \times 9 + 10 \times 9 \times 1$ (with *exactly* two identical digits, either the first two or the last two digits).

Proposition 1.1.3. *If k objects are taken, at random and without replacement, among n distinct objects and if the order in which the objects were drawn does not matter, then the number of different* **combinations** *that can be obtained is given, for $k = 0, 1, \ldots, n$, by*

$$\frac{n \times (n-1) \times \ldots \times [n-(k-1)]}{k!} = \frac{n!}{k!(n-k)!} := \binom{n}{k} \equiv C_k^n \tag{1.9}$$

Remark. If the objects are taken *with* replacement, then the number of different combinations is C_k^{n+k-1}. In this case, k may take any value in $\mathbb{N}^0 := \{0, 1, \ldots\}$.

Remark. In the preceding example, according to the common use, the word "combinations" was used for a padlock. However, they were indeed "permutations."

Example 1.1.5. In a given lottery, 6 balls are drawn at random and without replacement among 49 balls numbered from 1 to 49. We win a prize if the combination that we chose has at least three correct numbers. Let

F = we win a prize

and

F_k = our combination has exactly k correct numbers, for $k = 0, 1, \ldots, 6$.

We have

$$P[F] = 1 - \sum_{k=0}^{2} P[F_k] = 1 - \sum_{k=0}^{2} \frac{\binom{6}{k}\binom{43}{6-k}}{\binom{49}{6}}$$

$$= 1 - \frac{(1 \cdot 6,096,454) + (6 \cdot 962,598) + (15 \cdot 123,410)}{13,983,816}$$

$$= 1 - \frac{13,723,192}{13,983,816} \simeq 1 - 0.9814 = 0.0186$$

Notation. The expression $P[A \mid B]$ denotes the probability of the event A, *given that* (or *knowing that*, or simply *if*) the event B has occurred.

Definition 1.1.5. *We set*

$$P[A \mid B] = \frac{P[A \cap B]}{P[B]} \quad \text{if } P[B] > 0 \tag{1.10}$$

Proposition 1.1.4. (Multiplication rule) *We have*

$$P[A \cap B] = P[A \mid B] \times P[B] = P[B \mid A] \times P[A] \quad \text{if } P[A]P[B] > 0 \tag{1.11}$$

Example 1.1.6. In the preceding example, let

F_k = the number of the kth ball that is drawn is part of our combination.

Generalizing the multiplication rule, we may write that

$$P[F_1 \cap F_2 \cap F_3] = P[F_3 \mid F_1 \cap F_2]P[F_2 \mid F_1]P[F_1]$$

$$= \frac{4}{47} \times \frac{5}{48} \times \frac{6}{49} = \frac{120}{110,544} \simeq 0.0011$$

Definition 1.1.6. *The events $B_1, B_2, \ldots, B_n$ constitute a* **partition** *of the sample space S if*

i) $B_i \cap B_j = \emptyset \; \forall \; i \neq j$,

ii) $\bigcup_{k=1}^{n} B_k = S$,

iii) $P[B_k] > 0$, *for $k = 1, 2, \ldots, n$.*

If $B_1, B_2, \ldots, B_n$ is a partition of S, then we may write, for any event A, that

$$A = (A \cap B_1) \cup (A \cap B_2) \cup \ldots \cup (A \cap B_n) \tag{1.12}$$

where $(A \cap B_i) \cap (A \cap B_j) = \emptyset \; \forall \; i \neq j$. Making use of Axiom III in the definition of the function P, p. 2, we obtain the following result.

Proposition 1.1.5. (Total probability rule) *If $A \subset S$ and the events B_1, $B_2, \ldots, B_n$ form a partition of S, then*

$$P[A] = \sum_{k=1}^{n} P[A \cap B_k] = \sum_{k=1}^{n} P[A \mid B_k]P[B_k] \tag{1.13}$$

Finally, we deduce from the total probability rule and from the formula

$$P[A \mid B] = \frac{P[B \mid A]P[A]}{P[B]} \quad \text{if } P[A]P[B] > 0 \tag{1.14}$$

the result known as *Bayes'* [3] *rule* (or *formula*, or also *theorem*).

Proposition 1.1.6. (Bayes' rule) *If $P[A] > 0$, then*

$$P[B_j \mid A] = \frac{P[A \mid B_j]P[B_j]}{\sum_{k=1}^{n} P[A \mid B_k]P[B_k]} \quad \text{for } j = 1, 2, \ldots, n \tag{1.15}$$

where $B_1, B_2, \ldots, B_n$ is a partition of S.

Example 1.1.7. In a certain institution, 80% of the teaching staff are men. Moreover, 80% of the male teachers hold a Ph.D. and 90% of the female teachers hold a Ph.D. A teacher from this institution is taken at random. Let
$F =$ this teacher is a woman
and
$D =$ this teacher holds a Ph.D.
We may write, by the total probability rule, that

$$P[D] = P[D \mid F]P[F] + P[D \mid F^c]P[F^c] = 0.9 \times 0.2 + 0.8 \times 0.8 = 0.82$$

Furthermore, we have

$$P[F \mid D] = \frac{P[D \mid F]P[F]}{P[D]} = \frac{0.9 \times 0.2}{0.82} \simeq 0.2195$$

---

[3] The Reverend Thomas Bayes, 1702–1761, was born and died in England. His works on probability theory were published in a posthumous scientific paper in 1764.

Definition 1.1.7. *Two events, A and B, are said to be* **independent** *if*

$$P[A \cap B] = P[A]P[B] \tag{1.16}$$

Remark. Let C be an event such that $P[C] > 0$. We say that A and B are *conditionally independent with respect to C* if

$$P[A \cap B \mid C] = P[A \mid C]P[B \mid C] \tag{1.17}$$

If, in particular, A and C, or B and C, are incompatible, then A and B are conditionally independent with respect to C, whether they are independent or not. Moreover, A and B can be independent, but not conditionally independent with respect to C. For instance, this may be the case if A and C, and B and C, are not incompatible, but $A \cap B \cap C = \emptyset$.

For events A and B such that $P[A] \times P[B] > 0$, the next proposition could serve as a more intuitive definition of independence.

Proposition 1.1.7. *Two events, A and B, having a positive probability are independent if and only if*

$$P[A \mid B] = P[A] \quad or \quad P[B \mid A] = P[B] \tag{1.18}$$

Proposition 1.1.8. *If A and B are independent, then so are A^c and B, A and B^c, and A^c and B^c.*

Remark. The preceding proposition is obviously false if we replace the word "independent" by "incompatible" (and if A and B are not the sample space S).

All that remains to do is to generalize the notion of independence to any number of events.

Definition 1.1.8. *The events $A_1, A_2, \ldots, A_n$ are said to be* **independent** *if we may write that*

$$P[A_{i_1} \cap A_{i_2} \cap \ldots \cap A_{i_k}] = \prod_{j=1}^{k} P[A_{i_j}] \tag{1.19}$$

for $k = 2, 3, \ldots, n$, where the events $A_{i_j} \in \{A_1, \ldots, A_n\} \; \forall \, j$ are all different.

Remark. If the preceding definition is satisfied (at least) for $k = 2$, we say that the events $A_1, A_2, \ldots, A_n$ are *pairwise independent*.

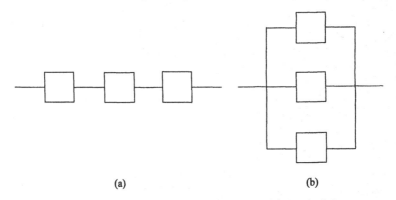

(a)   (b)

Fig. 1.2. Examples of (a) a series system and (b) a parallel system.

Example 1.1.8. A given system is made up of n components placed *in series* and that operate independently from one another [see Fig. 1.2 (a)]. Let
    F = the system is functioning at time t_0
and
    F_k = component k is functioning at time t_0, for $k = 1, \ldots, n$.
We have

$$P[F] = P[F_1 \cap F_2 \cap \ldots \cap F_n] \stackrel{\text{ind.}}{=} \prod_{k=1}^{n} P[F_k]$$

Remark. To help out the reader, the justification of the equality, as here *by independence* (abbreviated as *ind.*), is sometimes placed above the equality sign.

When the components are placed *in parallel*, we may write that

$$P[F] = 1 - P[F^c] = 1 - P[F_1^c \cap \ldots \cap F_n^c] \stackrel{\text{ind.}}{=} 1 - \prod_{k=1}^{n} (1 - P[F_k])$$

1.2 Random variables

Definition 1.2.1. *A **random variable** (r.v.) is a function X that associates a real number $X(s) = x$ with each element s of S, where S is a sample space associated to a random experiment E. We denote by S_X the set of all possible values of X (see Fig. 1.3).*

Remark. The reason for which we introduce the concept of a *random variable* is that the elements s of the sample space S can be anything, for example, a color or a brand of object. Since we prefer to work with real numbers, we transform (if needed) each s into a real number $x = X(s)$.

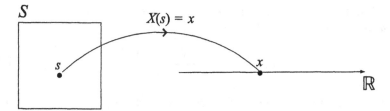

Fig. 1.3. Graphical representation of a random variable.

Example 1.2.1. Consider the random experiment E_3 in Example 1.1.1 that consists of taking a number at random in the interval $(0, 1)$. In this case, the elements of S are already real numbers, so that we can define the r.v. X that is simply the number obtained. That is, X is the *identity function* which with s associates the real number s.

We can define other random variables on the same sample space S, for example:

$$Y = \begin{cases} 1 \text{ if the number obtained is smaller than } 1/2 \\ 0 \text{ otherwise} \end{cases}$$

(called the *indicator variable* of the event A: the number obtained is smaller than $1/2$) and $Z(s) = \ln s$; that is, Z is the natural logarithm of the number taken at random in the interval $(0, 1)$.

We have

$$S_X \equiv S = (0, 1), \quad S_Y = \{0, 1\}, \quad \text{and} \quad S_Z = (-\infty, 0)$$

Definition 1.2.2. *The* **distribution function** *of the r.v. X is defined by*

$$F_X(x) = P[X \le x] \quad \forall \, x \in \mathbb{R} \tag{1.20}$$

Properties. i) $0 \le F_X(x) \le 1$.
ii) $\lim_{x \to -\infty} F_X(x) = 0$ and $\lim_{x \to \infty} F_X(x) = 1$.
iii) If $x_1 < x_2$, then $F_X(x_1) \le F_X(x_2)$.
iv) The function F_X is (at least) right-continuous:

$$F_X(x) = F_X(x^+) := \lim_{\epsilon \downarrow 0} F_X(x + \epsilon) \tag{1.21}$$

Proposition 1.2.1. *We have*
i) $P[a < X \le b] = F_X(b) - F_X(a)$,
ii) $P[X = x] = F_X(x) - F_X(x^-)$, *where* $F_X(x^-) := \lim_{\epsilon \downarrow 0} F_X(x - \epsilon)$.

Remark. Part ii) of the preceding proposition implies that $P[X = x] = 0$ for all x where the function $F_X(x)$ is *continuous*.

Definition 1.2.3. *The* **conditional distribution function** *of the r.v. X, given an event A for which P[A] > 0, is defined by*

$$F_X(x \mid A) = \frac{P[\{X \le x\} \cap A]}{P[A]} \tag{1.22}$$

Remark. A *marginal* distribution function $F_X(x)$ is simply the particular case of the preceding definition where the event A is the sample space S.

Example 1.2.2. The distribution function of the r.v. X in Example 1.2.1 is given by

$$F_X(x) \equiv P[X \le x] = \begin{cases} 0 \text{ if } x \le 0 \\ x \text{ if } 0 < x < 1 \\ 1 \text{ if } x \ge 1 \end{cases}$$

It is easy to check that this function possesses all the properties of a distribution function. In fact, it is continuous for all real x, so that we can state that the number that will be taken at random in the interval $(0, 1)$ had, a priori, a *zero* probability of being chosen, which might seem contradictory. However, there are so many real numbers in the interval $(0, 1)$ that if we assigned a positive probability to each of them, then the sum of all these probabilities would diverge.

Next, consider the event A: the number obtained is smaller than $1/2$. Since $P[A] = 1/2 > 0$, we calculate

$$F_X(x \mid A) \equiv P[X \le x \mid X < 1/2] = 2\,P[X \le x] \quad \text{if } 0 < x < 1/2$$

so that we may write that

$$F_X(x \mid A) = \begin{cases} 0 \text{ if } x \le 0 \\ 2x \text{ if } 0 < x < 1/2 \\ 1 \text{ if } x \ge 1/2 \end{cases}$$

Definition 1.2.4. *If the set S_X of values that the r.v. X can take is finite or countably infinite, we say that X is a* **discrete r.v.** *or an* **r.v. of discrete type.**

Definition 1.2.5. *The* **probability mass function** *of the discrete r.v. X is defined by*

$$p_X(x_k) = P[X = x_k] \quad \forall\, x_k \in S_X \tag{1.23}$$

Remarks. i) **Properties:** a) $p_X(x_k) \ge 0 \,\forall\, x_k$; b) $\sum_{x_k \in S_X} p_X(x_k) = 1$.
ii) We may write that

$$F_X(x) = \sum_{x_k \in S_X} p_X(x_k)\, u(x - x_k) \tag{1.24}$$

where $u(x)$ is the *Heaviside*[4] *function*, defined by

$$u(x) = \begin{cases} 0 \text{ if } x < 0 \\ 1 \text{ if } x \geq 0 \end{cases} \qquad (1.25)$$

iii) **Generalization**: the *conditional probability mass function* of X, given an event A having a positive probability, is defined by

$$p_X(x \mid A) = \frac{P[\{X = x\} \cap A]}{P[A]} \qquad (1.26)$$

Example 1.2.3. An urn contains five white balls and three red balls. We draw one ball at a time, at random and *without* replacement, until we obtain a white ball. Let X be the number of draws needed to end the random experiment. We find that

x	1	2	3	4	Σ
$p_X(x)$	$\frac{5}{8}$	$\left(\frac{3}{8}\right)\left(\frac{5}{7}\right)$	$\left(\frac{3}{8}\right)\left(\frac{2}{7}\right)\left(\frac{5}{6}\right)$	$\left(\frac{3}{8}\right)\left(\frac{2}{7}\right)\left(\frac{1}{6}\right)$	1

and

x	1	2	3	4
$F_X(x)$	$\frac{5}{8}$	$\left(\frac{5}{8}\right)+\left(\frac{3}{8}\right)\left(\frac{5}{7}\right)$	$\left(\frac{5}{8}\right)+\left(\frac{3}{8}\right)\left(\frac{5}{7}\right)+\left(\frac{3}{8}\right)\left(\frac{2}{7}\right)\left(\frac{1}{6}\right)$	1

Finally, let A: the first white ball is obtained after at most two draws. We have $P[A] = F_X(2) = \frac{5}{8} + \frac{3}{8} \times \frac{5}{7} = \frac{50}{56}$ and

x	1	2	Σ
$p_X(x \mid A)$	$\frac{7}{10}$	$\frac{3}{10}$	1

Important discrete random variables

i) **Bernoulli**[5] **distribution**: we say that X has a Bernoulli distribution with parameter p, where p is called the probability of a *success*, if

$$p_X(x) = p^x(1-p)^{1-x} \quad \text{for } x = 0 \text{ and } 1 \qquad (1.27)$$

---

[4] Oliver Heaviside, 1850–1925, who was born and died in England, was a physicist who worked in the field of electromagnetism. He invented *operational calculus* to solve ordinary differential equations.

[5] Jacob (or Jacques) Bernoulli, 1654–1705, was born and died in Switzerland. His important book on probability theory was published eight years after his death.

Remark. The term (probability) *distribution* is used to designate the set of possible values of a discrete random variable, along with their respective probabilities given by the probability mass function. By extension, the same term will be employed in the *continuous* case (see p. 13).

ii) **Binomial distribution** with parameters n and p: $S_X = \{0, 1, \ldots, n\}$ and

$$p_X(x) = \binom{n}{x} p^x (1-p)^{n-x} \tag{1.28}$$

We write $X \sim \mathrm{B}(n, p)$. Some values of its distribution function are given in Table 6.4, p. 358.

iii) **Geometric distribution** with parameter p: $S_X = \{1, 2, \ldots\}$ and

$$p_X(x) = (1-p)^{x-1} p \tag{1.29}$$

We write $X \sim \mathrm{Geom}(p)$.

iv) **Poisson**[6] **distribution** with parameter $\lambda > 0$: $S_X = \{0, 1, \ldots\}$ and

$$p_X(x) = e^{-\lambda} \frac{\lambda^x}{x!} \tag{1.30}$$

We write $X \sim \mathrm{Poi}(\lambda)$. Its distribution function is given, for some values of λ, in Table 6.4, p. 361.

Poisson approximation. If n is large enough (>20) and p is sufficiently small (<0.05), then we may write that

$$P[\mathrm{B}(n, p) = k] \simeq P[\mathrm{Poi}(\lambda = np) = k] \quad \text{for } k = 0, 1, \ldots, n \tag{1.31}$$

If the parameter p is greater than $1/2$, we proceed as follows:

$$\begin{aligned} P[\mathrm{B}(n, p) = k] &= P[\mathrm{B}(n, 1-p) = n-k] \\ &\simeq P[\mathrm{Poi}(\lambda = n(1-p)) = n-k] \end{aligned} \tag{1.32}$$

When $p > 1/2$, we also have

$$\begin{aligned} P[\mathrm{B}(n, p) \leq k] &= P[\mathrm{B}(n, 1-p) \geq n-k] \\ &\simeq P[\mathrm{Poi}(\lambda = n(1-p)) \geq n-k] \end{aligned} \tag{1.33}$$

---

[6] Siméon Denis Poisson, 1781–1840, was born and died in France. He first studied medicine and, from 1798, mathematics at the École Polytechnique de Paris, where he taught from 1802 to 1808. His professors at the École Polytechnique were, among others, Laplace and Lagrange. In mathematics, his main results were his papers on definite integrals and Fourier series. The Poisson distribution appeared in his important book on probability theory published in 1837. He also published works on mechanics, electricity, magnetism, and astronomy. His name is associated with numerous results in both mathematics and physics.

Example 1.2.4. Suppose that we repeat the random experiment in Example 1.2.3 20 times and that we count the number of times, which we denote by X, that the first white ball was obtained on the fourth draw. We may write that $X \sim B(n = 20, p = 1/56)$. We calculate

$$P[X \le 1] = \left(\frac{55}{56}\right)^{20} + 20 \left(\frac{1}{56}\right)\left(\frac{55}{56}\right)^{19} \simeq 0.9510$$

The approximation with a Poisson distribution gives

$$P[X \le 1] \simeq P[Y \le 1], \quad \text{where } Y \sim \text{Poi}(20/56)$$

$$= e^{-5/14} + e^{-5/14}\frac{5}{14} \simeq 0.9496$$

Definition 1.2.6. *A* **continuous random variable** *X is an r.v. that can take an uncountably infinite number of values and whose distribution function F_X is continuous.*

Definition 1.2.7. *The (probability)* **density function** *of the continuous r.v. X is defined (at all points where the derivative exists) by*

$$f_X(x) = \frac{d}{dx}F_X(x) \tag{1.34}$$

Remark. The function $f_X(x)$ is *not* the probability $P[X = x]$ for a continuous r.v. since $P[X = x] = 0 \; \forall \, x$ in this case. The interpretation that can be given to $f_X(x)$ is the following:

$$f_X(x) \simeq \frac{P\left[x - \frac{\epsilon}{2} \le X \le x + \frac{\epsilon}{2}\right]}{\epsilon} \tag{1.35}$$

where $\epsilon > 0$. The equality is obtained by taking the limit as $\epsilon \downarrow 0$.

Properties. i) $f_X(x) \ge 0$ [by the formula (1.35), or by the formula (1.34), because F_X is a *nondecreasing* function].

ii) We deduce from the formula (1.34) that

$$F_X(x) = \int_{-\infty}^{x} f_X(t)\, dt \tag{1.36}$$

It follows that

$$\int_{-\infty}^{\infty} f_X(x)\, dx = 1 \tag{1.37}$$

We also have

$$P[a < X \le b] = F_X(b) - F_X(a) = \int_{a}^{b} f_X(x)\, dx \tag{1.38}$$

Thus, the probability that X takes a value in the interval $(a, b]$ is given by the area under the curve $y = f_X(x)$ from a to b.

Definition 1.2.8. *The* **conditional density function** *of the continuous r.v.* *X, given an event A for which* $P[A] > 0$, *is given by*

$$f_X(x \mid A) = \frac{d}{dx} F_X(x \mid A) \tag{1.39}$$

Remark. We find that the function $f_X(x \mid A)$ may be expressed as follows:

$$f_X(x \mid A) = \frac{f_X(x)}{P[A]} \tag{1.40}$$

for all $x \in S_X$ for which the event A occurs. For example, if $S_X = [0, 1]$ and $A = \{X < 1/2\}$, then $S_Y = [0, 1/2)$, where $Y := X \mid A$.

Remark. If X is an r.v. that can take an uncountably infinite number of values, but the function F_X is *not* continuous, then X is an r.v. of *mixed type*. An example of an r.v. of mixed type is the quantity X (in inches) of rain or snow that will fall during a certain day in a given region. We certainly have the following: $P[X = 0] > 0$, so that X is not a continuous r.v. (since $P[X = 0] = 0$ for any continuous r.v.). It is not an r.v. of discrete type either, because it can (theoretically) take any value in the interval $[0, \infty)$.

Example 1.2.5. Suppose that

$$f_X(x) = \begin{cases} 1/4 & \text{if } -1 < x < 1 \\ 1/(2x) & \text{if } 1 \le x \le e \\ 0 & \text{elsewhere} \end{cases}$$

We can check that the function f_X is nonnegative and that its integral over $\mathbb{R}$ is indeed equal to 1.

We calculate

$$F_X(x) = \begin{cases} 0 & \text{if } x \le -1 \\ (x+1)/4 & \text{if } -1 < x < 1 \\ (1 + \ln x)/2 & \text{if } 1 \le x \le e \\ 1 & \text{if } x > e \end{cases}$$

Note that the density function $f_X(x)$ is discontinuous at $x = 1$ (and at $x = -1$ and $x = e$), which is allowed, whereas the distribution function F_X is a continuous function, as it should be (for a continuous random variable).

Next, we can calculate $F_X(x \mid X < 0)$ and differentiate this function to obtain $f_X(x \mid X < 0)$. It is, however, more efficient to simply calculate $P[X < 0] = 1/4$ and write that

$$f_X(x \mid X < 0) = \begin{cases} 1 & \text{if } -1 < x < 0 \\ 0 & \text{elsewhere} \end{cases}$$

Finally, note that the function $f_X(x \mid X < 0)$ also satisfies the two properties of probability density functions: it is nonnegative and its integral over $\mathbb{R}$ is equal to 1.

Important continuous random variables

i) **Uniform distribution** on the interval $[a, b]$:

$$f_X(x) = (b - a)^{-1} \quad \text{for } a \leq x \leq b \tag{1.41}$$

Notation: $X \sim \mathrm{U}[a, b]$.

ii) **Exponential distribution** with parameter $\lambda > 0$:

$$f_X(x) = \lambda e^{-\lambda x} \quad \text{for } x \geq 0 \tag{1.42}$$

Notation: $X \sim \mathrm{Exp}(\lambda)$.

iii) **Gamma distribution** with parameters $\alpha > 0$ and $\lambda > 0$:

$$f_X(x) = \frac{\lambda e^{-\lambda x}(\lambda x)^{\alpha-1}}{\Gamma(\alpha)} \quad \text{for } x \geq 0 \tag{1.43}$$

where $\Gamma(\alpha) = (\alpha - 1)!$ if $\alpha \in \mathbb{N}$ (see p. 115). Notation: $X \sim \mathrm{G}(\alpha, \lambda)$.

iv) **Gaussian**[7] distribution with parameters μ and σ^2, where $\sigma > 0$:

$$f_X(x) = \frac{1}{\sqrt{2\pi}\sigma} \exp\left\{-\frac{(x-\mu)^2}{2\sigma^2}\right\} \quad \text{for } x \in \mathbb{R} \tag{1.44}$$

Notation: $X \sim \mathrm{N}(\mu, \sigma^2)$.

Remarks. i) In the particular case where $\mu = 0$ and $\sigma = 1$ (see Fig. 1.4), X is called a *standard* Gaussian distribution. Its distribution function is denoted by Φ:

$$\Phi(z) := \int_{-\infty}^{z} \frac{1}{\sqrt{2\pi}} e^{-x^2/2} \, dx \tag{1.45}$$

The values of this function are presented in Table A.3, p. 370, for $z \geq 0$. By symmetry, we may write that $\Phi(-z) = 1 - \Phi(z)$.

ii) If we define $Y = aX + b$, where X has a Gaussian distribution with parameters μ and σ^2, then we find that $Y \sim \mathrm{N}(a\mu + b, a^2\sigma^2)$. In particular, $Z := (X - \mu)/\sigma \sim \mathrm{N}(0, 1)$.

Transformations. If X is a r.v., then any transformation $Y := g(X)$, where g is a real-valued function defined on $\mathbb{R}$, is also a random variable.

---

[7] Carl Friedrich Gauss, 1777–1855, was born and died in Germany. He carried out numerous works in astronomy and physics, in addition to his important mathematical discoveries. He was interested, in particular, in algebra and geometry. He introduced the *law of errors*, that now bears his name, as a model for the errors in astronomical observations.

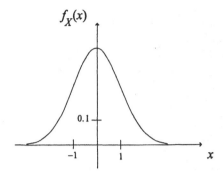

$f_X(x)$

0.1

-1   1 x

Fig. 1.4. Standard Gaussian distribution.

Proposition 1.2.2. *Suppose that the transformation* $y = g(x)$ *is* bijective. *Then the density function of* $Y := g(X)$ *is given by*

$$f_Y(y) = f_X\left(g^{-1}(y)\right)\left|\frac{d}{dy}g^{-1}(y)\right| \tag{1.46}$$

for $y \in [g(a), g(b)]$ *(respectively,* $[g(b), g(a)]$*) if* g *is a strictly* increasing *(resp., decreasing) function and* $S_X = [a, b]$.

Example 1.2.6. If $X \sim U(0, 1)$ and we define $Y = e^X$, then we obtain

$$f_Y(y) = f_X(\ln y)\left|\frac{d}{dy}\ln y\right| = \frac{1}{y}$$

for $y \in (1, e)$.

Definition 1.2.9. *The* **mathematical expectation** *(or the* **mean***)* $E[X]$ *of the r.v.* X *is defined by*

$$E[X] = \sum_{k=1}^{\infty} x_k\, p_X(x_k) \quad \text{(discrete case)} \tag{1.47}$$

or

$$E[X] = \int_{-\infty}^{\infty} x\, f_X(x)\, dx \quad \text{(continuous case)} \tag{1.48}$$

Remarks. i) **Generalization**: we obtain the *conditional (mathematical) expectation* $E[X \mid A]$ of X, given an event A, by replacing $p_X(x)$ by $p_X(x \mid A)$ or $f_X(x)$ by $f_X(x \mid A)$ in the definition.

ii) The mathematical operator E is *linear*.

Proposition 1.2.3. *The mathematical expectation of* $g(X)$ *is given by*

$$E[g(X)] = \sum_{k=1}^{\infty} g(x_k) p_X(x_k) \quad (discrete\ case) \tag{1.49}$$

or

$$E[g(X)] = \int_{-\infty}^{\infty} g(x) f_X(x)\, dx \quad (continuous\ case) \tag{1.50}$$

Remark. We can calculate the mathematical expectation of $g(X)$ by *conditioning*, as follows:

$$E[g(X)] = \sum_{i=1}^{n} E[g(X) \mid B_i] P[B_i] \tag{1.51}$$

where $B_1, \dots, B_n$ is a partition of a sample space S.

The next two definitions are particular cases of mathematical expectations of transformations $g(X)$ of the r.v. X.

Definition 1.2.10. *The* **kth** **moment** *(or* **moment of order** *k) of the r.v. X about the origin is given by $E[X^k]$, for $k = 0, 1, 2, \dots$.*

Definition 1.2.11. *The* **variance** *of the r.v. X is the nonnegative quantity*

$$V[X] = E[(X - E[X])^2] \tag{1.52}$$

Remarks. i) The *standard deviation* of X is defined by $\mathrm{STD}[X] = (V[X])^{1/2}$. The r.v. X and $\mathrm{STD}[X]$ have the same units of measure.

ii) We can also calculate the variance of X by conditioning with respect to a partition of a sample space S, together with the formula (1.52):

$$V[X] = \sum_{i=1}^{n} E[(X - E[X])^2 \mid B_i] P[B_i]$$

iii) **Generalization**: the *conditional variance* of X, given an event A, is defined by

$$V[X \mid A] = E[(X - E[X \mid A])^2 \mid A] \tag{1.53}$$

Proposition 1.2.4. *i)* $V[aX + b] = a^2 V[X] \quad \forall\, a, b \in \mathbb{R}.$
ii) $V[X] = E[X^2] - (E[X])^2.$

Example 1.2.7. The mean (or the *expected value*) of the r.v. Y in Example 1.2.6 is given by

$$E[Y] = \int_1^e y \frac{1}{y}\, dy = e - 1$$

We also have

$$E[Y^2] = \int_1^e y^2 \frac{1}{y}\, dy = \frac{y^2}{2}\Big|_1^e = \frac{e^2-1}{2}$$

It follows that

$$V[Y] = \frac{e^2-1}{2} - (e-1)^2 = \frac{-e^2+4e-3}{2} \simeq 0.2420$$

Now, let $Z := \ln Y$. We have

$$E[Z] = \int_1^e \ln y\, \frac{1}{y}\, dy = \frac{(\ln y)^2}{2}\Big|_1^e = \frac{1}{2}$$

Note that Z is identical to the r.v. X in Example 1.2.6 and that the mean of $X \sim U(0,1)$ is indeed equal to $1/2$.

A very important special case of the mathematical expectation $E[g(X)]$ occurs when $g(X) = e^{j\omega X}$.

Definition 1.2.12. *The function*

$$C_X(\omega) = E[e^{j\omega X}] \tag{1.54}$$

*where $j = \sqrt{-1}$, is called the **characteristic function** of the r.v. X.*

If X is a continuous r.v., then $C_X(\omega)$ is the *Fourier*[8] *transform* of the density function $f_X(x)$:

$$C_X(\omega) = \int_{-\infty}^{\infty} e^{j\omega x} f_X(x)\, dx \tag{1.55}$$

We can invert this Fourier transform and write that

$$f_X(x) = \frac{1}{2\pi} \int_{-\infty}^{\infty} e^{-j\omega x} C_X(\omega)\, d\omega \tag{1.56}$$

Since the Fourier transform is *unique*, the function $C_X(\omega)$ *characterizes* entirely the r.v. X. For instance, there is only the standard Gaussian distribution that possesses the characteristic function $C_X(\omega) = e^{-\omega^2/2}$.

We can also use the function $C_X(\omega)$ to obtain the moments of order n of the r.v. X, generally more easily than from the definition of $E[X^n]$ (for $n \in \{2,3,\dots\}$).

Proposition 1.2.5. *If the mathematical expectation $E[X^n]$ exists and is finite for all $n \in \{1,2,\dots\}$, then*

$$E[X^n] = (-j)^n \frac{d^n}{d\omega^n}\, C_X(\omega)\big|_{\omega=0} \tag{1.57}$$

---

[8] Joseph (Baron) Fourier, 1768–1830, was born and died in France. He taught at the Collège de France and at the École Polytechnique. In his main work, the *Théorie Analytique de la Chaleur*, published in 1822, he made wide use of the series that now bears his name, but that he did not invent.

Table 1.1. Means, Variances, and Characteristic Functions of the Main Random Variables (with $q := 1 - p$)

Distribution	Mean	Variance	Characteristic function
Bernoulli	p	pq	$pe^{j\omega} + q$
$B(n,p)$	np	npq	$(pe^{j\omega} + q)^n$
Geom(p)	$1/p$	q/p^2	$\dfrac{pe^{j\omega}}{1 - qe^{j\omega}}$
Poi(λ)	λ	λ	$\exp\{\lambda(e^{j\omega} - 1)\}$
$U[a,b]$	$\dfrac{a+b}{2}$	$\dfrac{(b-a)^2}{12}$	$\dfrac{e^{j\omega b} - e^{j\omega a}}{j\omega(b-a)}$
Exp(λ)	$\dfrac{1}{\lambda}$	$\dfrac{1}{\lambda^2}$	$\dfrac{\lambda}{\lambda - j\omega}$
$G(\alpha,\lambda)$	$\dfrac{\alpha}{\lambda}$	$\dfrac{\alpha}{\lambda^2}$	$\left(\dfrac{\lambda}{\lambda - j\omega}\right)^\alpha$
$N(\mu,\sigma^2)$	μ	σ^2	$\exp\{j\omega\mu - \frac{1}{2}\omega^2\sigma^2\}$

Remark. Table 1.1 gives the mean, the variance, and the characteristic function of all the discrete and continuous random variables mentioned previously.

Many authors prefer to work with the following function, which, as its name indicates, also enables us to calculate the moments of a random variable.

Definition 1.2.13. *The* **moment-generating function** *of the r.v. X is defined, if the mathematical expectation exists, by $M_X(t) = E[e^{tX}]$.*

Remarks. i) When X is a continuous and nonnegative r.v., $M_X(t)$ is the Laplace[9] *transform* of the function $f_X(x)$.

ii) Corresponding to the formula (1.57), we find that

$$E[X^n] = \frac{d^n}{dt^n} M_X(t)\big|_{t=0} \quad \text{for } n = 1, 2, \ldots \tag{1.58}$$

---

[9] Pierre Simon (Marquis de) Laplace, 1749–1827, was born and died in France. In addition to being a mathematician and an astronomer, he was also a minister and a senator. He was made a count by Napoléon and marquis by Louis XVIII. He participated in the organization of the École Polytechnique of Paris. His main works were on astronomy and on the calculus of probabilities: the *Traité de Mécanique Céleste,* published in five volumes, from 1799, and the *Théorie Analytique des Probabilités,* whose first edition appeared in 1812. Many mathematical formulas bear his name.

Example 1.2.8. If $X \sim \text{Poi}(\lambda)$, we calculate

$$M_X(t) = \sum_{k=0}^{\infty} e^{tk} e^{-\lambda} \frac{\lambda^k}{k!} = e^{-\lambda} \sum_{k=0}^{\infty} \frac{(e^t\lambda)^k}{k!} = e^{-\lambda} \exp(e^t\lambda)$$

We deduce from this formula and from Eq. (1.58) that

$$E[X] = M'_X(0) = \lambda \quad \text{and} \quad E[X^2] = \lambda^2 + \lambda$$

so that $V[X] = \lambda^2 + \lambda - (\lambda)^2 = \lambda$.

Note that to obtain $E[X^2]$, we can proceed as follows:

$$E[X^2] = \sum_{k=0}^{\infty} k^2 e^{-\lambda} \frac{\lambda^k}{k!} = e^{-\lambda} \sum_{k=1}^{\infty} k \frac{\lambda^k}{(k-1)!} = e^{-\lambda}\lambda \sum_{k=1}^{\infty} \frac{d}{d\lambda} \frac{\lambda^k}{(k-1)!}$$

$$= e^{-\lambda}\lambda \frac{d}{d\lambda} \sum_{k=1}^{\infty} \frac{\lambda^k}{(k-1)!} = e^{-\lambda}\lambda \frac{d}{d\lambda}(\lambda e^\lambda) = \lambda + \lambda^2$$

It is clear that it is easier to differentiate twice the function $e^{-\lambda} \exp(e^t\lambda)$ than to evaluate the infinite sum above.

When we do not know the distribution of the r.v. X, we can use the following inequalities to obtain bounds for the probability of certain events.

Proposition 1.2.6. *a)* (**Markov's**[10] **inequality**) *If X is an r.v. that takes only nonnegative values, then*

$$P[X \geq c] \leq \frac{E[X]}{c} \quad \forall\, c > 0 \tag{1.59}$$

b) (**Chebyshev's**[11] **inequality**) *If $E[Y]$ and $V[Y]$ exist, then we have*

$$P[|Y - E[Y]| \geq c] \leq \frac{V[Y]}{c^2} \quad \forall\, c > 0 \tag{1.60}$$

---

[10] Andrei Andreyevich Markov, 1856–1922, who was born and died in Russia, was a professor at St. Petersburg University. His first works were on number theory and mathematical analysis. He proved the *central limit theorem* under quite general conditions. His study of what is now called *Markov chains* initiated the theory of stochastic processes. He was also interested in poetry.

[11] Pafnuty Lvovich Chebyshev, 1821–1894, was born and died in Russia. By using the inequality that bears his name, he gave a simple proof of the *law of large numbers*. He also worked intensively on the *central limit theorem*.

1.3 Random vectors

Definition 1.3.1. *An n-dimensional* **random vector** *is a function* $\mathbf{X} = (X_1, \ldots, X_n)$ *that associates a vector* $(X_1(s), \ldots, X_n(s))$ *of real numbers with each element s of a sample space S of a random experiment E. Each component X_k of the vector is a random variable. We denote by $S_{\mathbf{X}}$ ($\subset \mathbb{R}^n$) the set of all possible values of* $\mathbf{X}$.

Remark. As in the case of the random variables, we will use the abbreviation r.v., for random vector, since there is no risk of confusion between a random *variable* and a random *vector*.

Two-dimensional random vectors

Definition 1.3.2. *The* **joint distribution function** *of the r.v.* (X, Y) *is defined, for all points* $(x, y) \in \mathbb{R}^2$, *by*

$$F_{X,Y}(x, y) = P[\{X \le x\} \cap \{Y \le y\}] \equiv P[X \le x, Y \le y] \qquad (1.61)$$

Properties. i) $F_{X,Y}(-\infty, y) = F_{X,Y}(x, -\infty) = 0$ and $F_{X,Y}(\infty, \infty) = 1$.

ii) $F_{X,Y}(x_1, y_1) \le F_{X,Y}(x_2, y_2)$ if $x_1 \le x_2$ and $y_1 \le y_2$.

iii) We have

$$\lim_{\epsilon \downarrow 0} F_{X,Y}(x + \epsilon, y) = \lim_{\epsilon \downarrow 0} F_{X,Y}(x, y + \epsilon) = F_{X,Y}(x, y) \qquad (1.62)$$

Remark. We can show that

$$P[a < X \le b, c < Y \le d] = F_{X,Y}(b, d) - F_{X,Y}(b, c) - F_{X,Y}(a, d) + F_{X,Y}(a, c) \qquad (1.63)$$

where $a, b, c,$ and d are constants.

It is easy to obtain the *marginal* distribution function of X when the function $F_{X,Y}$ is known. Indeed, we may write that

$$F_X(x) = P[X \le x, Y < \infty] = F_{X,Y}(x, \infty) \qquad (1.64)$$

Definition 1.3.3. *A two-dimensional r.v.,* $\mathbf{Z} = (X, Y)$, *is of* **discrete type** *if $S_{\mathbf{Z}}$ is a finite or countably infinite set of points in $\mathbb{R}^2$:*

$$S_{\mathbf{Z}} \equiv S_{X \times Y} = \{(x_j, y_k), j = 1, 2, \ldots; k = 1, 2, \ldots\} \qquad (1.65)$$

Definition 1.3.4. *The* **joint probability mass function** *of the discrete r.v.* (X, Y) *is defined by*

$$p_{X,Y}(x_j, y_k) = P[X = x_j, Y = y_k] \qquad (1.66)$$

for $j, k = 1, 2, \ldots$.

To obtain the *marginal* probability mass function of X, from $p_{X,Y}$, we make use of the *total probability rule*:

$$p_X(x_j) \equiv P[X = x_j] = \sum_{k=1}^{\infty} P[\{X = x_j\} \cap \{Y = y_k\}] = \sum_{k=1}^{\infty} p_{X,Y}(x_j, y_k)$$

(1.67)

Example 1.3.1. The generalization of the binomial distribution to the two-dimensional case is the joint probability mass function given by

$$p_{X_1,X_2}(x_1, x_2) = \frac{m!}{x_1! x_2! (m - x_1 - x_2)!} p_1^{x_1} p_2^{x_2} (1 - p_1 - p_2)^{m-x_1-x_2} \quad (1.68)$$

where $x_1, x_2 \in \{0, 1, \ldots, m\}$ and $x_1 + x_2 \leq m \in \mathbb{N}$. We say that (X_1, X_2) has a *trinomial distribution* with parameters m, p_1, and p_2, where $0 < p_k < 1$, for $k = 1, 2$. We can generalize further the binomial distribution and obtain the *multinomial distribution* (in n dimensions).

Definition 1.3.5. *A two-dimensional r.v.,* $\mathbf{Z} = (X, Y)$, *is of* **continuous type** *if $S_{\mathbf{Z}}$ is an uncountably infinite subset of $\mathbb{R}^2$. (We assume that X and Y are two continuous random variables.)*

Remark. We will not consider in this book the case of random vectors with at least one component being a random variable of mixed type.

Definition 1.3.6. *The* **joint (probability) density function** *of the continuous r.v.* $\mathbf{Z} = (X, Y)$ *is defined by*

$$f_{X,Y}(x, y) = \frac{\partial^2}{\partial x \partial y} F_{X,Y}(x, y) \qquad (1.69)$$

for any point where the derivative exists.

Remarks. i) Corresponding to the formula (1.67), the **marginal (probability) density function** of X can be obtained as follows:

$$f_X(x) = \int_{-\infty}^{\infty} f_{X,Y}(x, y) \, dy \qquad (1.70)$$

where we integrate in practice over all the values that Y can take when $X = x$.
ii) The probability of the event $\{\mathbf{Z} \in A\}$, where $A \subset S_{\mathbf{Z}}$, can be calculated as follows:

$$P[\mathbf{Z} \in A] = \int_A \int f_{X,Y}(x, y) \, dx dy \qquad (1.71)$$

iii) The distribution function of the continuous r.v. (X, Y) is given by

$$F_{X,Y}(x, y) = \int_{-\infty}^{y} \int_{-\infty}^{x} f_{X,Y}(u, v) \, du dv \tag{1.72}$$

In the discrete case, the formula above becomes

$$F_{X,Y}(x, y) = \sum_{x_j \leq x} \sum_{y_k \leq y} p_{X,Y}(x_j, y_k) \tag{1.73}$$

Example 1.3.2. The continuous r.v. (X, Y) has the joint density function

$$f_{X,Y}(x, y) = \begin{cases} \dfrac{\ln x}{x} & \text{if } 1 \leq x \leq e, \, 0 < y < x \\ 0 & \text{elsewhere} \end{cases}$$

We calculate

$$f_X(x) = \int_0^x \frac{\ln x}{x} \, dy = \ln x \quad \text{if } 1 \leq x \leq e$$

and

$$f_Y(y) = \begin{cases} \displaystyle\int_1^e \frac{\ln x}{x} \, dx = \frac{\ln^2 x}{2} \Big|_1^e = \frac{1}{2} & \text{if } 0 < y < 1 \\[2mm] \displaystyle\int_y^e \frac{\ln x}{x} \, dx = \frac{\ln^2 x}{2} \Big|_y^e = \frac{1 - \ln^2 y}{2} & \text{if } 1 \leq y < e \\[2mm] 0 & \text{elsewhere} \end{cases}$$

Example 1.3.3. Let

$$f_{X,Y}(x, y) = \begin{cases} 2y \, e^{-x} & \text{if } x \geq 0, \, 0 < y < 1 \\ 0 & \text{elsewhere} \end{cases}$$

First, we calculate

$$f_X(x) = \int_0^1 2y \, e^{-x} \, dy = e^{-x} \quad \text{if } x \geq 0$$

and

$$f_Y(y) = \int_0^\infty 2y \, e^{-x} \, dx = 2y \quad \text{if } 0 < y < 1$$

Moreover, we find that

$$F_{X,Y}(x, y) = \begin{cases} 0 & \text{if } x < 0 \text{ or } y \leq 0 \\ (1 - e^{-x})y^2 & \text{if } x \geq 0 \text{ and } 0 < y < 1 \\ e^{-x} & \text{if } x \geq 0 \text{ and } y \geq 1 \end{cases}$$

Finally, we calculate

$$P[X + Y > 1] = 1 - P[X + Y \leq 1] = 1 - \int_0^1 \int_0^{1-y} 2y \, e^{-x} \, dx \, dy$$

$$= 1 - \int_0^1 2y \, (1 - e^{y-1}) \, dy = 1 - (1 - 2e^{-1}) = 2e^{-1}$$

because

$$\int_0^1 y e^y \, dy = y e^y \big|_0^1 - \int_0^1 e^y \, dy = e - (e - 1) = 1$$

Definition 1.3.7. *Let (X, Y) be a random vector. We say that X and Y are* **independent** *random variables if*

$$p_{X,Y}(x_j, y_k) \equiv p_X(x_j) p_Y(y_k) \quad \text{if } (X, Y) \text{ is discrete} \qquad (1.74)$$

or

$$f_{X,Y}(x, y) \equiv f_X(x) f_Y(y) \quad \text{if } (X, Y) \text{ is continuous} \qquad (1.75)$$

Remarks. i) More generally, X and Y are independent if (and only if)

$$P[X \in A, Y \in B] = P[X \in A] P[Y \in B] \qquad (1.76)$$

where A (respectively, B) is any event that involves only X (resp., Y). In particular, we must have

$$F_{X,Y}(x, y) \equiv P[X \leq x, Y \leq y] = F_X(x) F_Y(y) \quad \forall \, (x, y) \qquad (1.77)$$

ii) If X and Y are independent r.v.s, then so are $g(X)$ and $h(Y)$.

iii) Let $Z := X + Y$, where X and Y are two independent r.v.s. Then [see the formula (1.102)],

$$M_Z(t) \equiv E[e^{tZ}] = E[e^{t(X+Y)}] = E[e^{tX}] E[e^{tY}] = M_X(t) M_Y(t) \qquad (1.78)$$

Similarly, $C_Z(\omega) = C_X(\omega) C_Y(\omega)$.

Example 1.3.4. We deduce from Eq. (1.75) that the r.v.s X and Y in Example 1.3.2 are *not* independent, whereas those in Example 1.3.3 are.

Definition 1.3.8. *If Y is a discrete r.v., then the* **conditional distribution function** *of X, given that $Y = y_k$, is defined by*

$$F_{X|Y}(x \mid y_k) = \frac{P[X \leq x, Y = y_k]}{P[Y = y_k]} \quad \text{if } P[Y = y_k] > 0 \qquad (1.79)$$

Remark. In theory, Y can be a continuous r.v. in the preceding definition. However, in practice, most of the time the two random variables in the pair (X, Y) are of the same type.

Definition 1.3.9. *If (X, Y) is a discrete r.v., then the* **conditional probability mass function** *of X, given that $Y = y_k$, is defined by*

$$p_{X|Y}(x_j \mid y_k) = \frac{p_{X,Y}(x_j, y_k)}{p_Y(y_k)} = \frac{P[X = x_j, Y = y_k]}{P[Y = y_k]} \quad \text{if } P[Y = y_k] > 0 \tag{1.80}$$

Remark. The conditional functions possess the same properties as the corresponding *marginal* functions.

When Y is a continuous r.v., we cannot condition on the event $\{Y = y\}$ directly, because $P[Y = y] = 0$ for all y. We must rather take the limit as dy decreases to zero of the functions defined by conditioning on the event $\{y < Y \leq y + dy\}$. We then obtain the following proposition.

Proposition 1.3.1. *If (X, Y) is a continuous r.v. and $f_Y(y) > 0$, then the* **conditional distribution function** *and the* **conditional density function** *of X, given that $Y = y$, are given, respectively, by*

$$F_{X|Y}(x \mid y) = \frac{\int_{-\infty}^{x} f_{X,Y}(u, y) \, du}{f_Y(y)} \quad \text{and} \quad f_{X|Y}(x \mid y) = \frac{f_{X,Y}(x, y)}{f_Y(y)} \tag{1.81}$$

Example 1.3.5. The conditional density function of Y, given that $X = x$, in Example 1.3.2 is

$$f_{Y|X}(y \mid x) = \frac{f_{X,Y}(x, y)}{f_X(x)} = \frac{1}{x} \quad \text{if } 0 < y < x$$

That is, $Y \mid \{X = x\}$ has a uniform distribution on the interval $(0, x)$. Hence, we easily find that $F_{Y|X}(y \mid x) = 0$ if $y \leq 0$,

$$F_{Y|X}(y \mid x) = \frac{y}{x} \quad \text{if } 0 < y < x$$

and $F_{Y|X}(y \mid x) = 1$ if $y \geq x$.

Proposition 1.3.2. *The r.v.s X and Y are independent if and only if the conditional distribution function, the conditional probability mass function, or the conditional density function of X, given that $Y = y$, is identical to the corresponding marginal function.*

Example 1.3.6. We say that the continuous random variables X and Y have a *binormal (or bivariate normal) distribution* with parameters $\mu_X \in \mathbb{R}$, $\mu_Y \in \mathbb{R}$, $\sigma_X^2 > 0$, $\sigma_Y^2 > 0$, and $\rho \in (-1, 1)$, and we write that $(X, Y) \sim N(\mu_X, \mu_Y; \sigma_X^2, \sigma_Y^2; \rho)$, if their joint density function is

$$f_{X,Y}(x, y) = \frac{1}{2\pi\sigma_X\sigma_Y(1 - \rho^2)^{1/2}}$$

$$\times \exp\left\{-\frac{1}{2(1-\rho^2)}\left[\left(\frac{x-\mu_X}{\sigma_X}\right)^2+\left(\frac{y-\mu_Y}{\sigma_Y}\right)^2\right.\right.$$

$$\left.\left.-2\rho\frac{(x-\mu_X)(y-\mu_Y)}{\sigma_X\sigma_Y}\right]\right\} \qquad (1.82)$$

for $(x, y) \in \mathbb{R}^2$.

We easily find that $X \sim N(\mu_X, \sigma_X^2)$ and $Y \sim N(\mu_Y, \sigma_Y^2)$. It follows that

$$X \mid \{Y = y\} \sim N(\mu_X + \rho(\sigma_X/\sigma_Y)(y - \mu_Y), \sigma_X^2(1 - \rho^2)) \qquad (1.83)$$

Since X and $X \mid \{Y = y\}$ have the same distribution if $\rho = 0$, we can state that X and Y are independent r.v.s if the parameter ρ is equal to zero. This parameter is actually the *correlation coefficient* of X and Y.

Definition 1.3.10. *The* **conditional expectation** *of X, given that $Y = y$, is defined by*

$$E[X \mid Y = y] = \sum_{j=1}^{\infty} x_j \, p_{X|Y}(x_j \mid y) \quad \text{(discrete case)} \qquad (1.84)$$

or

$$E[X \mid Y = y] = \int_{-\infty}^{\infty} x \, f_{X|Y}(x \mid y) \, dx \quad \text{(continuous case)} \qquad (1.85)$$

The mean $E[g(X)]$ of a transformation g of a random variable X is a real constant, while $E[g(X) \mid Y = y]$ is a function of y, where y is a particular value taken by the r.v. Y. We now consider $E[g(X) \mid Y]$. It is a function of the r.v. Y that takes the value $E[g(X) \mid Y = y]$ when $Y = y$. Consequently, $E[g(X) \mid Y]$ is a random variable, whose mean can be calculated. We then obtain the following important proposition.

Proposition 1.3.3. *We have*

$$E[g(X)] = E[E[g(X) \mid Y]] \qquad (1.86)$$

Remarks. i) We deduce from the preceding proposition that

$$E[X] = E[E[X \mid Y]] = \begin{cases} \sum_{k=1}^{\infty} E[X \mid Y = y_k] \, p_Y(y_k) \text{ (discrete case)} \\ \\ \int_{-\infty}^{\infty} E[X \mid Y = y] \, f_Y(y) \, dy \text{ (continuous case)} \end{cases}$$

$$(1.87)$$

ii) We can calculate the variance of X by conditioning on another r.v. Y as follows:

$$V[X] = E[E[X^2 \mid Y]] - (E[E[X \mid Y]])^2 \tag{1.88}$$

iii) Let $X_1, X_2, \ldots$ be r.v.s that possess the same distribution as the r.v. X, so that $E[X_k] = E[X]$ and $V[X_k] = V[X]$, for $k = 1, 2, \ldots$, and let N be an r.v. independent of the X_k's and taking its values in the set $\{1, 2, \ldots\}$. By making use of the formula (1.86), we can show (see p. 254) that

$$E\left[\sum_{k=1}^{N} X_k\right] = E[N]E[X] \tag{1.89}$$

If the r.v.s X_k are independent among themselves, we also have (see p. 254)

$$V\left[\sum_{k=1}^{N} X_k\right] = E[N]V[X] + V[N](E[X])^2 \tag{1.90}$$

iv) Suppose that we wish to *estimate* a random variable X by using another r.v. Y. It can be shown that the function $g(Y)$ that minimizes the *mean-square error* (MSE)

$$\text{MSE} := E[(X - g(Y))^2] \tag{1.91}$$

is $g(Y) = E[X \mid Y]$. If we look for a function of the form $g(Y) = \alpha Y + \beta$, we can show that the constants α and β that minimize MSE are

$$\hat{\alpha} = \frac{E[XY] - E[X]E[Y]}{V[Y]} \quad \text{and} \quad \hat{\beta} = E[X] - \hat{\alpha}E[Y] \tag{1.92}$$

Finally, if $g(Y) \equiv c$, we easily find that the constant c that yields the smallest MSE is $c = E[X]$.

The function $g(Y) = E[X \mid Y]$ is the *best estimator* of X, in terms of Y, while $g(Y) = \hat{\alpha}Y + \hat{\beta}$ is the *best linear estimator* of X, in terms of Y. If X and Y both have a Gaussian distribution, then the two estimators are equal (see Ex. 1.3.6).

Proposition 1.3.3 also enables us to calculate the probability of the event $\{X \in A\}$ by conditioning on the possible values of an r.v. Y. We only have to define the r.v. W such that $W = 1$ if $X \in A$ and $W = 0$ if $X \notin A$, and use the fact that $E[W] = P[X \in A]$. We can then show the following proposition, which is the equivalent of the *total probability rule* for random variables. This proposition and Proposition 1.3.3 will be very useful in the next chapters.

Proposition 1.3.4. *We may write that*

$$P[X \in A] = \begin{cases} \sum_{k=1}^{\infty} P[X \in A \mid Y = y_k]\, p_Y(y_k) & \text{(discrete case)} \\ \int_{-\infty}^{\infty} P[X \in A \mid Y = y]\, f_Y(y)\, dy & \text{(continuous case)} \end{cases} \tag{1.93}$$

Corresponding to the definition of the conditional variance $V[X \mid A]$ that was given in the preceding section, we now have the following definition.

Definition 1.3.11. *The conditional variance of X, given the r.v. Y, is defined by*

$$V[X \mid Y] = E[(X - E[X \mid Y])^2 \mid Y] \tag{1.94}$$

Remarks. i) We find that

$$V[X \mid Y] = E[X^2 \mid Y] - (E[X \mid Y])^2 \tag{1.95}$$

ii) We can show the following useful result:

$$V[X] = E[V[X \mid Y]] + V[E[X \mid Y]] \tag{1.96}$$

Example 1.3.7. Instead of calculating the variance of the r.v. Y in Example 1.3.2 from the density function f_Y that we obtained and from the definition of $V[Y]$, we can use the fact (see Ex. 1.3.5) that $Y \mid X \sim U(0, X)$. It follows that $E[Y \mid X] = X/2$ and $V[Y \mid X] = X^2/12$, and then, by the formula (1.96),

$$V[Y] = \frac{1}{12}E[X^2] + \frac{1}{4}V[X] = \frac{1}{3}E[X^2] - \frac{1}{4}(E[X])^2 \simeq 0.4252$$

because

$$E[X] = \int_1^e x \ln x \, dx = \frac{1}{4}(e^2 + 1) \quad \text{and} \quad E[X^2] = \int_1^e x^2 \ln x \, dx = \frac{1}{9}(2e^3 + 1)$$

Remark. We can check that

$$E[Y] = \frac{1}{8}(e^2 + 1) \quad \text{and} \quad E[Y^2] = \frac{1}{27}(2e^3 + 1)$$

Remark. We can consider conditional expectations $E[X \mid A]$, or conditional density functions $f_{X,Y}(x, y \mid A)$, etc., with respect to more general events A, like $Y \le y$, $Y > 0$, etc., and also with respect to events A that involve both random variables, X and Y. For instance, let X_1 and X_2 be independent random variables having a $U(0, 1)$ distribution. We have
$P[X_1 \le x_1 \mid X_1 < X_2]$

$$= \frac{P[X_1 \le x_1, X_1 < X_2]}{P[X_1 < X_2]} = 2 \int_0^1 P[X_1 \le x_1, X_1 < X_2 \mid X_2 = x_2] \cdot 1 \, dx_2$$

$$\stackrel{\text{ind.}}{=} 2 \int_0^1 P[X_1 \le x_1, X_1 < x_2] \, dx_2 = 2 \left\{ \int_0^{x_1} x_2 \, dx_2 + \int_{x_1}^1 x_1 \, dx_2 \right\}$$

$$= x_1(2 - x_1)$$

so that

$$f_{X_1}(x_1 \mid X_1 < X_2) = 2(1 - x_1) \quad \text{for } 0 < x_1 < 1$$

Next, we have

$$E[X_1 \mid X_1 < X_2] = \int_0^1 x_1 \cdot 2(1 - x_1) \, dx_1 = \frac{1}{3}$$

Actually, if we are only looking for the mean of X_1, given that $X_1 < X_2$, we can directly write that

$$E[X_1 \mid X_1 < X_2] = \int_0^1 \int_{x_1}^1 x_1 \cdot \frac{1}{1/2} \, dx_2 dx_1 = \frac{1}{3}$$

where we used the formula

$$f_{X_1,X_2}(x_1, x_2 \mid X_1 < X_2) = \frac{f_{X_1,X_2}(x_1, x_2)}{P[X_1 < X_2]} \quad \text{for } 0 < x_1 < x_2 < 1$$

The following proposition is the two-dimensional version of Proposition 1.2.2.

Proposition 1.3.5. *Let* $W := g_1(X, Y)$ *and* $Z := g_2(X, Y)$, *where* X *and* Y *are two continuous r.v.s. If*
1) the system $w = g_1(x, y)$, $z = g_2(x, y)$ *has the* unique *solution* $x = h_1(w, z)$, $y = h_2(w, z)$
and
2) the functions g_1 *and* g_2 *have continuous partial derivatives* $\forall \, (x, y)$ *and the Jacobian* $J(x, y)$ *of the transformation is such that*

$$J(x, y) := \frac{\partial g_1}{\partial x} \frac{\partial g_2}{\partial y} - \frac{\partial g_2}{\partial x} \frac{\partial g_1}{\partial y} \neq 0 \quad \forall \, (x, y) \tag{1.97}$$

then

$$f_{W,Z}(w, z) = f_{X,Y}(h_1(w, z), h_2(w, z)) \, |J(h_1(w, z), h_2(w, z))|^{-1} \tag{1.98}$$

Remarks. i) The proposition can be easily generalized to the n-dimensional case, where $n \in \{3, 4, \dots\}$.

ii) In the particular case where X and Y are independent and $Z := X + Y$, we could use the proposition to obtain the density function of Z. We must first define an appropriate *auxiliary* variable W, then calculate the joint density function of the r.v. (W, Z), and finally integrate this joint density function with respect to w to obtain $f_Z(z)$. We can also proceed as follows:

$$F_Z(z) = \int_{-\infty}^{\infty} \int_{-\infty}^{z-u} f_X(u) f_Y(v) \, dv \, du \implies f_Z(z) = \int_{-\infty}^{\infty} f_X(u) f_Y(z - u) \, du \tag{1.99}$$

Note that the density function of Z is the *convolution product* of the density functions of X and Y.

Proposition 1.3.6. *The mathematical expectation of the random variable* $Z := g(X, Y)$ *is given by*

$$E[Z] = \begin{cases} \sum_{j=1}^{\infty} \sum_{k=1}^{\infty} g(x_j, y_k) \, p_{X,Y}(x_j, y_k) & \text{(discrete case)} \\ \int_{-\infty}^{\infty} \int_{-\infty}^{\infty} g(x, y) \, f_{X,Y}(x, y) \, dx \, dy & \text{(continuous case)} \end{cases} \tag{1.100}$$

Remark. If the mathematical expectations $E[X]$ and $E[Y]$ exist, we have

$$E[aX + bY] = aE[X] + bE[Y] \quad \forall \, a, b \in \mathbb{R} \tag{1.101}$$

Moreover, if X and Y are independent r.v.s and $g(X, Y) = g_1(X)g_2(Y)$, then

$$E[g(X, Y)] = E[g_1(X)]E[g_2(Y)] \tag{1.102}$$

Definition 1.3.12. *The* **covariance** *of X and Y is defined by*

$$\text{Cov}[X, Y] \equiv E[(X - E[X])(Y - E[Y])] = E[XY] - E[X]E[Y] \tag{1.103}$$

Remarks. i) The covariance generalizes the variance, since $\text{Cov}[X, X] = V[X]$, but the covariance $\text{Cov}[X, Y]$ can be negative. For example, if $Y = -X$, then we have

$$\text{Cov}[X, Y] = \text{Cov}[X, -X] = E[X(-X)] - E[X]E[-X] = -V[X] \le 0 \tag{1.104}$$

ii) We deduce from Eq. (1.102) that if X and Y are independent, then $\text{Cov}[X, Y] = 0$. However, the converse is not always true.

iii) We also define the *correlation coefficient* of X and Y by

$$\rho_{X,Y} = \frac{\text{Cov}[X, Y]}{\text{STD}[X]\text{STD}[Y]} \tag{1.105}$$

We then deduce from Example 1.3.6 that, in the case of the bivariate normal distribution, the r.v.s X and Y are independent *if and only if* their correlation coefficient is equal to zero.

An important particular case of transformations of random vectors is the one where the random variable $Z := g(X_1, \ldots, X_n)$ is a *linear combination* of the r.v.s $X_1, \ldots, X_n$:

$$Z = a_0 + a_1 X_1 + \cdots + a_n X_n \tag{1.106}$$

where the a_k's are real constants $\forall \, k$. We can show the following proposition.

Proposition 1.3.7. *Let Z be a linear combination of the r.v.s $X_1, \ldots, X_n$. We can write (if the mathematical expectations exist) that*

$$E[Z] = a_0 + a_1 E[X_1] + \cdots + a_n E[X_n] \qquad (1.107)$$

and

$$V[Z] = \sum_{k=1}^{n} a_k^2 V[X_k] + 2 \underbrace{\sum_{j=1}^{n} \sum_{k=1}^{n}}_{j<k} a_j a_k \text{Cov}[X_j, X_k] \qquad (1.108)$$

Example 1.3.8. If X and Y are two independent r.v.s having a uniform distribution on the interval $[0, 1]$ and $Z := X + Y$, then $S_Z = [0, 2]$ and

$$f_Z(z) = \int_{-\infty}^{\infty} f_X(u) f_Y(z - u) \, du = \int_0^1 f_Y(z - u) \, du$$

Since

$$f_Y(z - u) = \begin{cases} 1 \text{ if } z - 1 \leq u \leq z \\ 0 \text{ elsewhere} \end{cases}$$

we may write that

$$f_Z(z) = \begin{cases} \displaystyle\int_0^z 1 \, du = z & \text{if } 0 \leq z \leq 1 \\[2mm] \displaystyle\int_{z-1}^1 1 \, du = 2 - z & \text{if } 1 < z \leq 2 \\[2mm] 0 & \text{elsewhere} \end{cases}$$

Remark. If we define the auxiliary variable $W = X$, then we find that

$$f_{W,Z}(w, z) = 1 \quad \text{if } 0 \leq w \leq 1, \, 0 \leq z \leq 2, \, w \leq z \leq w + 1$$

Integrating $f_{W,Z}(w, z)$ with respect to w, we retrieve the function $f_Z(z)$ above.

Next, if $Z := X - \frac{1}{2}$ and $W := Z^2$, we calculate

$$\text{Cov}[Z, W] = E[ZW] - E[Z]E[W] = E[Z^3] - 0 = 0$$

because $E[Z^{2k+1}] = 0$, for all $k \in \{0, 1, \ldots\}$. However, Z and W are *not* independent, since $Z = 0 \Rightarrow W = 0$, in particular.

Finally, Eq. (1.108) enables us to write that

$$V[Z - 3W] = V[Z] + 9 V[W] - 6 \text{Cov}[Z, W] = V\left[X - \frac{1}{2}\right] + 9 V\left[\left(X - \frac{1}{2}\right)^2\right]$$

$$= V[X] + 9\,V[T^2], \quad \text{where } T \sim U\left[-\frac{1}{2}, \frac{1}{2}\right]$$

$$= \frac{1}{12} + 9\left(\frac{1}{180}\right) = \frac{2}{15}$$

because

$$V[T^2] = E[T^4] - (E[T^2])^2 = \int_{-1/2}^{1/2} t^4 \cdot 1\,dt - \left(\int_{-1/2}^{1/2} t^2 \cdot 1\,dt\right)^2$$

$$= \frac{1}{5 \times 16} - \left(\frac{1}{3 \times 4}\right)^2 = \frac{64}{80 \times 144} = \frac{1}{180}$$

We continue with limit theorems that will be used in the subsequent chapters.

Proposition 1.3.8. *Let $X_1, X_2, \ldots$ be an infinite sequence of independent and identically distributed (i.i.d.) r.v.s, and let $S_n := X_1 + \cdots + X_n$.*

a) (**Weak law of large numbers**) *If $E[X_1] = \mu \in \mathbb{R}$, then*

$$\lim_{n \to \infty} P\left[\left|\frac{S_n}{n} - \mu\right| < c\right] = 1. \quad \forall \ c > 0 \tag{1.109}$$

b) (**Strong law of large numbers**) *If $E[X_1^2] < \infty$, then we may write that*

$$P\left[\lim_{n \to \infty} \frac{S_n}{n} = \mu\right] = 1 \tag{1.110}$$

c) (**Central limit theorem**) *If $E[X_1] = \mu \in \mathbb{R}$ and $V[X_1] = \sigma^2 \in (0, \infty)$, then we have*

$$\lim_{n \to \infty} P\left[\frac{S_n - n\mu}{\sqrt{n}\sigma} \leq z\right] = P[N(0,1) \leq z] \tag{1.111}$$

Remarks. i) Actually, the condition $E[X_1^2] < \infty$ is a *sufficient* condition for the strong law of large numbers to hold. It may be replaced by the weaker condition $E[|X_1|] < \infty$, which reduces to $E[X_1] < \infty$ in the case when $X_1 \geq 0$.

ii) The central limit theorem (CLT) implies that

$$S_n \approx N(n\mu, n\sigma^2) \quad \text{and} \quad \frac{S_n}{n} \approx N(\mu, \sigma^2/n) \tag{1.112}$$

In general, from $n = 30$, the approximation by the Gaussian distribution should be rather good.

iii) An application of the CLT is the *Moivre*[12]*–Laplace Gaussian approxima-tion* to a binomial distribution:

$$P[B(n,p) = k] \simeq f_Z(k), \quad \text{where } Z \sim N(np, np(1-p)) \qquad (1.113)$$

For the approximation to be good, the minimum between np and $n(1-p)$ should be at least equal to 5.

Example 1.3.9. If $X_k \sim \text{Exp}(1)$, for $k = 1, \ldots, 30$, and if the X_k's are inde-pendent r.v.s, then we can show that

$$S_{30} := X_1 + \ldots + X_{30} \sim G(30, 1)$$

Making use of the formula

$$P[G(n, \lambda) \leq x] = P[\text{Poi}(\lambda x) \geq n] \qquad (1.114)$$

we obtain (from a table of the distribution function of the Poisson distribution) that

$$P[S_{30} \leq 30] = P[\text{Poi}(30) \geq 30] \simeq 0.5243$$

The approximation by the CLT yields

$$P[S_{30} \leq 30] \simeq P[N(30, 30) \leq 30] = 0.5$$

Example 1.3.10. Suppose that 1% of the tires manufactured by a certain com-pany do not conform to the norms (or are defective). What is the probability that among 1000 tires, there are exactly 10 that do not conform to the norms?

Solution. Let X be the number of tires that do not conform to the norms among the 1000 tires. If we assume that the tires are independent, then X has a binomial distribution with parameters $n = 1000$ and $p = 0.01$. We seek

$$P[X = 10] \simeq f_Z(10), \quad \text{where } Z \sim N(10, 9.9)$$
$$= \frac{1}{\sqrt{2\pi}\sqrt{9.9}} \exp\left\{-\frac{1}{2}\frac{(10-10)^2}{9.9}\right\} \simeq 0.1268$$

Remarks. i) In fact, we obtain that $P[X = 10] \simeq 0.1257$. By using the Poisson approximation (see p. 12), we find that

---

[12] One of the pioneers of the calculus of probabilities, Abraham de Moivre, 1667–1754, was born in France and died in England. The definition of independence of two events can be found in his book *The Doctrine of Chance* published in 1718. The formula attributed to Stirling appeared in a book that he published in 1730. He later used this formula to prove the Gaussian approximation to the binomial distribution.

$$P[X = 10] \simeq P[\text{Poi}(10) = 10] \simeq 0.1251$$

In this example, the Poisson approximation is slightly more accurate. However, if we increase the value of the probability p, the Moivre–Laplace approximation should be better.

ii) To calculate approximately a probability like $P[5 \leq X \leq 10]$, we would rather use the distribution function of the Gaussian distribution. It is then recommended to make a *continuity correction* to improve the approximation. That is, we write that

$$P[5 \leq X \leq 10] = P[4.5 \leq X \leq 10.5] \simeq P[4.5 \leq Z \leq 10.5]$$

1.4 Exercises

Section 1.1

Question no. 1

In urn A, there are four red balls and two white balls, while urn B contains two red balls and four white balls. We throw, *only once*, a coin for which the probability of "tails" is equal to p $(0 < p < 1)$. If we get "tails," then we will draw balls from urn A; otherwise, urn B will be used.

(a) What is the probability of obtaining a red ball on any draw?

(b) If we obtained a red ball on each of the first two draws, what is the probability of obtaining a red ball on the third draw?

(c) If we obtained a red ball on each of the first n draws, what is the probability that we are using urn A?

Question no. 2

Box 1 contains 1000 transistors, of which 100 are defective, and box 2 contains 2000 transistors, of which 100 are also defective. A box is taken at random and two transistors are drawn from it, at random and without replacement.

(a) Calculate the probability that both transistors are defective.

(b) Given that both transistors are defective, what is the probability that they come from box 1?

Question no. 3

Assume that there is a leap year every four years. How many (independent) persons must be in a room, at minimum, if we want the probability that at least one of these persons was born on February 29 to be greater than $1/2$?

Question no. 4

Object O_1 moves on the x-axis between 0 and 2, while object O_2 moves on the y-axis between 0 and 1. Suppose that the position of each object is completely random. What is the probability that the distance between the two objects is greater than 1?

Question no. 5

A certain user of the public transport system can take bus no. N_1 or bus no. N_2 to go to work. A bus no. N_1 runs near his home every hour, from 6:00 a.m., while a bus no. N_2 runs there d minutes after the hour, where $d \in (0, 30]$. The user arrives at the bus stop at a completely random time between 7:45 a.m. and 8:15 a.m. What is the value of d if he takes a bus no. N_1 thrice more often than a bus no. N_2?

Question no. 6

Two sport teams play a series of (independent) games to win a trophy. The first team that wins four games gets the trophy. There are no draws. What is the probability that a team having, for each game, only a one-in-three chance of winning gets the trophy?

Question no. 7

In how many ways can we permute the numbers $1, 2, \ldots, n$ if we do not want a single number to remain in its original position?

Question no. 8

In the dice game called *craps*, the player tosses two (fair) dice simultaneously. If the sum of the two numbers that show up is equal to 7 or 11, the player wins. If the sum is equal to 2, 3, or 12, he loses. When the sum is a number x different from the preceding numbers, the player must toss the two dice again until he gets a sum equal to x or 7. If x is obtained first, the player wins; otherwise, he loses. What is the probability that the player wins?

Question no. 9

A says that B told her that C has lied. If the three persons tell the truth and lie with probability $p \in (0, 1)$, independently from one another, what is the probability that C has indeed lied?

Question no. 10

A man takes part in a television game show. At the end, he is presented with three doors and is asked to choose one among them. The grand prize is hidden, at random, behind one of the doors, while there is nothing behind the other two doors. The game show host knows where the grand prize has been hidden. Suppose that the man has chosen door no. 1 and that the host tells him that he did well in not choosing door no. 3, because there was nothing behind it. He then offers the man the opportunity to change his choice and, therefore, to select door no. 2 instead. What is the probability that the man will win the grand prize if he decides to stick with door no. 1?

Section 1.2

Question no. 11
   Boxes I and II both contain n transistors. At each step, a fair coin is tossed. If "heads" (respectively, "tails") is obtained, we take, at random and without replacement, a transistor in box I (resp., II). We repeat this experiment until one of the two boxes is empty. Let N be the number of transistors that remain in the other box at that moment. If we assume that the repeated trials are independent, what is the probability mass function of N?

Question no. 12
   Let X be a continuous random variable whose density function is given by

$$f_X(x) = c^2 x e^{-cx} \quad \text{for } x > 0$$

where c is a positive constant. Calculate $E[X \mid X < 1]$.

Question no. 13
   A mathematician hesitates between three methods to solve a certain problem. With the first (respectively, second) method he will work in vain for two (resp., three) hours, while the third method will give him the solution at once. If we assume that at each step the mathematician uses a method taken at random among those that he still has not tried, what is the variance of the number of hours that he will have to work to solve his problem?

Question no. 14
   Let X be a random variable whose moment-generating function, $M_X(t)$, exists for $t \in (-c, c)$. Show that

$$P[X \geq a] \leq e^{-at} M_X(t) \quad \text{for } 0 < t < c$$

and

$$P[X \leq a] \leq e^{-at} M_X(t) \quad \text{for } -c < t < 0$$

where a is a real constant.

Question no. 15
   Suppose that the moment-generating function of X exists for every real value of t and is given by

$$M_X(t) = \begin{cases} \dfrac{e^t - e^{-t}}{2t} & \text{if } t \neq 0 \\ 1 & \text{if } t = 0 \end{cases}$$

Use the results of the preceding question to show that

$$P[X \geq 1] = 0 \quad \text{and} \quad P[X \leq -1] = 0$$

Question no. 16
Let X be a continuous random variable whose set of possible values is the interval $[a, b]$. We define $Y = g(X)$.

(a) Calculate the probability density function of Y if $g(x) = 1 - F_X(x)$.

Indication. The inverse distribution function F_X^{-1} exists.

(b) Find a transformation $g(x)$ such that

$$f_Y(y) = \frac{1}{2} \quad \text{for } 1 \leq y \leq 3$$

Question no. 17
Calculate $E[X \mid X > 1]$, where X is a random variable having a standard Gaussian distribution.

Question no. 18
Two players, X and Y, take turns at tossing a fair coin. The first one that gets "tails" wins. Calculate, assuming that X starts,

(a) the probability that X wins,

(b) the probability that X wins, given that she did not obtain "tails" on her first two trials,

(c) the average number of tosses needed to end the game, given that X lost.

Question no. 19
Suppose that the probability that a family has exactly n children is given by

$$P_n = c\, p^n \quad \text{for } n = 1, 2, \ldots$$

where $c > 0$ and $0 < p < 1$, and $P_0 = 1 - \sum_{n=1}^{\infty} P_n$. Suppose also that every child is equally likely to be a male or a female.

(a) Calculate the probability that a family with n children has exactly k male children, for $k = 0, 1, \ldots, n$.

(b) Find the probability that a family has no male children.

(c) What is the average number of male children per family?

Question no. 20
A box contains 200 brand A and 10 brand B transistors. Twenty transistors are taken at random. Let X be the number of brand A transistors obtained.

(a) Calculate $P[X = 20]$, assuming that the transistors are taken without replacement.

(b) Calculate $P[X = 19]$ if the transistors are taken with replacement.

(c) Use a Poisson distribution to calculate approximately $P[X = 18]$ when the transistors are taken with replacement.

Question no. 21

The density function of the random variable X is given by

$$f_X(x) = \begin{cases} k(1-x^2) & \text{if } 0 \le x \le 1 \\ 0 & \text{elsewhere} \end{cases}$$

where k is a positive constant.

(a) Calculate $f_X(x \mid X^2 \le 1/4)$.

(b) Find the constant b that minimizes $E[(X-b)^2]$.

(c) Find the constant c that minimizes $E[|X-c|]$.

Indication. The value x_m for which $F_X(x_m) = 1/2$ is called the *median* of the continuous r.v. X. It can be shown that, for any real constant a,

$$E[|x-a|] = E[|x-x_m|] + 2 \int_a^{x_m} (x-a) f_X(x)\, dx$$

Question no. 22

We say that the continuous random variable X, whose set of possible values is the interval $[0, \infty)$, has a *Pareto*[13] *distribution* with parameter $\theta > 0$ if its density function is of the form

$$f_X(x) = \begin{cases} \dfrac{\theta}{(1+x)^{\theta+1}} & \text{if } x \ge 0 \\ 0 & \text{elsewhere} \end{cases}$$

In economics, the Pareto distribution is used to describe the (unequal) distribution of wealth. Suppose that, in a given country, the wealth X of an individual (in thousands of dollars) has a Pareto distribution with parameter $\theta = 1.2$.

(a) Calculate $f_X(2 \mid 1 < X \le 3)$.

(b) What is the median wealth (see Question no. 21) in this country?

(c) We find that about 11.65% of the population has a personal wealth of at least $5000, which is the average wealth in this population. What percentage of the total wealth of this country does this 11.65% of the population own?

Question no. 23

Let

$$f_X(x) = \begin{cases} k\, x^2 e^{-x^2/2} & \text{if } x \ge 0 \\ 0 & \text{elsewhere} \end{cases}$$

---

[13] Vilfredo Pareto, 1848–1923, born in France and died in Switzerland, was an economist and sociologist. He observed that 20% of the Italian population owned 80% of the wealth of the country, which was generalized by the concept of Pareto distribution.

where k is a positive constant.

(a) Calculate $f_X(x \mid X < 1.282)$.

Indication. We have $P[N(0,1) \leq 1.282] \simeq 0.9$.

(b) Find the value of x_0 for which $F_X(x_0) \simeq 0.35$.

Remark. The random variable X defined above actually has a *Maxwell*[14] *distribution* with parameter $\alpha = 1$. In the general case where $\alpha > 0$, we may write that

$$f_X(x) = \frac{\sqrt{2/\pi}}{\alpha^3} x^2 e^{-x^2/(2\alpha^2)} \quad \text{if } x \geq 0$$

We find that $E[X] = 2\alpha\sqrt{2/\pi}$ and $V[X] = \alpha^2[3 - (8/\pi)]$. This distribution is used in statistical mechanics, in particular, to describe the velocity of molecules in thermal equilibrium.

Question no. 24

Let X be a continuous random variable having the density function

$$f_X(x) = \begin{cases} \dfrac{x}{\theta^2} e^{-x^2/(2\theta^2)} & \text{if } x > 0 \\ 0 & \text{elsewhere} \end{cases}$$

We say that X has a *Rayleigh*[15] *distribution* with parameter $\theta > 0$.

(a) Show that $E[X] = \theta\sqrt{\pi/2}$ and $V[X] = \theta^2[2 - (\pi/2)]$.

(b) Let $Y := \ln X$, where X has a Rayleigh distribution with parameter $\theta = 1$. Calculate (i) $f_Y(1)$ and (ii) the moment-generating function of Y at $t = 2$.

(c) We define $Z = 1/X$. Calculate the mathematical expectation of Z if $\theta = 1$ as in (b).

Section 1.3

Question no. 25

The lifetime X (in days) of a device has an exponential distribution with parameter λ. Moreover, the fraction of time during which the device is used each day has a uniform distribution on the interval $[0, 1]$, independently from one day to another. Let N be the number of complete days during which the device is in a working state.

(a) Show that $P[N \geq n] = (1 - e^{-\lambda})^n/\lambda^n$, for $n = 1, 2, \ldots$.

---

[14] James Clerk Maxwell, 1831–1879, was born in Scotland and died in England. He was a physicist and mathematician who worked in the fields of electricity and magnetism.

[15] John William Strutt Rayleigh, 1842–1919, was born and died in England. He won the Nobel Prize for physics in 1905. The distribution that bears his name is associated with the phenomenon known as *Rayleigh fading* in communication theory.

Indication. Because of the *memoryless property* of the exponential distribution, that is,

$$P[X > s + t \mid X > t] = P[X > s] \quad \forall \, s, t \geq 0$$

it is as if we started anew every day.

(b) Calculate $E[N \mid N \leq 2]$ if $\lambda = 1/10$.

Question no. 26

Let X_1 and X_2 be two independent $N(\mu, \sigma^2)$ random variables. We set $Y_1 = X_1 + X_2$ and $Y_2 = X_1 + 2X_2$.

(a) What is the joint density function of Y_1 and Y_2?

(b) What is the covariance of Y_1 and Y_2?

Question no. 27

Let X_1 and X_2 be two independent random variables. If X_1 has a gamma distribution with parameters $n/2$ and $1/2$, and $Y := X_1 + X_2$ has a gamma distribution with parameters $m/2$ and $1/2$, where $m > n$, what is the distribution of X_2?

Indication. If X has a gamma distribution with parameters α and λ, then (see Table 1.1, p. 19)

$$C_X(\omega) = \frac{\lambda^\alpha}{(\lambda - j\omega)^\alpha}$$

Question no. 28

Show that if $E[(X - Y)^2] = 0$, then $P[X = Y] = 1$, where X and Y are arbitrary random variables.

Question no. 29

The conditional variance of X, given the random variable Y, has been defined (see p. 28) by

$$V[X \mid Y] = E[[X - E[X \mid Y]]^2 \mid Y]$$

Prove the formula (1.96):

$$V[X] = E[V[X \mid Y]] + V[E[X \mid Y]]$$

Question no. 30

Let X be a random variable having a Poisson distribution with parameter Y, where Y is an exponential r.v. with mean equal to 1. Show that $W := X + 1$ has a geometric distribution with parameter $1/2$. That is,

$$p_W(n) = (1/2)^n \quad \text{for } n = 1, 2, \ldots$$

Question no. 31

Let X_1 and X_2 be two independent random variables, both having a standard Gaussian distribution.

(a) Calculate the joint density function of $Y_1 := X_1^2 + X_2^2$ and $Y_2 := X_2$.

(b) What is the marginal density function of Y_1?

Question no. 32

Let X and Y be independent and identically distributed random variables.

(a) Show that $h(z) := E[X \mid X + Y = z] = z/2$.

(b) Evaluate $E[(X - h(Z))^2]$ in terms of $V[X]$.

Question no. 33

A company found out that the quantity X of a certain product it sells during a given time period has the following conditional density function:

$$f_{X|Y}(x \mid y) = \frac{4}{y^2} x e^{-2x/y} \quad \text{if } x > 0$$

where Y is a random variable whose reciprocal $Z := 1/Y$ has a gamma distribution. That is,

$$f_Z(z) = \lambda e^{-\lambda z} \frac{(\lambda z)^{\alpha - 1}}{\Gamma(\alpha)} \quad \text{for } z \geq 0$$

(a) Obtain the marginal density function of X.

(b) Calculate $E[Z \mid X = x]$.

Question no. 34

Let X and Y be continuous and independent random variables. Express the conditional density function of $Z := X + Y$, given that $X = x$, in terms of f_Y.

Question no. 35

Show that for continuous random variables X and Y, we have

$$E[Y \mid X \leq x] = \frac{1}{F_X(x)} \int_{-\infty}^{x} E[Y \mid X = u] f_X(u) \, du$$

if $F_X(x) > 0$.

Question no. 36

Let $X_1, \ldots, X_n$ be independent random variables such that

$$f_{X_k}(x) = \frac{a_k/\pi}{x^2 + a_k^2}$$

for $-\infty < x < \infty$ and $k = 1, 2, \ldots, n$, where $a_k > 0 \; \forall \, k$.

(a) Calculate the density function of the sum $Z := X_1 + \ldots + X_n$.

Indication. The characteristic function of X_k is given by

$$C_{X_k}(\omega) = e^{-a_k|\omega|}$$

(b) Assuming that $a_k = a_1$, for $k = 2, 3, \ldots$, can we state that $f_Z(z)$ tends toward a Gaussian probability density? Justify.

Question no. 37

Suppose that $X \sim N(0, 1)$ and $Y \sim N(1, 1)$ are random variables such that $\rho_{X,Y} = \rho$, where $\rho_{X,Y}$ is the *correlation coefficient* of X and Y. Calculate $E[X^2 Y^2]$.

Question no. 38

Let

$$f_{X,Y}(x, y) = \begin{cases} e^{-y} & \text{if } x > 0, \, y > x \\ 0 & \text{elsewhere} \end{cases}$$

be the joint density function of the random vector (X, Y).

(a) Find the estimator $g(Y)$ of X, in terms of Y, that minimizes the mean-square error MSE $:= E[(X - g(Y))^2]$.

(b) Calculate the minimum mean-square error.

Question no. 39

The joint density function of the random vector (X, Y) is given by

$$f_{X,Y}(x, y) = \begin{cases} \frac{3}{8}(x^2 + xy + y^2) & \text{if } -1 < x < 1 \text{ and } -1 < y < 1 \\ 0 & \text{elsewhere} \end{cases}$$

Calculate (a) $E[X \mid Y = 1/2]$, (b) $E[Y \mid X]$, (c) the mean-square error MSE $:= E[(Y - g(X))^2]$ made by using $g(X) := E[Y \mid X]$ to estimate the random variable Y.

Indication. It can be shown that

$$\text{MSE} = E[Y^2] - E[g^2(X)] \quad \text{if } g(X) = E[Y \mid X]$$

Question no. 40

A number X is taken at random in the interval $[0, 1]$, and then a number Y is taken at random in the interval $[0, X]$. Finally, a number Z is taken at random in the interval $[0, Y]$. Calculate (a) $E[Z]$, (b) $V[Y]$, (c) $P[Z < 1/2]$.

Question no. 41

An angle A is taken at random in the interval $[0, \pi/2]$, so that

$$f_A(a) = \frac{2}{\pi} \quad \text{for } 0 \le a \le \pi/2$$

Let $X := \cos A$ and $Y := \sin A$. Calculate
(a) $P[X = 1 \mid Y = 0]$,   (b) $E[Y \mid X]$,   (c) $E[Y]$,   (d) $E[X \mid X + Y]$,
(e) $E[X^2 \mid X + Y]$,   (f) $E[X \mid A]$,   (g) $P[X = 0 \mid \{X = 0\} \cup \{X = \sqrt{3}/2\}]$.

Indication. We have

$$\frac{d}{dx}\arccos x = -\frac{1}{\sqrt{1-x^2}} \quad \text{for } -1 < x < 1$$

(h) $V[X \mid Y]$ if the angle A is taken at random in the interval $(0, \pi)$.

Question no. 42
Let

$$f_{X|Y}(x \mid y) = ye^{-xy} \quad \text{for } x > 0, 0 < y < 1$$

Calculate, assuming that Y has a uniform distribution on the interval $(0, 1)$,
(a) $f_X(x)$, (b) $P[XY > 1]$, (c) $V[X \mid Y]$, (d) $E[X]$.

Question no. 43
We suppose that the (random) number N of customers that arrive at an automatic teller machine to withdraw money, during a given hour, has a Poisson distribution with parameter $\lambda = 5$. Moreover, the amount X of money withdrawn by an arbitrary customer is a discrete random variable whose probability mass function is given by

$$p_X(x) = \frac{1}{5} \quad \text{if } x = 20, 40, 60, 80, \text{ or } 100$$

Finally, we assume that N and X are independent random variables. Let Y be the total amount of money withdrawn over a one-hour period. Calculate
(a) $E[Y \mid N > 0]$, (b) $P[Y = 60]$, (c) $P[N = 3 \mid Y = 60]$.

Question no. 44
Let X_1 and X_2 be two independent random variables having a uniform distribution on the interval $(0, 1)$. We define $Y = \max\{X_1, X_2\}$. Calculate (a) $F_{Y|X_1}(y \mid x_1)$, (b) $E[Y \mid X_1 = x_1]$, (c) $V[E[Y \mid X_1]]$, (d) $E[V[Y \mid X_1]]$.
Indication. If X is a nonnegative continuous (or mixed type) random variable, then

$$E[X] = \int_0^\infty [1 - F_X(x)]\, dx$$

Question no. 45
We consider a system made up of two components placed in parallel and operating independently. That is, both components operate at the same time, but the system functions if at least one of them is operational. Let T_i be the lifetime of component i, for $i = 1, 2$, and let T be the lifetime of the system. We suppose that $T_i \sim \text{Exp}(1/2)$, for $i = 1, 2$. Calculate
(a) $E[T \mid T_1 = 1]$, (b) $E[T \mid T_1 > 1]$, (c) $E[T \mid \{T_1 > 1\} \cup \{T_2 > 1\}]$.
Indication. In (c), we can use the formula

$$E[X \mid A \cup B] = E[X \mid A]\, P[A \mid A \cup B] + E[X \mid B]\, P[B \mid A \cup B]$$
$$- E[X \mid A \cap B]\, P[A \cap B \mid A \cup B]$$

Question no. 46

Let $X_1 \sim U[-1,1]$, $X_2 \sim U[0,1]$, and $X_3 \sim U[0,2]$ be independent random variables. Calculate

(a) $P[X_1 < X_2]$,   (b) $P[X_1 < X_2 < X_3]$,   (c) $E\left[\dfrac{X_1}{X_2+1}\right]$,   (d) $E[YZ \mid X_1]$,

where $Y := X_1 + X_2$ and $Z := X_1 + X_3$.

Question no. 47

Suppose that X_1 and X_2 are independent random variables such that

$$f_{X_i}(x_i) = \tfrac{1}{2}\lambda_i e^{-\lambda_i |x_i|} \quad \text{for } x_i \in \mathbb{R}$$

where λ_i is a positive constant, for $i = 1, 2$. Calculate

(a) $P[X_1 < X_2]$,   (b) $V[X_1 \mid X_1 > 0]$,   (c) $E[|X_1| \mid |X_1| > 1]$,
(d) $E[X_1 + X_2 \mid X_1 < X_2]$ if $\lambda_1 = \lambda_2$.

Question no. 48

Calculate $P[Y \geq X]$ if $X \sim B(2, 1/2)$ and $Y \sim \mathrm{Poi}(1)$ are two independent random variables.

Question no. 49

Use the central limit theorem to calculate (approximately)

$$P[X_1 + \ldots + X_{40} < X_{41} + \ldots + X_{100}]$$

where $X_1, \ldots, X_{100}$ are independent random variables, each having a $U[0,1]$ distribution.

Question no. 50

Suppose that the random variables $X_1, \ldots, X_{30}$ are independent and all have the probability density function

$$f_X(x) = \frac{1}{x} \quad \text{for } 1 \leq x \leq e$$

What is the approximate density function of the product $X_1 X_2 \cdots X_{30}$?

Question no. 51

Let (X, Y) be a random vector having a bivariate normal distribution.

(a) Calculate $P[XY < 0]$ if $\mu_X = 0$, $\mu_Y = 0$, $\sigma_X^2 = 1$, $\sigma_Y^2 = 4$, and $\rho = 0$.
(b) What is the best estimator of X^2 in terms of Y when $\mu_X = 0$, $\mu_Y = 0$, $\sigma_X^2 = 1$, $\sigma_Y^2 = 1$, and $\rho = 0$?
(c) Calculate $E[XY]$ when $\mu_X = 1$, $\mu_Y = 2$, $\sigma_X^2 = 1$, $\sigma_Y^2 = 4$, and $\rho = 1/2$.

Question no. 52

Let X and Y be two random variables, and let g and h be real-valued functions. Show that

(a) $E[g(X) \mid X] = g(X)$,

(b) $E[g(X)h(Y)] = E[h(Y)E[g(X) \mid Y]]$.

Question no. 53

Letters are generated at random (among the 26 letters of the alphabet) until the word "me" has been formed, in this order, with the two most recent letters. Let N be the total number of letters that will have to be generated to end the random experiment, and let X_k be the kth generated letter. It can be shown that $E[N] = 676$ and $V[N] = 454{,}948$. Calculate (a) $E[N \mid X_2 = e]$, (b) $E[N \mid X_1 = m]$, and (c) $E[N^2 \mid X_1 = m]$.

Remark. The variable X_k is not a random variable in the strict sense of the term, because its possible values are not real numbers. We can say that it is an example of a *qualitative* (rather than *quantitative*) variable. It could easily be transformed into a *real* random variable by defining X_k instead to be equal to j if the kth generated letter is the jth letter of the alphabet, for $j = 1, \ldots, 26$.

Question no. 54

Let X_i, for $i = 1, 2, 3$, be independent random variables, each having a uniform distribution on the interval $(0, 1)$. Calculate

(a) $E[X_1 + X_2 + X_3 \mid X_1 + X_2]$,

(b) $E\left[X_1 + \frac{1}{2}X_2 \mid X_1 + X_2 + X_3\right]$,

(c) $E[V[X_1 \mid X_1 + X_2 + X_3]]$.

2

Stochastic Processes

2.1 Introduction and definitions

Definition 2.1.1. *Suppose that with each element s of a sample space S of some random experiment E, we associate a function $X(t, s)$, where t belongs to $T \subset \mathbb{R}$. The set $\{X(t, s), t \in T\}$ is called a* **stochastic** *(or* **random***)* **process**.

Remarks. i) The function $X(t, s)$ is a random variable for any particular value of t.

ii) In this book, the set T will generally be the set $\mathbb{N}^0 = \{0, 1, \dots\}$ or the interval $[0, \infty)$.

Classification of the stochastic processes

We consider the case when T is either a countably infinite set or an uncountably infinite set. Moreover, the set of possible values of the random variables $X(t, s)$ can be *discrete* (that is, finite or countably infinite) or *continuous* (that is, uncountably infinite). Consequently, there are four different types of stochastic processes (s.p.).

Definition 2.1.2. *If T is a countably infinite set (respectively, an interval or a set of intervals), then $\{X(t, s), t \in T\}$ is said to be a* **discrete-time** *(resp.,* **continuous-time***) stochastic process.*

Remarks. i) Except in Section 2.3, it will not be necessary to write explicitly the argument s of the function $X(t, s)$. Thus, the stochastic process will be denoted by $\{X(t), t \in T\}$. However, in the discrete case, it is customary to write $\{X_n, n \in T\}$.

ii) We will not consider in this book the case when T is the union of a set of points and of an uncountably infinite set.

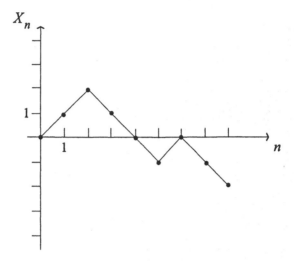

Fig. 2.1. Example of a random walk.

Example 2.1.1. A classic example of a stochastic process is the one where we consider a particle that, at time 0, is at the origin. At each time unit, a coin is tossed. If "tails" (respectively, "heads") is obtained, the particle moves one unit to the right (resp., left) (see Fig. 2.1). Thus, the random variable X_n denotes the position of the particle after n tosses of the coin, and the s.p. $\{X_n, n = 0, 1, \ldots\}$ is a particular *random walk* (see Chapter 3). Note that here the index n can simply denote the toss number (or the number of times the coin has been tossed) and it is not necessary to introduce the notion of *time* in this example.

Example 2.1.2. An elementary continuous-time s.p., $\{X(t), t \geq 0\}$, is obtained by defining

$$X(t) = Yt \quad \text{for } t \geq 0$$

where Y is a random variable having an arbitrary distribution.

Definition 2.1.3. *The set $S_{X(t)}$ of values that the r.v.s $X(t)$ can take is called the* **state space** *of the stochastic process $\{X(t), t \in T\}$. If $S_{X(t)}$ is finite or countably infinite (respectively, uncountably infinite), $\{X(t), t \in T\}$ is said to be a* **discrete-state** *(resp.,* **continuous-state***) process.*

Example 2.1.3. The random walk in Example 2.1.1 is a discrete-time and discrete-state s.p., since $S_{X_n} = \{0, \pm1, \pm2, \ldots\}$. For the continuous time s.p. in Example 2.1.2, it is a continuous-state process, unless Y takes on the value 0, because $S_{X(t)} = [0, \infty)$ if $Y > 0$ and $S_{X(t)} = (-\infty, 0]$ if $Y < 0$.

As we mentioned above, for any fixed value of t, we obtain a random variable $X(t)$ $(= X(t, s))$. Although many authors use the notation $X(t)$ to designate the stochastic process itself, we prefer to use the notation $\{X(t), t \in T\}$ to avoid the possible confusion between the s.p. and the random variable.

Definition 2.1.4. *The* **distribution function of order** k *of the stochastic process* $\{X(t), t \in T\}$ *is the joint distribution function of the random vector* $(X(t_1), \dots, X(t_k))$:

$$F(x_1, \dots, x_k; t_1, \dots, t_k) = P[X(t_1) \leq x_1, \dots, X(t_k) \leq x_k] \qquad (2.1)$$

Similarly, we define the **probability mass and density functions of order** k *of an s.p.:*

$$p(x_1, \dots, x_k; n_1, \dots, n_k) = P[X_{n_1} = x_1, \dots, X_{n_k} = x_k] \qquad (2.2)$$

and (where the derivative exists)

$$f(x_1, \dots, x_k; t_1, \dots, t_k) = \frac{\partial^k}{\partial x_1 \dots \partial x_k} F(x_1, \dots, x_k; t_1, \dots, t_k) \qquad (2.3)$$

Remark. When $k = 1$ or 2, the preceding definitions are in fact only new notations for functions already defined in Sections 1.2 and 1.3.

Example 2.1.4. If the tosses of the coin are independent in Example 2.1.1, then we may write, with $p := P[\{\text{Tails}\}]$, that the first-order probability mass function (or probability mass function of order 1) of the process at time $n = 2$ is given by

$$p(x; n = 2) \equiv P[X_2 = x] = \begin{cases} 2p(1-p) & \text{if } x = 0 \\ p^2 & \text{if } x = 2 \\ (1-p)^2 & \text{if } x = -2 \\ 0 & \text{otherwise} \end{cases}$$

First- and second-order moments of stochastic processes

Just like the means, variances, and covariances enable us to characterize, at least partially, random variables and vectors, we can also characterize a stochastic process with the help of its moments.

Definition 2.1.5. *The* **mean** $E[X(t)]$ *of an s.p.* $\{X(t), t \in T\}$ *at time t is denoted by* $m_X(t)$. *Moreover, the* **autocorrelation function** *and the* **autocovariance function** *of the process at the point* (t_1, t_2) *are defined, respectively, by*

$$R_X(t_1, t_2) = E[X(t_1)X(t_2)] \qquad (2.4)$$

and

$$C_X(t_1, t_2) = R_X(t_1, t_2) - m_X(t_1)m_X(t_2) \qquad (2.5)$$

Finally, the **correlation coefficient** *of the s.p. at the point* (t_1, t_2) *is*

$$\rho_X(t_1, t_2) = \frac{C_X(t_1, t_2)}{[C_X(t_1, t_1)C_X(t_2, t_2)]^{1/2}} \qquad (2.6)$$

Remarks. i) In the case of pairs of random variables, the quantity $E[XY]$ is called the *correlation* of the vector (X, Y). Here, we use the prefix "auto" because the function is calculated for two values of the same stochastic process $\{X(t), t \in T\}$. The function $R_{X,Y}(t_1, t_2) := E[X(t_1)Y(t_2)]$, where $\{Y(t), t \in T^*\}$ is another s.p., is named the *cross-correlation function*, etc. In fact, we could simply use the term *correlation function* in the case of the function $R_X(t_1, t_2)$.

ii) The function $R_X(t, t) = E[X^2(t)]$ is called the *average power* of the stochastic process $\{X(t), t \in T\}$. Furthermore, the *variance* of the process at time t is

$$V[X(t)] = C_X(t, t) \qquad (2.7)$$

Since $V[X(t)] \geq 0$, we then deduce from Eq. (2.6) that $\rho_X(t, t) = 1$.

Two properties of stochastic processes that will be assumed to hold true in the definition of the Wiener[1] process (see Chapter 4) and of the Poisson[2] process (see Chapter 5), in particular, are given in the following definitions.

Definition 2.1.6. *If the random variables $X(t_4) - X(t_3)$ and $X(t_2) - X(t_1)$ are* independent $\forall \ t_1 < t_2 \leq t_3 < t_4$, *we say that the stochastic process $\{X(t), t \in T\}$ is a process with* **independent increments**.

Definition 2.1.7. *If the random variables $X(t_2 + s) - X(t_1 + s)$ and $X(t_2) - X(t_1)$ have the same distribution function for all s, $\{X(t), t \in T\}$ is said to be a process with* **stationary increments**.

Remarks. i) The random variables $X(t_2 + s) - X(t_1 + s)$ and $X(t_2) - X(t_1)$ in the preceding definition are *identically distributed*. However, in general, they are *not* equal.

ii) The Poisson process is a process that counts the number of *events*, for instance, the arrival of customers or of phone calls, that occurred in the interval $[0, t]$. By assuming that this process possesses these two properties, we take for granted that the r.v.s that designate the number of events in disjoint intervals are independent, and that the distribution of the number of events in a given interval depends only on the length of this interval. In practice, we can doubt these assertions. For example, the fact of having had many (or

---

[1] Norbert Wiener, 1894–1964, was born in the United States and died in Sweden. He obtained his Ph.D. in philosophy from Harvard University at the age of 18. His research subject was *mathematical logic*. After a stay in Europe to study mathematics, he started working at the Massachusetts Institute of Technology, where he did some research on *Brownian motion*. He contributed, in particular, to communication theory and to control. In fact, he is the inventor of *cybernetics*, which is the "science of communication and control in the animal and the machine."

[2] See p. 12.

very few) customers on a given morning should give us some indication about how the rest of the day will unfold. Similarly, the *arrival rate* of customers at a store is generally not constant in time. There are rush periods and slack periods that occur at about the same hours day after day. Nevertheless, these simplifying assumptions enable us, for instance, to obtain explicit answers to problems in the theory of queues. Without these assumptions, it would be very difficult to calculate many quantities of interest exactly.

Example 2.1.5. *Independent* trials for which the probability of *success* is the same for each of these trials are called *Bernoulli trials*. For example, we can roll some die independently an indefinite number of times and define a success as being the rolling of a "6."

A *Bernoulli process* is a sequence $X_1, X_2, \ldots$ of Bernoulli r.v.s associated with Bernoulli trials. That is, $X_k = 1$ if the kth trial is a success and $X_k = 0$ otherwise. We easily calculate

$$E[X_k] = p \quad \forall \, k \in \{1, 2, \ldots\}$$

where p is the probability of a success, and

$$R_X(k_1, k_2) \equiv E[X_{k_1} X_{k_2}] = \begin{cases} p^2 & \text{if } k_1 \neq k_2 \\ p & \text{if } k_1 = k_2 \end{cases}$$

It follows that $C_X(k_1, k_2) = 0$ if $k_1 \neq k_2$ and $C_X(k_1, k_2) = p(1-p)$ if $k_1 = k_2$.

Example 2.1.6. Let Y be a random variable having a $U(0, 1)$ distribution. We define the stochastic process $\{X(t), t \geq 0\}$ by

$$X(t) = e^Y t \quad \text{for } t \geq 0$$

The first-order density function of the process can be obtained by using Proposition 1.2.2 (see Example 1.2.6):

$$f(x; t) \equiv f_{X(t)}(x) = f_Y(\ln(x/t)) \left| \frac{d \ln(x/t)}{dx} \right|$$

$$= 1 \cdot \left| \frac{1}{x} \right| = \frac{1}{x} \quad \text{if } x \in (t, te)$$

Next, the mean $E[X(t)]$ of the process at time $t \geq 0$ is given by

$$E[X(t)] = \int_0^1 e^y t \cdot 1 \, dy = t(e - 1) \quad \text{for } t \geq 0$$

or, equivalently, by

$$E[X(t)] = \int_t^{te} x \cdot \frac{1}{x} \, dx = te - t = t(e - 1) \quad \text{for } t \geq 0$$

Finally, we have

$$X(t)X(t+s) = e^Y t \cdot e^Y \, (t+s) = e^{2Y} \, t(t+s)$$

It follows that

$$R_X(t, t+s) \equiv E[X(t)X(t+s)] = E[e^{2Y} t(t+s)] = t(t+s)\frac{e^2 - 1}{2} \quad \forall \; s, t \geq 0$$

2.2 Stationarity

Definition 2.2.1. *We say that the stochastic process* $\{X(t), t \in T\}$ *is* **stationary**, *or* **strict-sense stationary** *(SSS), if its distribution function of order n is invariant under any change of origin:*

$$F(x_1, \ldots, x_n; t_1, \ldots, t_n) = F(x_1, \ldots, x_n; t_1 + s, \ldots, t_n + s) \tag{2.8}$$

for all s, n, *and* $t_1, \ldots, t_n$.

Remark. The value of s in the preceding definition must be chosen so that $t_k + s \in T$, for $k = 1, \ldots, n$. So, if $T = [0, \infty)$, for instance, then $t_k + s$ must be nonnegative for all k.

In practice, it is difficult to show that a given stochastic process is stationary in the strict sense (except in the case of *Gaussian* processes, as will be seen in Section 2.4). Consequently, we often satisfy ourselves with a weaker version of the notion of stationarity, by considering only the cases where $n = 1$ and $n = 2$ in Definition 2.2.1.

If $\{X(t), t \in T\}$ is a (continuous) SSS process, then we may write that

$$f(x; t) = f(x; t+s) \quad \forall \; s, t \tag{2.9}$$

and

$$f(x_1, x_2; t_1, t_2) = f(x_1, x_2; t_1 + s, t_2 + s) \quad \forall \; s, t_1, t_2 \tag{2.10}$$

We deduce from Eq. (2.9) that the first-order density function of the process must actually be independent of t:

$$f(x; t) = f(x) \quad \forall \; t \tag{2.11}$$

Moreover, Eq. (2.10) implies that it is not necessary to know explicitly the values of t_1 and t_2 to be able to evaluate $f(x_1, x_2; t_1, t_2)$. It is sufficient to know the difference $t_2 - t_1$:

$$f(x_1, x_2; t_1, t_2) = f(x_1, x_2; t_2 - t_1) \quad \forall \; t_1, t_2 \tag{2.12}$$

In terms of the moments of the process, Eqs. (2.11) and (2.12) imply that $m_X(t)$ is a constant and that the autocorrelation function $R_X(t_1, t_2)$ is, in fact, a function R_X^* of a single variable: $R_X(t_1, t_2) = R_X^*(t_2 - t_1)$. By abuse of notation, we simply write that $R_X(t_1, t_2) = R_X(t_2 - t_1)$.

Definition 2.2.2. *We say that the stochastic process $\{X(t), t \in T\}$ is* **wide-sense stationary** *(WSS) if $m_X(t) \equiv m$ and*

$$R_X(t_1, t_2) = R_X(t_2 - t_1) \quad \forall\, t_1, t_2 \in T \tag{2.13}$$

Remarks. i) Since $m_X(t) \equiv m$ if $\{X(t), t \in T\}$ is wide-sense stationary, we can also write that

$$C_X(t_1, t_2) \equiv R_X(t_1, t_2) - m_X(t_1)m_X(t_2) = R_X(t_2 - t_1) - m^2 = C_X(t_2 - t_1) \tag{2.14}$$

$\forall\, t_1, t_2 \in T$. Similarly, we have

$$\rho_X(t_1, t_2) = \rho_X(t_2 - t_1) = \frac{C_X(t_2 - t_1)}{C_X(0)} \tag{2.15}$$

ii) By choosing $t_1 = t_2 = t$, we obtain that $E[X^2(t)] = R_X(t, t) = R_X(0)$, for all $t \in T$. Therefore, the *average power* of a WSS s.p. does *not* depend on t.

iii) We often take $t_1 = t$ and $t_2 = t + s$ when we calculate the function R_X (or C_X). If the process considered is WSS, then the function obtained depends only on s.

iv) It is clear that an SSS stochastic process is also WSS. We will see in Section 2.4 that, in the case of *Gaussian* processes, the converse is true as well.

Example 2.2.1. The most important continuous-time and continuous-state stochastic process is the *Wiener process*, $\{W(t), t \geq 0\}$, which will be the subject (in part) of Chapter 4. We will see that $E[W(t)] \equiv 0$ and that

$$C_W(t, t + s) = R_W(t, t + s) = \sigma^2 t$$

where $\sigma > 0$ is a constant and $s, t \geq 0$. Since the function $R_W(t, t+s)$ depends on t (rather than on s), the Wiener process is not wide-sense stationary.

Example 2.2.2. The Poisson process, that we denote by $\{N(t), t \geq 0\}$ and that will be studied in detail in Chapter 5, possesses the following characteristics:

$$E[N(t)] = \lambda t \quad \text{and} \quad R_N(t_1, t_2) = \lambda \min\{t_1, t_2\}$$

for all t, t_1, and $t_2 \geq 0$, where λ is a positive constant. It is therefore not stationary (not even in the wide sense), because its mean depends on t. If we define the stochastic process $\{X(t), t > 0\}$ by

$$X(t) = \frac{N(t)}{t} \quad \text{for } t > 0$$

then the mean of the process is a constant. However, we find that we cannot write that $R_X(t_1, t_2) = R_X(t_2 - t_1)$. Consequently, this process is not stationary either.

Remark. By definition, the Wiener and Poisson processes have *stationary increments*. However, as we have just seen, they are not even wide-sense stationary. Therefore, these two notions must not be confused.

Example 2.2.3. An elementary example of a strict-sense stationary stochastic process is obtained by setting

$$X(t) = Y \quad \text{for } t \geq 0$$

where Y is an arbitrary random variable. Since $X(t)$ does not depend on the variable t, the process $\{X(t), t \geq 0\}$ necessarily satisfies Eq. (2.8) in Definition 2.2.1.

Definition 2.2.3. *The* **spectral density** *of a wide-sense stationary stochastic process,* $\{X(t), t \in T\}$, *is the* Fourier transform $S_X(\omega)$ *of its autocorrelation function:*

$$S_X(\omega) = \int_{-\infty}^{\infty} e^{-j\omega s} R_X(s) \, ds \tag{2.16}$$

Remarks. i) Inverting the Fourier transform, we obtain that

$$R_X(s) = \frac{1}{2\pi} \int_{-\infty}^{\infty} e^{j\omega s} S_X(\omega) \, d\omega \tag{2.17}$$

ii) Since the autocorrelation function of a WSS process is an even function (that is, $R_X(-s) = R_X(s)$), the spectral density $S_X(\omega)$ is a *real* and *even* function. We can then write that

$$S_X(\omega) = 2 \int_0^{\infty} R_X(s) \cos \omega s \, ds \quad \text{and} \quad R_X(s) = \frac{1}{\pi} \int_0^{\infty} S_X(\omega) \cos \omega s \, d\omega \tag{2.18}$$

iii) It can be shown (the *Wiener–Khintchin*[3] *theorem*) that the spectral density $S_X(\omega)$ is a *nonnegative* function. Actually, a function $S_X(\omega)$ is a spectral density if and only if it is nonnegative.

Suppose now that the following relation holds between the processes $\{X(t), t \in T\}$ and $\{Y(t), t \in T\}$:

$$Y(t) = X(t) * h(t) \equiv \int_{-\infty}^{\infty} X(t-s)h(s) \, ds \tag{2.19}$$

---

[3] Aleksandr Yakovlevich Khintchin, 1894–1959, was born and died in Russia. He contributed in a very important way to the development of the theory of stochastic processes. He was also interested in statistical mechanics and in information theory.

Remark. We can interpret the process $\{Y(t), t \in T\}$ as being the *output* of a *linear system* whose *input* is the process $\{X(t), t \in T\}$. We write $Y(t) = L[X(t)]$.

We assume again that $\{X(t), t \in T\}$ is stationary in the wide sense. If $E[X(t)] \equiv 0$, we can show that $\{Y(t), t \in T\}$ is a WSS process with zero mean and such that

$$S_Y(\omega) = S_X(\omega)|H(\omega)|^2 \tag{2.20}$$

where

$$H(\omega) := \int_{-\infty}^{\infty} e^{-j\omega s} h(s)\, ds \tag{2.21}$$

We also have

$$E[Y^2(t)] = R_Y(0) = \frac{1}{2\pi} \int_{-\infty}^{\infty} S_X(\omega)|H(\omega)|^2\, d\omega \quad (\geq 0) \tag{2.22}$$

2.3 Ergodicity

In *statistics*, to estimate an unknown parameter of a distribution function, for example, the parameter λ of an r.v. X having a Poi(λ) distribution, we draw a *random sample* of X. That is, we take n *observations*, $X_1, \ldots, X_n$, of X and we assume that the X_k's have the same distribution function as X and are independent. Next, we write that the estimator $\hat{\lambda}$ of λ (which is the mean of the distribution) is the arithmetic mean of the observations. Similarly, to estimate the mean $m_X(t)$ of a stochastic process $\{X(t), t \in T\}$ at time t, we must first take *observations* $X(t, s_k)$ of the process. Next, we define

$$\widehat{m_X(t)} = \frac{1}{n} \sum_{k=1}^{n} X(t, s_k) \tag{2.23}$$

Thus, we estimate the mean $m_X(t)$ of the s.p. by the mean of a random sample taken at time t. Of course, the more observations of the process at time t we have, the more precise the estimator $\widehat{m_X(t)}$ should be. Suppose, however, that we only have a *single* observation, $X(t, s_1)$, of $X(t)$. Since we cannot estimate $m_X(t)$ in a reasonable way from a single observation, we would like to use the values of the process for the other values of t to estimate $m_X(t)$. For this to be possible, it is necessary (but not sufficient) that the mean $m_X(t)$ be independent of t.

Definition 2.3.1. *The* **temporal mean** *of the s.p.* $\{X(t), t \in \mathbb{R}\}$ *is defined by*

$$\langle X(t) \rangle_S = \frac{1}{2S} \int_{-S}^{S} X(t, s)\, dt \tag{2.24}$$

Remarks. i) In this section, we will suppose that the set T is the entire real line. If $T = [0, \infty)$, for example, we can modify the definition above. Moreover, in the discrete case, the integral is replaced by a sum. Thus, when $T = \{0, \pm 1, \pm 2, \dots\}$, we can write that

$$\langle X_n \rangle_N := \frac{1}{2N+1} \sum_{k=-N}^{N} X_{n_k}(s) \tag{2.25}$$

where N is a natural number.

ii) We call a *realization* or *trajectory* of the process $\{X(t), t \in T\}$ the graph of $X(t, s)$ as a function of t, for a fixed value of s.

Definition 2.3.2. *The stochastic process $\{X(t), t \in T\}$ is said to be* **ergodic** *if any characteristic of the process can be obtained, with probability 1, from a single realization $X(t, s)$ of the process.*

A stochastic process can be, in particular, *mean ergodic, distribution ergodic* (that is, ergodic with respect to the distribution function), *correlation ergodic* (with respect to the correlation function), etc. In this book, we will limit ourselves to the most important case, namely, the one where the process $\{X(t), t \in T\}$ is mean ergodic (see also p. 72).

Definition 2.3.3. *An s.p. $\{X(t), t \in T\}$ for which $m_X(t) = m \; \forall \; t \in T$ is called* **mean ergodic** *if*

$$P\left[\lim_{S \to \infty} \langle X(t) \rangle_S = m \right] = 1 \tag{2.26}$$

Now, the temporal mean, $\langle X(t) \rangle_S$, is a random variable. Since

$$E[\langle X(t) \rangle_S] = \frac{1}{2S} \int_{-S}^{S} E[X(t, s)]\, dt = \frac{1}{2S} \int_{-S}^{S} m\, dt = m \tag{2.27}$$

we can state that Eq. (2.26) in the definition above will be satisfied if the variance $V[\langle X(t) \rangle_S]$ of the temporal mean decreases to 0 when S tends to infinity. Indeed, if $\lim_{S \to \infty} V[\langle X(t) \rangle_S] = 0$, then $\langle X(t) \rangle_S$ converges to its mean value when S tends to infinity. To calculate the variance $V[\langle X(t) \rangle_S]$, we can use the following proposition.

Proposition 2.3.1. *The variance of the temporal mean of the stochastic process $\{X(t), t \in T\}$ is given by*

$$V[\langle X(t) \rangle_S] = \frac{1}{4S^2} \int_{-S}^{S} \int_{-S}^{S} C_X(t_1, t_2)\, dt_1 dt_2 \tag{2.28}$$

Corollary 2.3.1. *If the process* $\{X(t), t \in T\}$ *is wide-sense stationary, then we may write that*

$$V[\langle X(t)\rangle_S] = \frac{1}{2S} \int_{-2S}^{2S} C_X(s) \left[1 - \frac{|s|}{2S}\right] ds \qquad (2.29)$$

Often, it is not necessary to calculate $V[\langle X(t)\rangle_S]$ to determine whether the stochastic process $\{X(t), t \in T\}$ is mean ergodic. For example, when the process is WSS, we can use either of the sufficient conditions given in the following proposition.

Proposition 2.3.2. *The WSS s.p.* $\{X(t), t \in T\}$ *is mean ergodic if*

$$C_X(0) < \infty \quad and \quad \lim_{|s|\to\infty} C_X(s) = 0 \qquad (2.30)$$

or if its autocovariance function $C_X(s)$ *is absolutely integrable, that is, if*

$$\int_{-\infty}^{\infty} |C_X(s)| \, ds < \infty \qquad (2.31)$$

Example 2.3.1. The elementary stochastic process defined in Example 2.2.3 is strict-sense stationary. However, it is not mean ergodic. Indeed, we have

$$\langle X(t)\rangle_S = \frac{1}{2S} \int_{-S}^{S} Y \, dt = Y$$

Thus, we may write that $V[\langle X(t)\rangle_S] = V[Y]$. Now, if Y is not a constant, the variance $V[Y]$ is strictly positive and does not decrease to 0 when S tends to infinity (since $V[Y]$ does not depend on S). Therefore, an arbitrary stochastic process can be mean ergodic without even being WSS (provided that $m_X(t)$ is a constant), and a strict-sense stationary process is not necessarily mean ergodic.

Example 2.3.2. As will be seen in Chapter 5, the s.p. $\{X(t), t \in T\}$ called the *random telegraph signal*, which is defined from a Poisson process, is zero mean and its autocovariance function is given by

$$C_X(s) = e^{-2\lambda|s|}$$

where λ is a positive constant. Using the conditions in Eq. (2.30), we can state that the process is mean ergodic. Indeed, we have

$$C_X(0) = 1 < \infty \quad and \quad \lim_{|s|\to\infty} C_X(s) = \lim_{|s|\to\infty} e^{-2\lambda|s|} = 0$$

Actually, we also have

$$\int_{-\infty}^{\infty} |C_X(s)|\, ds = 2 \int_{0}^{\infty} e^{-2\lambda s}\, ds = \frac{1}{\lambda} < \infty$$

Remarks. i) It is important to remember that the conditions in Proposition 2.3.2 are *sufficient*, but not necessary, conditions. Consequently, if we cannot show that the process considered is mean ergodic by making use of this proposition, then we must calculate the variance of the temporal mean and check whether it decreases to 0 or not with $S \to \infty$.

ii) It can be shown that the random telegraph signal is an example of a strict-sense stationary stochastic process. Therefore, we can calculate the variance $V[\langle X(t)\rangle_S]$ by using the formula (2.29):

$$V[\langle X(t)\rangle_S] = \frac{2}{2S} \int_{0}^{2S} e^{-2\lambda s} \left[1 - \frac{s}{2S}\right] ds$$

The integral above is not difficult to evaluate. However, here, it is not even necessary to calculate it explicitly. It is sufficient to replace the expression between the square brackets by 1, because this expression is comprised of values between 0 and 1 when s varies from 0 to $2S$. It follows that

$$V[\langle X(t)\rangle_S] < \frac{1 - e^{-4\lambda S}}{2\lambda S}$$

Finally, we have

$$\lim_{S\to\infty} V[\langle X(t)\rangle_S] \leq \lim_{S\to\infty} \frac{1 - e^{-4\lambda S}}{2\lambda S} = 0$$

Since the variance $V[\langle X(t)\rangle_S]$ is nonnegative, we can conclude that

$$\lim_{S\to\infty} V[\langle X(t)\rangle_S] = 0,$$

which confirms the fact that the random telegraph signal is mean ergodic.

2.4 Gaussian and Markovian processes

The *bivariate normal distribution* was defined in Example 1.3.6. The generalization of this distribution to the n-dimensional case is named the *multinormal distribution*.

Definition 2.4.1. *We say that the random vector $(X_1, \ldots, X_n)$ has a **multinormal distribution** if each random variable X_k can be expressed as a linear combination of independent random variables $Z_1, \ldots, Z_m$, where $Z_j \sim N(0,1)$, for $j = 1, \ldots, m$. That is, if*

$$X_k = \mu_k + \sum_{j=1}^{m} c_{kj} Z_j \quad \text{for } k = 1, \ldots, n \tag{2.32}$$

where μ_k is a real constant, for all k.

Just as a Gaussian distribution $N(\mu, \sigma^2)$ is completely determined by its mean μ and its variance σ^2, and a bivariate normal distribution by its parameters μ_X, μ_Y, σ_X^2, σ_Y^2 and ρ, the joint density function of the random vector $\mathbf{X} = (X_1, \ldots, X_n)$ is completely determined by the vector of means $\mathbf{m} := (\mu_{X_1}, \ldots, \mu_{X_n})$ and the *covariance matrix* $\mathbf{K}$, where

$$
\mathbf{K} := \begin{bmatrix} V[X_1] & \mathrm{Cov}[X_1, X_2] & \ldots & \mathrm{Cov}[X_1, X_n] \\ \mathrm{Cov}[X_2, X_1] & V[X_2] & \ldots & \mathrm{Cov}[X_2, X_n] \\ \ldots & & \ldots & \ldots \\ \mathrm{Cov}[X_n, X_1] & \mathrm{Cov}[X_n, X_2] & \ldots & V[X_n] \end{bmatrix} \tag{2.33}
$$

By analogy with the one-dimensional case, we write that $\mathbf{X} \sim N(\mathbf{m}, \mathbf{K})$.

The matrix $\mathbf{K}$ is symmetrical, because $\mathrm{Cov}[X, Y] = \mathrm{Cov}[Y, X]$, and non-negative definite:

$$
\sum_{i=1}^{n} \sum_{k=1}^{n} c_i c_k \mathrm{Cov}[X_i, X_k] \geq 0 \quad \forall \, c_i, c_k \in \mathbb{R} \tag{2.34}
$$

If, in addition, it is nonsingular, then we may write that

$$
f_{\mathbf{X}}(\mathbf{x}) = \frac{1}{(2\pi)^{n/2}} \frac{1}{(\det \mathbf{K})^{1/2}} \exp \left\{ -\frac{1}{2}(\mathbf{x} - \mathbf{m}) \mathbf{K}^{-1} (\mathbf{x}^T - \mathbf{m}^T) \right\} \tag{2.35}
$$

for $\mathbf{x} := (x_1, \ldots, x_n) \in \mathbb{R}^n$, where "det" denotes the determinant and T denotes the transpose of the vector.

Proposition 2.4.1. *Let* $\mathbf{X} \sim N(\mathbf{m}, \mathbf{K})$. *The* joint characteristic function *of the r.v.* $\mathbf{X}$:

$$
\phi_{\mathbf{X}}(\omega_1, \ldots, \omega_n) := E\left[\exp\{j(\omega_1 X_1 + \ldots + \omega_n X_n)\}\right] \tag{2.36}
$$

is given by

$$
\phi_{\mathbf{X}}(\omega) = \exp\left\{ j \sum_{i=1}^{n} \mu_{X_i} \omega_i - \frac{1}{2} \sum_{i=1}^{n} \sum_{k=1}^{n} \sigma_{ik} \omega_i \omega_k \right\} = \exp\left\{ j\mathbf{m}\omega^T - \frac{1}{2}\omega\mathbf{K}\omega^T \right\} \tag{2.37}
$$

where $\sigma_{ik} := \mathrm{Cov}[X_i, X_k]$ *and* $\omega := (\omega_1, \ldots, \omega_n)$.

Proof. We use the fact that any linear combination of Gaussian random variables also has a Gaussian distribution. More precisely, we can write that

$$
Y := \omega_1 X_1 + \ldots + \omega_n X_n \sim N(\mu_Y, \sigma_Y^2) \tag{2.38}
$$

where

$$\mu_Y := \sum_{i=1}^{n} w_i \mu_{X_i} \quad \text{and} \quad \sigma_Y^2 := \sum_{i=1}^{n} \sum_{k=1}^{n} w_i w_k \sigma_{ik} \qquad (2.39)$$

We obtain the formula (2.37) by observing (see Table 1.1, p. 19) that

$$\phi_{\mathbf{X}}(\omega) = E[e^{jY}] = \phi_Y(1) = \exp\left(j\mu_Y - \frac{1}{2}\sigma_Y^2\right) \quad \square \qquad (2.40)$$

Properties. i) If $\text{Cov}[X_i, X_k] = 0$, then the random variables X_i and X_k are *independent*.

ii) If Y_i, for $i = 1, \dots, m$, is a linear combination of the random variables X_k of a random vector $(X_1, \dots, X_n)$ having a multinormal distribution, then the r.v. $(Y_1, \dots, Y_m)$ also has a multinormal distribution.

Example 2.4.1. Let $\mathbf{X} = (X_1, \dots, X_n)$ be a random vector having a multinormal distribution $N(\mathbf{0}, \mathbf{I}_n)$, where $\mathbf{0} := (0, \dots, 0)$ and $\mathbf{I}_n$ is the *identity matrix* of order n. Thus, all the random variables X_k have a standard Gaussian distribution and are independent (because $\sigma_{ij} = 0 \ \forall \ i \neq j$). It follows that the mathematical expectation of the square of the distance of the vector $\mathbf{X}$ from the origin is

$$E[X_1^2 + X_2^2 + \dots + X_n^2] = \sum_{k=1}^{n} E[X_k^2] \stackrel{\text{i.d.}}{=} \sum_{k=1}^{n} 1 = n$$

and the variance of the squared distance is

$$V[X_1^2 + X_2^2 + \dots + X_n^2] \stackrel{\text{ind.}}{=} \sum_{k=1}^{n} V[X_k^2] \stackrel{\text{i.d.}}{=} \sum_{k=1}^{n} 2 = 2n$$

because if $Z \sim N(0,1)$, then we have

$$E[Z^4] = (-j)^4 \frac{d^4}{d\omega^4} e^{-\omega/2}\Big|_{\omega=0} = 3$$

so that $V[Z^2] = 3 - 1^2 = 2$.

Definition 2.4.2. *A stochastic process $\{X(t), t \in T\}$ is said to be a **Gaussian** process if the random vector $(X(t_1), \dots, X(t_n))$ has a multinormal distribution, for any n and for all $t_1, \dots, t_n$.*

Remark. Let X be a random variable whose distribution is $N(\mu_X, \sigma_X^2)$. Any *affine transformation* of X also has a Gaussian distribution:

$$Y := aX + b \quad \Longrightarrow \quad Y \sim N(a\mu_X + b, a^2\sigma_X^2) \qquad (2.41)$$

Similarly, any affine transformation of a Gaussian process remains a Gaussian process. For example, if $\{X(t), t \in T\}$ is a Gaussian process, then the s.p. $\{Y(t), t \in T\}$ defined by

$$Y(t) = 2X(t) - 1 \quad \text{or} \quad Y(t) = X(t^2) \tag{2.42}$$

is Gaussian as well. We can also show that $\{Y(t), t \in T\}$ is a Gaussian process if

$$Y(t) = \int_0^t X(s)\, ds \tag{2.43}$$

because an integral is the limit of a sum. However, the process is *not* Gaussian if

$$Y(t) = X^2(t) \quad \text{or} \quad Y(t) = e^{X(t)} \tag{2.44}$$

etc.

Proposition 2.4.2. *If a Gaussian process $\{X(t), t \in T\}$ is such that its mean $m_X(t)$ is a constant m_X and if its autocovariance function $C_X(t, t+s)$ depends only on s, then it is stationary (in the strict sense).*

Proof. Since the nth-order characteristic function of the s.p. $\{X(t), t \in T\}$ is given by [see Eq. (2.37)]

$$E\left[\exp\left\{j\sum_{i=1}^{n} \omega_i X(t_i)\right\}\right] = \exp\left\{jm_X \sum_{i=1}^{n} \omega_i - \frac{1}{2}\sum_{i=1}^{n}\sum_{k=1}^{n} C_X(t_i - t_k)\omega_i\omega_k\right\} \tag{2.45}$$

we can assert that the statistical characteristics of a Gaussian process depend only on its mean and its autocovariance function. Now, we see that the function above is *invariant* under any change of time origin. $\square$

Remark. The preceding proposition means that a wide-sense stationary Gaussian process is also *strict-sense* stationary.

Definition 2.4.3. *An s.p. $\{X(t), t \in T\}$ is said to be* **Markovian** *if*

$$P[X(t_n) \leq x_n \mid X(t), t \leq t_{n-1}] = P[X(t_n) \leq x_n \mid X(t_{n-1})] \tag{2.46}$$

where $t_{n-1} < t_n$.

Remarks. i) We say that the *future*, given the *present* and the *past*, depends only on the present.

ii) If $\{X(t), t \in T\}$ is a discrete-state process, we can write the formula (2.46) as follows:

$$P[X_{t+s} = j \mid X_t = i, X_{t_n} = i_n, \ldots, X_{t_1} = i_1] = P[X_{t+s} = j \mid X_t = i] \quad (2.47)$$

for all states $i_1, \ldots, i_n, i, j$, and for all time instants $t_1 < \ldots < t_n < t$ and $s \geq 0$.

iii) The *random walk* considered in Example 2.1.1 is a typical example of a Markovian process, which follows directly from the fact that we assume that the tosses of the coin are independent.

A Markovian, continuous-time, and continuous-state stochastic process, $\{X(t), t \in T\}$, is completely determined by its first-order density function:

$$f(x; t) := \frac{\partial}{\partial x} P[X(t) \leq x] \quad (2.48)$$

and by its *conditional transition density function*, defined by

$$
\begin{aligned}
p(x, x_0; t, t_0) &\equiv f_{X(t)|X(t_0)}(x \mid x_0) \quad &(2.49)\\
&= \lim_{dx \downarrow 0} \frac{P[X(t) \in (x, x + dx] \mid X(t_0) = x_0]}{dx} \quad \text{for } t > t_0
\end{aligned}
$$

Since the process must be somewhere at time t, we have

$$\int_{-\infty}^{\infty} f(x; t) \, dx = 1 \quad \text{and} \quad \int_{-\infty}^{\infty} p(x, x_0; t, t_0) \, dx = 1 \quad (2.50)$$

Moreover, by conditioning on all possible initial states, we may write that

$$f(x; t) = \int_{-\infty}^{\infty} f(x_0; t_0) p(x, x_0; t, t_0) \, dx_0 \quad (2.51)$$

Similarly, we deduce from the *Chapman[4]–Kolmogorov[5] equations* (see Chapter 3) that

$$p(x, x_0; t, t_0) = \int_{-\infty}^{\infty} p(x, x_1; t, t_1) p(x_1, x_0; t_1, t_0) \, dx_1 \quad (2.52)$$

where $t_0 < t_1 < t$. Finally, since at the initial time the distribution of the process is *deterministic*, we also have

---

[4] Sydney Chapman, 1888–1970, was born in England and died in the United States. He is especially known for his work in geophysics. One of the craters of the moon is named after him.

[5] Andrei Nikolaevich Kolmogorov, 1903–1987, was born and died in Russia. He was a great mathematician who, before getting his Ph.D., had already published 18 scientific papers, many of which were written during his undergraduate studies. His work on Markov processes in continuous time and with continuous-state space is the basis of the theory of *diffusion processes*. His book on theoretical probability, published in 1933, marks the beginning of *modern* probability theory. He also contributed in an important way to many other domains of mathematics, notably to the theory of dynamical systems.

$$\lim_{t \downarrow t_0} p(x, x_0; t, t_0) = \delta(x - x_0) \tag{2.53}$$

where $\delta(\cdot)$ is the *Dirac*[6] *delta function* defined by

$$\delta(x) = \begin{cases} 0 & \text{if } x \neq 0 \\ \infty & \text{if } x = 0 \end{cases} \tag{2.54}$$

so that

$$\int_{-\infty}^{\infty} \delta(x) \, dx = 1 \tag{2.55}$$

Definition 2.4.4. *The* **infinitesimal mean** $m(x;t)$ *and the* **infinitesimal variance** $v(x;t)$ *of the continuous-time and continuous-state stochastic process* $\{X(t), t \in T\}$ *are defined, respectively, by*

$$m(x;t) = \lim_{\epsilon \downarrow 0} \frac{1}{\epsilon} E[X(t + \epsilon) - X(t) \mid X(t) = x] \tag{2.56}$$

and

$$v(x;t) = \lim_{\epsilon \downarrow 0} \frac{1}{\epsilon} E[(X(t + \epsilon) - X(t))^2 \mid X(t) = x] \tag{2.57}$$

Remarks. i) We can also obtain $m(x_0; t_0)$ and $v(x_0; t_0)$ as follows:

$$m(x_0; t_0) = \lim_{t \downarrow t_0} \frac{\partial}{\partial t} E[X(t) \mid X(t_0) = x_0] \tag{2.58}$$

and

$$v(x_0; t_0) = \lim_{t \downarrow t_0} \frac{\partial}{\partial t} V[X(t) \mid X(t_0) = x_0] \tag{2.59}$$

ii) Suppose that the process $\{X(t), t \in T\}$ has infinitesimal moments $m(x;t) = m(x) \,\forall\, t$ and $v(x;t) \equiv v(x)$ and that its state space is the interval $[a,b]$ (or $[a,b)$, etc.). Let

$$Y(t) := g[X(t)] \quad \text{for } t \in T \tag{2.60}$$

If the function g is strictly increasing or decreasing on the interval $[a, b]$ and if the second derivative $g''(x)$ exists and is continuous, for $a < x < b$, then

---

[6] Paul Adrien Maurice Dirac, 1902–1984, was born in England and died in the United States. He was a physicist whose father was a French-speaking Swiss. He won the Nobel Prize for physics in 1933 for his work on quantum mechanics. He was a professor at the University of Cambridge for 37 years.

we can show that the infinitesimal moments of the process $\{Y(t), t \in T\}$ are given by

$$m_Y(y) = m(x)g'(x) + \frac{1}{2}v(x)g''(x) \tag{2.61}$$

and

$$v_Y(y) = v(x)[g'(x)]^2 \tag{2.62}$$

where $x = g^{-1}(y)$. Moreover, the state space of the process is the interval $[g(a), g(b)]$ (respectively, $[g(b), g(a)]$) if g is strictly increasing (resp., decreasing).

It can be shown that the function $p(x, x_0; t, t_0)$ satisfies the following partial differential equations:

$$\frac{\partial p}{\partial t} + \frac{\partial}{\partial x}[m(x; t)\, p] - \frac{1}{2}\frac{\partial^2}{\partial x^2}[v(x; t)\, p] = 0 \tag{2.63}$$

and

$$\frac{\partial p}{\partial t_0} + m(x_0; t_0)\frac{\partial p}{\partial x_0} + \frac{1}{2}v(x_0; t_0)\frac{\partial^2 p}{\partial x_0^2} = 0 \tag{2.64}$$

These equations are called the *Kolmogorov equations* or the *diffusion equations*. The first one is the Kolmogorov *forward equation* (or *Fokker[7]–Planck[8] equation*), and the second one is the Kolmogorov *backward equation*.

Definition 2.4.5. *If the function $p(x, x_0; t, t_0)$ depends only on $(x, x_0$ and) the difference $t - t_0$, the stochastic process $\{X(t), t \in T\}$ is said to be* **time-homogeneous**.

Remarks. i) If the s.p. $\{X(t), t \in T\}$ is time-homogeneous, then the functions $m(x; t)$ and $v(x; t)$ do not depend on t.

ii) Note that a time-homogeneous s.p. is not necessarily (wide-sense) stationary, because the function $f(x; t)$ may depend on the variable t. On the other hand, if $f(x; t) \equiv f(x)$, then the process is *strict-sense* stationary, because it is completely determined by $f(x; t)$ and $p(x, x_0; t, t_0)$.

Definition 2.4.6. *A stochastic process $\{B(t), t \geq 0\}$ is called a* **white noise (process)** *if its mean is equal to zero and if its autocovariance function is of the form*

$$C_B(t_1, t_2) = q(t_1)\,\delta(t_2 - t_1) \tag{2.65}$$

where $q(t_1) \geq 0$ and $\delta(\cdot)$ is the Dirac delta function (see p. 63).

---

[7] Adriaan Daniël Fokker, 1887–1972, was born in Indonesia and died in the Netherlands. He was a physicist and musician. He proved the equation in question in 1913 in his Ph.D. thesis.

[8] Max Karl Ernst Ludwig Planck, 1858–1947, born and died in Germany, was a renowned physicist famous for the development of quantum theory.

Remark. Actually, $\{B(t), t \geq 0\}$ is not a stochastic process in the proper sense of the term.

Finally, the following important result can be shown.

Proposition 2.4.3. *A Gaussian and stationary process, $\{X(t), t \in T\}$, is Markovian if and only if its autocorrelation function is of the form*

$$R_X(s) = \sigma^2 e^{-\alpha|s|} \tag{2.66}$$

where α and σ are positive constants.

Example 2.4.2. The Wiener process $\{W(t), t \geq 0\}$ (see Example 2.2.1) is such that $E[W(t) \mid W(t_0) = w_0] \equiv w_0$ and $V[W(t) \mid W(t_0) = w_0] = \sigma^2(t - t_0)$. It follows that its infinitesimal mean and variance are given by $m(w; t) \equiv 0$ and $v(w; t) \equiv \sigma^2$. Therefore, the conditional transition density function $p(w, w_0; t, t_0)$ satisfies the Kolmogorov forward equation

$$\frac{\partial p}{\partial t} - \frac{1}{2} \frac{\partial^2}{\partial w^2}(\sigma^2 p) = 0$$

We can check that $W(t) \mid \{W(t_0) = w_0\}$ has a Gaussian distribution with parameters w_0 and $\sigma^2(t - t_0)$. That is, the function $p(w, w_0; t, t_0)$ given by

$$p(w, w_0; t, t_0) = \frac{1}{\sqrt{2\pi\sigma^2(t - t_0)}} \exp\left\{-\frac{1}{2}\frac{(w - w_0)^2}{\sigma^2(t - t_0)}\right\} \quad \text{for } t > t_0$$

is a solution of the partial differential equation above. Moreover, we have

$$\lim_{t \downarrow t_0} p(w, w_0; t, t_0) = \delta(w - w_0)$$

as required.

2.5 Exercises

Section 2.1

Question no. 1

We define the stochastic process $\{X(t), t > 0\}$ by

$$X(t) = e^{-Yt} \quad \text{for } t > 0$$

where Y is a random variable having a uniform distribution on the interval $(0, 1)$. Calculate

(a) the first-order density function of the process $\{X(t), t > 0\}$,

(b) $E[X(t)]$, for $t > 0$,

(c) $C_X(t, t + s)$, where $s, t > 0$.

Question no. 2

Let $\{X_n, n = 1, 2, \dots\}$ be a Bernoulli process. That is, the random variables $X_1, X_2, \dots$ are independent and all have a Bernoulli distribution with parameter p. Calculate

(a) the particular case $p(0, 1; n_1 = 0, n_2 = 1)$ of the second-order probability mass function of the process,

(b) the correlation coefficient $\rho_X(n, m)$ of the process if $p = 1/2$ and $n, m \in \{1, 2, \dots\}$.

Question no. 3

Calculate the first-order density function of the s.p. $\{X(t), t \geq 0\}$ defined by

$$X(t) = tY + 1$$

where Y is a random variable having a $U(0, 1)$ distribution.

Question no. 4

We define the stochastic process $\{X(t), t > 0\}$ by

$$X(t) = \frac{t}{Y} \quad \text{for } t > 0$$

where Y has a uniform distribution on the interval $(0, 2)$. Calculate the function $f(x; t)$, for $x > t/2$.

Question no. 5

We consider the process $\{X(t), t \geq 0\}$ defined by $X(t) = (\tan Y)t$, for $t \geq 0$, where Y is a random variable having a uniform distribution on the interval $(-\pi/2, \pi/2)$. Calculate the probability $P[\exists \, t \in (0, 1): X(t) \notin [0, 1]]$. In other words, calculate the probability that the process $\{X(t), t \geq 0\}$ leaves the interval $[0, 1]$ between 0 and 1.

Indication. It can be shown that the r.v. $W := \tan Y$ has the following density function:

$$f_W(w) = \frac{1}{\pi(w^2 + 1)} \quad \text{for } w \in \mathbb{R}$$

That is, W has a *Cauchy*[9] *distribution*, or a *Student*[10] *distribution* with one degree of freedom.

---

[9] Augustin Louis Cauchy, 1789–1857, was born and died in France. He is considered the father of mathematical analysis and the inventor of the theory of functions of a complex variable.

[10] Pseudonym of William Sealy Gosset, 1876–1937, who was born and died in England. He worked as a chemist for the Guinness brewery, in Ireland, where he invented a *statistical test* for the control of the quality of beer. This test uses the distribution that bears his name.

Question no. 6

Are the increments of the stochastic process $\{X(t), t > 0\}$ defined in Example 2.1.6 independent? stationary? Justify.

Question no. 7

(a) Find the autocorrelation function of the process $\{X(t), t > 0\}$ defined by $X(t) = 1$ if $Y \geq t$ and $X(t) = 0$ if $Y < t$, where Y is a random variable having a $U(0, c)$ distribution.

(b) Calculate $R_X(t_1, t_2)$ in (a) if $X(t) = \delta(Y - t)$ instead, where $\delta(\cdot)$ is the Dirac delta function.

Question no. 8

We consider the process $\{X(t), t > 0\}$ defined by $X(t) = e^{-Yt}$, for $t > 0$, where Y is a continuous random variable whose density function is $f_Y(y)$, for $y \geq 0$.

(a) Find $f(x; t)$ in terms of $f_Y(y)$.

(b) Calculate $E[X(t)]$ and $R_X(t_1, t_2)$ when Y has an exponential distribution with parameter 1.

Question no. 9

Let

$$X(t) = \int_0^t \left[\int_0^s B(\tau) \, d\tau \right] ds \quad \text{for } t \geq 0$$

where $\{B(t), t \geq 0\}$ is the white noise process defined on p. 64. What is the average power of the stochastic process $\{X(t), t \geq 0\}$?

Section 2.2

Question no. 10

Is the stochastic process $\{X(t), t > 0\}$ defined in Question no. 1 wide-sense stationary? Justify.

Question no. 11

Let $\{X(t), t \geq 0\}$ be a stochastic process whose autocorrelation and auto-covariance functions are

$$R_X(t_1, t_2) = e^{-|t_1 - t_2|} + 1 \quad \text{and} \quad C_X(t_1, t_2) = e^{-|t_1 - t_2|}$$

Is the process wide-sense stationary? Justify.

Question no. 12

Let $\{X(t), t \geq 0\}$ be a wide-sense stationary stochastic process, with zero mean and autocorrelation function given by $R_X(s) = e^{-|s|}$. We define

$$Y(t) = tX^2(1/t) \quad \text{for } t > 0$$

Is the stochastic process $\{Y(t), t > 0\}$ wide-sense stationary? Justify.

Question no. 13

We consider a stochastic process $\{X(t), t \geq 0\}$ for which

$$C_X(t_1, t_2) = \frac{1}{|t_2 - t_1| + 1} \quad \text{and} \quad R_X(t_1, t_2) = \frac{1}{|t_2 - t_1| + 1} + 4$$

Is the process wide-sense stationary? Justify.

Question no. 14

Let Y be a random variable having a uniform distribution on the interval $(-1, 1)$. We consider the stochastic process $\{X(t), t \geq 0\}$ defined by

$$X(t) = Y^3 t \quad \text{for } t \geq 0$$

Is the process stationary? Justify.

Question no. 15

Let Y be a random variable such that $\phi_Y(1) = 2\phi_Y(-1)$ and $\phi_Y(2) = 4\phi_Y(-2)$, where $\phi_Y(\cdot)$ is the characteristic function of Y. We set

$$X(t) = \cos(\omega t + Y) \quad \text{for } t \geq 0$$

Show that the stochastic process $\{X(t), t \geq 0\}$ is wide-sense stationary.

Indication. We have the following trigonometric identity:

$$\cos(\omega t_1 + s) \cos(\omega t_2 + s) = \frac{1}{2} \{\cos \omega(t_1 - t_2) + \cos(\omega t_1 + \omega t_2 + 2s)\}$$

Question no. 16

We set

$$Y(t) = X(t+1) - X(t) \quad \text{for } t \in \mathbb{R}$$

where $\{X(t), t \in \mathbb{R}\}$ is a wide-sense stationary stochastic process. Find the spectral density of the process $\{Y(t), t \in \mathbb{R}\}$ in terms of $S_X(\omega)$.

Question no. 17

Consider the wide-sense stationary process $\{X(t), t \in \mathbb{R}\}$ whose spectral density is $S_X(\omega) = 2/(1 + \omega^2)$. We set

$$Y(t) = \int_{t-1}^{t+1} X(s) \, ds \quad \text{for } t \in \mathbb{R}$$

Calculate $S_Y(\omega)$.

Indication. Write $Y(t)$ in the form

$$Y(t) = \int_{-\infty}^{\infty} h(t - s) X(s) \, ds$$

for an appropriately chosen function h.

Question no. 18

We consider the random variable $Y(t)$ defined by Eq. (2.19). Calculate the mean of $X(t)Y(t)$ if $\{X(t), t \in T\}$ is a white noise process.

Section 2.3

Question no. 19

Are the stochastic processes defined in Questions nos. 11, 12, and 13 mean ergodic? Justify.

Question no. 20

Is the Bernoulli process defined in Example 2.1.5 mean ergodic? Justify.

Question no. 21

We define $X(t) = m + B(t)$, for $t \geq 0$, where m is a constant and $\{B(t), t \geq 0\}$ is a white noise process, so that $E[B(t)] \equiv 0$ and

$$C_B(t_1, t_2) = q(t_1)\,\delta(t_2 - t_1)$$

where $\delta(\cdot)$ is the Dirac delta function. Show that the process $\{X(t), t \geq 0\}$ is mean ergodic if the function q is bounded.

Question no. 22

Let $\{X(t), t \geq 0\}$ be a wide-sense stationary stochastic process, with zero mean and for which

$$R_X(s) = e^{-\alpha|s|} \cos 2\pi\omega s$$

where α and ω are positive constants. Is this process mean ergodic? Justify.

Question no. 23

The stochastic process $\{X(t), t \geq 0\}$ is defined by

$$X(t) = Y + B(t)$$

where $Y \sim N(0, 1)$ and $\{B(t), t \geq 0\}$ is a white noise process whose autocorrelation function is $R_B(s) = c\,\delta(s)$, with $c > 0$. We assume that the random variable Y and the white noise are independent. Is the process $\{X(t), t \geq 0\}$

(a) wide-sense stationary?

(b) mean ergodic?

Question no. 24

Let $\{B(t), t \geq 0\}$ be a white noise process for which the function $q(t_1)$ is a constant $c > 0$. Calculate the variance of the random variable

$$I_T := \frac{1}{2T} \int_{-T}^{T} B(t)\, dt$$

Question no. 25

The *standard Brownian motion* is the particular case of the Wiener process $\{W(t), t \geq 0\}$ (see Example 2.2.1) for which $E[W(t)] \equiv 0$ and $C_W(t_1, t_2) = \min\{t_1, t_2\}$.

(a) Calculate $V[\langle W(t)\rangle_S]$.

(b) Is the process mean ergodic? Justify.

Section 2.4

Question no. 26

Let $(X_1, \ldots, X_n)$ be a random vector having a multinormal distribution for which $\mathbf{m} = (0, \ldots, 0)$ and the covariance matrix $\mathbf{K}$ is the identity matrix of order n (≥ 4).

(a) Calculate the characteristic function of $Y := X_1 + X_2$.

Reminder. The characteristic function of a random variable X having a Gaussian $N(\mu, \sigma^2)$ distribution is given by $\phi_X(\omega) = \exp\left(j\mu\omega - \frac{1}{2}\sigma^2\omega^2\right)$.

(b) Does the random vector (Y, Z), where $Z := X_3 - X_4$, have a bivariate normal distribution? If it does, give its five parameters; otherwise, justify.

Question no. 27

Let $\{X(t), t \geq 0\}$ be a Gaussian process such that $E[X(t)] = \mu t$, for $t \geq 0$, where μ is a nonzero constant, and

$$R_X(t, t+s) = 2t + \mu^2 t(t+s) \quad \text{for } s, t \geq 0$$

We define $Y(t) = X(t) - \mu t$, for $t \geq 0$. Is the process $\{Y(t), t \geq 0\}$ stationary? Justify.

Question no. 28

We consider a Gaussian process $\{X(t), t \geq 0\}$ for which $E[X(t)] \equiv 0$ and whose autocovariance function is

$$C_X(t, t+s) = e^{-t} \quad \text{for } s, t \geq 0$$

Let $Y(t) := X^2(t)$, for $t \geq 0$. Is the stochastic process $\{Y(t), t \geq 0\}$ (a) stationary? (b) mean ergodic? Justify.

Question no. 29

Let $\{X_n, n = 1, 2, \ldots\}$ be a discrete-time and discrete-state process such that

$$P[X_{n+1} = j \mid X_n = i, X_{n-1} = i_{n-1}, \ldots, X_1 = i_1]$$

$$= P[X_{n+1} = j \mid X_n = i, X_{n-1} = i_{n-1}]$$

for all states $i_1, \ldots, i_{n-1}, i, j$ and for any time instant n. This process is *not* Markovian. Transform the state space so that the process thus obtained is Markovian.

Question no. 30

Suppose that $\{X(t), t \geq 0\}$ is a Gaussian process such that

$$E[X(t) \mid X(t_0) = x_0] \equiv x_0 \quad \text{and} \quad V[X(t) \mid X(t_0) = x_0] = t \quad \forall t \geq 0$$

Let

$$Y(t) := e^{X(t)} \quad \text{for } t \geq 0$$

(a) Calculate the infinitesimal parameters of the process $\{Y(t), t \geq 0\}$ by using the formulas (2.56) and (2.57).

(b) Check your results with the help of the formulas (2.58) and (2.59).

Question no. 31

Find a solution of the Fokker–Planck equation

$$\frac{\partial p}{\partial t} + \frac{\partial}{\partial x}(\mu p) - \frac{1}{2}\frac{\partial^2}{\partial x^2}(\sigma^2 p) = 0 \quad (\text{for } x \in \mathbb{R} \text{ and } t \geq t_0 \geq 0)$$

where $\mu \in \mathbb{R}$ and $\sigma > 0$ are constants, for which

$$\lim_{t \downarrow t_0} p(x, x_0; t, t_0) = \delta(x - x_0)$$

Indications. i) Try the density function of a Gaussian distribution as a solution.

ii) We can take the *Fourier transform* (with respect to x) of the equation and then invert the solution of the ordinary differential equation obtained. We have

$$\lim_{x \to \pm\infty} p(x, x_0; t, t_0) = 0$$

which implies that

$$\lim_{x \to \pm\infty} \frac{\partial}{\partial x} p(x, x_0; t, t_0) = 0$$

as well.

Question no. 32

Let $\{X(t), t \in \mathbb{R}\}$ be a wide-sense stationary Gaussian process, with zero mean and $R_X(s) = 2e^{-4|s|}$. We define $Y = X(t+s)$ and $Z = X(t-s)$, where $s \geq 0$.

(a) What is the distribution of the random variable $X(t)$?

(b) Calculate the mean of YZ.

(c) Calculate the variance of $Y + Z$.

Question no. 33

The Gaussian process $\{X(t), t \geq 0\}$ is such that $E[X(t)] = 0 \; \forall \, t$ and

$$R_X(t_1, t_2) = c_1 e^{-c_2|t_1 - t_2|}$$

where $c_1, c_2 > 0$.

(a) Show that the stochastic process $\{X(t), t \geq 0\}$ is stationary and mean ergodic.

(b) We define the process $\{Y(t), t \geq 0\}$ by

$$Y(t) = \begin{cases} 1 \text{ if } X(t) \leq x \\ 0 \text{ if } X(t) > x \end{cases}$$

We then have $E[Y(t)] = F(x;t)$. If the process $\{X(t), t \geq 0\}$ is stationary, we may write that $F(x;t) \equiv F(x)$. We say that the s.p. $\{X(t), t \geq 0\}$ is *distribution ergodic* if $\{Y(t), t \geq 0\}$ is mean ergodic. Is the stochastic process $\{X(t), t \geq 0\}$ distribution ergodic?

(c) When the process $\{X(t), t \geq 0\}$ is stationary, we also define the process $\{Z_s(t), t \geq 0\}$ by $Z_s(t) = X(t)X(t+s)$, where $s \geq 0$. It follows that $E[Z_s(t)] = R_X(s)$. We say that the s.p. $\{X(t), t \geq 0\}$ is *correlation ergodic* if $\{Z_s(t), t \geq 0\}$ is mean ergodic. Can we assert that $\{X(t), t \geq 0\}$ is correlation ergodic?

Question no. 34

Suppose that $\{X(t), t \in \mathbb{R}\}$ is a stationary Gaussian process such that

$$C_X(s) = R_X(s) = e^{-2|s|} \quad \text{for } s \in \mathbb{R}$$

(a) What is the distribution of $X(t)$? Justify.

(b) We define the random variable

$$Y(t) = \Phi[X(t)] + 1$$

where $\Phi(\cdot)$ is the distribution function of the $N(0,1)$ distribution. That is, $\Phi(z) := P[N(0,1) \leq z]$. What is the distribution of $Y(t)$? Justify.

Indication. The inverse function Φ^{-1} exists.

Question no. 35

Let $\{X(t), t \in \mathbb{R}\}$ be a stationary Gaussian process whose autocovariance function is

$$C_X(s) = \begin{cases} 4[1 - (|s|/2)] \text{ if } |s| \leq 2 \\ 0 \qquad\qquad \text{if } |s| > 2 \end{cases}$$

Is the stochastic process $\{X(t), t \in \mathbb{R}\}$

(a) mean ergodic?

(b) distribution ergodic (see Question no. 33)? Justify.

3

Markov Chains

3.1 Introduction

The notion of a *Markovian process* was seen in Section 2.4. In the general case, the stochastic process $\{X(t), t \in T\}$ is said to be Markovian if

$$P[X(t_{n+1}) \in A \mid X(t) = x_t, t \le t_n] = P[X(t_{n+1}) \in A \mid X(t_n) = x_{t_n}] \quad (3.1)$$

for all events A and for all time instants $t_n < t_{n+1}$.

Equation (3.1) means that the probability that the process moves from state x_{t_n}, where it is at time t_n, to a state included in A at time t_{n+1} does not depend on the way the process reached x_{t_n} from x_{t_0}, where t_0 is the initial time (that is, does not depend on the path followed by the process from x_{t_0} to x_{t_n}).

In this chapter, we will consider the cases where $\{X(t), t \in T\}$ is a discrete-time process and where it is a continuous-time and discrete-state process. Actually, we will only mention briefly, within the framework of an example, the case of discrete-time and continuous-state processes.

When $\{X_n, n = 0, 1, \dots\}$ is a discrete-time and discrete-state process, the *Markov property* implies that

$$P[X_{n+1} = j \mid X_n = i, X_{n-1} = i_{n-1}, \dots, X_0 = i_0] = P[X_{n+1} = j \mid X_n = i] \tag{3.2}$$

for all states $i_0, \dots, i_{n-1}, i, j$, and for any time n. We also have

$$P[X_{n+1} = j \mid X_{n-1} = i_{n-1}, X_{n-2} = i_{n-2}, \dots, X_0 = i_0]$$
$$= P[X_{n+1} = j \mid X_{n-1} = i_{n-1}] \tag{3.3}$$

etc., which means that the *transition probabilities* depend only on the most recent information about the process that is available.

Remark. We call a **Markov chain** a discrete-time process that possesses the Markov property. Originally, this expression denoted discrete-time and discrete-state Markovian processes. However, we can accept the fact that the state space of the process may be continuous. By extension, we will call a **continuous-time Markov chain** a *discrete*-state (and continuous-time) Markovian stochastic process. Discrete-time and continuous-time Markov chains will be studied in Sections 3.2 and 3.3, respectively.

We can generalize Eq. (3.2) as follows.

Proposition 3.1.1. *If the discrete-time and discrete-state stochastic process* $\{X_n, n = 0, 1, \ldots\}$ *possesses the Markov property, then*

$$P[(X_{n+1}, X_{n+2}, \ldots) \in B \mid X_n = i_n, X_{n-1} = i_{n-1}, \ldots, X_0 = i_0]$$
$$= P[(X_{n+1}, X_{n+2}, \ldots) \in B \mid X_n = i_n] \tag{3.4}$$

where B is an infinite-dimensional event.

Remark. For example, by using the fact that

$$P[A \cap B \mid C] = \frac{P[A \cap B \cap C]}{P[C]} = \frac{P[A \cap B \cap C]}{P[B \cap C]} \frac{P[B \cap C]}{P[C]}$$
$$= P[A \mid B \cap C] P[B \mid C] \tag{3.5}$$

we can easily show that

$$P[(X_{n+1}, X_{n+2}) = (i_{n+1}, i_{n+2}) \mid X_n = i_n, X_{n-1} = i_{n-1}, \ldots, X_0 = i_0]$$
$$= P[(X_{n+1}, X_{n+2}) = (i_{n+1}, i_{n+2}) \mid X_n = i_n] \tag{3.6}$$

We now give several examples of discrete-time and continuous-time Markov chains.

Example 3.1.1. One of the simplest, but not trivial, examples of a Markovian process is that of the *random walk* (see p. 48). We can generalize the random walk as follows: if the particle is in state i at time n, then it will be in state j at time $n + 1$ with probability

$$P[X_{n+1} = j \mid X_n = i] = \begin{cases} p & \text{if } j = i + 1 \\ q & \text{if } j = i - 1 \\ r & \text{if } j = i \\ 0 & \text{otherwise} \end{cases}$$

where p, q, and r are nonnegative constants such that $p + q + r = 1$. That is, we add the possibility that the particle does not move from the position it occupies at time n.

Another way of generalizing the random walk is to assume that the probabilities p and q (and r in the case above) may depend on the state i where the particle is. Thus, we have

$$P[X_{n+1} = j \mid X_n = i] = \begin{cases} p_i & \text{if } j = i+1 \\ q_i & \text{if } j = i-1 \\ 0 & \text{otherwise} \end{cases} \tag{3.7}$$

and $p_i + q_i = 1$, for any $i \in \{0, \pm 1, \pm 2, \dots\}$.

Finally, we can generalize the random walk to the d-dimensional case, where $d \in \mathbb{N}$. For example, a *two-dimensional* ($d = 2$) random walk is a stochastic process $\{(X_n, Y_n), n = 0, 1, \dots\}$ for which the state space is the set

$$S_{(X_n, Y_n)} := \{(i, j) : i, j \in \{0, \pm 1, \pm 2, \dots\}\}$$

and such that

$$P[(X_{n+1}, Y_{n+1}) = (i_{n+1}, j_{n+1}) \mid (X_n, Y_n) = (i_n, j_n)]$$

$$= \begin{cases} p_1 & \text{if } i_{n+1} = i_n + 1, j_{n+1} = j_n \\ p_2 & \text{if } i_{n+1} = i_n, j_{n+1} = j_n + 1 \\ q_1 & \text{if } i_{n+1} = i_n - 1, j_{n+1} = j_n \\ q_2 & \text{if } i_{n+1} = i_n, j_{n+1} = j_n - 1 \\ 0 & \text{otherwise} \end{cases}$$

where $p_1 + p_2 + q_1 + q_2 = 1$ (and $p_i \geq 0$ and $q_i \geq 0$, for $i = 1, 2$). Thus, the particle can only move to one of its four nearest neighbors. In three dimensions, the particle can move from a given state to one of its six nearest neighbors [the neighbors of the origin being the triplets $(\pm 1, 0, 0)$, $(0, \pm 1, 0)$ and $(0, 0, \pm 1)$], etc.

Example 3.1.2. An important particular case of the random walk having the transition probabilities given in Eq. (3.7) is the one where

$$p_i = \frac{N-i}{N} \quad \text{and} \quad q_i = \frac{i}{N}$$

If we suppose that the state space is the finite set $\{0, 1, \dots, N\}$, then we can give the following interpretation to this random walk: we consider two urns that contain N balls in all. At each time unit, a ball is taken at random from all the balls and is placed in the other urn. Let X_n be the number of balls in urn I after n shifts. This model was used by Paul and Tatiana Ehrenfest[1] to study the transfer of heat between the molecules in a gas.

Example 3.1.3. Suppose that we observe the number of customers that are standing in line in front of an automated teller machine. Let X_n be the number of customers in the queue at time $n \in \{0, 1, \dots\}$. That is, we observe the system at deterministic time instants separated by one *time unit*, for example,

---

[1] Paul Ehrenfest, 1880–1933, born in Austria and died in the Netherlands, was a physicist who studied statistical mechanics and quantum theory. His wife, of Russian origin, was Tatiana Ehrenfest-Afanaseva, 1876–1964. She was interested in the same subjects.

every hour, or every 10 minutes (a time unit may be an arbitrary number of minutes), etc. Suppose next that if a customer is using the teller machine at time n, then the probability that he is finished before time $n+1$ is $q \in (0,1)$. Finally, suppose that the probability that k customers arrive during a time unit is given by

$$P[Y_n = k] = e^{-\lambda} \frac{\lambda^k}{k!}$$

for $k = 0, 1, \ldots$ and for any $n \in \{0, 1, \ldots\}$. That is, the number Y_n of arrivals in the system in the interval $[n, n+1)$ has a Poisson distribution with parameter $\lambda\ \forall n$. Thus, we may write that $X_{n+1} = X_n + Y_n$ with probability $p = 1 - q$ and $X_{n+1} = (X_n - 1) + Y_n$ with probability q. We find that $\{X_n, n = 0, 1, \ldots\}$ is a (discrete-time and discrete-state) Markov chain for which the transition probabilities are

$$P[X_{n+1} = j \mid X_n = i] = \begin{cases} p\, e^{-\lambda} \dfrac{\lambda^k}{k!} + q\, e^{-\lambda} \dfrac{\lambda^{k+1}}{(k+1)!} & \text{if } j = i + k \text{ and } i \in \mathbb{N} \\ q\, e^{-\lambda} & \text{if } j = i - 1 \text{ and } i \in \mathbb{N} \\ e^{-\lambda} \dfrac{\lambda^k}{k!} & \text{if } j = k \text{ and } i = 0 \\ 0 & \text{otherwise} \end{cases}$$
$$(3.8)$$

This is an example of a queueing model in discrete-time.

Example 3.1.4. An example of a discrete-time, but continuous-state, Markov chain is obtained from a problem in *optimal control*: suppose that an object moves on the real line. Let X_n be its position at time $n \in \{0, 1, \ldots\}$. We suppose that

$$X_{n+1} = a_n X_n + b_n u_n + \epsilon_n$$

where a_n and b_n are nonzero constants, u_n is the *control* (or *command*) variable, and ϵ_n is the term that corresponds to the *noise* in the system. We assume that $\epsilon_n \sim N(0, \sigma^2)$, for all n and that the random variables X_n and ϵ_n are independent. The objective is to bring the object close to the origin. To do so, we choose

$$u_n = -\frac{a_n}{b_n} X_n \quad \forall\, n \tag{3.9}$$

Remark. In stochastic optimal control, the variable u_n must be chosen so as to minimize the mathematical expectation of a certain cost function J. If the function in question is of the form

$$J(X_n, u_n) = \sum_{n=0}^{\infty} \left(k X_n^2 + \frac{1}{2} c\, u_n^2 \right)$$

where $c > 0$ and k are constants, then the optimal solution is indeed obtained by using the formula (3.9) above.

The s.p. $\{X_n, n = 0, 1, \dots\}$ is a Markov chain. If $f_{X_{n+1}|X_n}(y \mid x)$ denotes the conditional density function of the random variable X_{n+1}, given that $X_n = x$, we may write that

$$f_{X_{n+1}|X_n}(y \mid x) = \frac{1}{\sqrt{2\pi}\sigma} \exp\left\{-\frac{y^2}{2\sigma^2}\right\} \quad \text{for } y \in \mathbb{R}$$

That is, $X_{n+1} \mid X_n \sim N(0, \sigma^2)$.

We can also suppose that the noise ϵ_n has a variance depending on X_n. For instance, $V[\epsilon_n] = \sigma^2 X_n^2$. In this case, we have that $X_{n+1} \mid X_n \sim N(0, \sigma^2 X_n^2)$.

Example 3.1.5. Suppose that $X(t)$ denotes the number of customers who arrived in the interval $[0, t]$ in the example of a queueing system in discrete-time (see Example 3.1.3). If we assume that the time τ_n elapsed until the arrival of a new customer into the system, when there have been n arrivals since the initial time, has an exponential distribution with parameter λ, for all n, and that the random variables τ_0, $\tau_1, \dots$ are independent, then we find that $\{X(t), t \geq 0\}$ is a continuous-time and discrete-state Markovian process called a *Poisson process*. In this case, the number of arrivals in the system in the interval $[0, t]$ has a Poisson distribution with parameter λt. Therefore, during *one* time unit, the number of arrivals indeed has a Poi(λ) distribution.

In general, if the random variable τ_k has an exponential distribution with parameter λ_k, then $\{X(t), t \geq 0\}$ is a continuous-time Markov chain.

Example 3.1.6. In the preceding example, we considered only the arrivals in the system. Suppose that the time an arbitrary customer spends using the automated teller machine is an exponential random variable with parameter μ. Now, let $X(t)$ be the number of customers *in the system* at time $t \geq 0$. If the customers arrive one at a time, then the process $\{X(t), t \geq 0\}$ is an example of a continuous-time Markov chain called a *birth and death process*. It is also the basic model in the theory of queues, as will be seen in Chapter 6. It is denoted by the symbol $M/M/1$. When there are two *servers* (for example, two automated teller machines) rather than a single one, we write $M/M/2$, etc.

Remark. Actually, the Poisson process is a particular case of the so-called pure birth (or growth) processes.

3.2 Discrete-time Markov chains

3.2.1 Definitions and notations

Definition 3.2.1. *A stochastic process* $\{X_n, n = 0, 1, \dots\}$ *whose state space* S_{X_n} *is finite or countably infinite is a* **stationary** *(or* **time-homogeneous***)* **Markov chain** *if*

$$P[X_{n+1} = j \mid X_n = i, X_{n-1} = i_{n-1}, \dots, X_0 = i_0]$$
$$= P[X_{n+1} = j \mid X_n = i] = p_{i,j} \tag{3.10}$$

for all states $i_0, \dots, i_{n-1}, i, j$ and for any $n \geq 0$.

Remarks. i) In the general case, we can denote the conditional probability $P[X_{n+1} = j \mid X_n = i]$ by $p_{i,j}(n)$. Moreover, when there is no risk of confusion, we may also write $p_{i,j}$ simply as p_{ij}.

ii) When the states of the Markov chain are identified by a coding system, we will use, by convention, the set $\mathbb{N}^0 := \{0, 1, \dots\}$ as the set space. For example, suppose that we wish to model the flow of a river as a Markov chain and that we use three adjectives to describe the flow X_n during the nth day of the year: *low, average,* or *high,* rather than considering the exact flow. In this case, we would denote the state *low flow* by state 0, the state *average flow* by 1, and the state *high flow* by 2.

iii) Note that, in the example given above, the conditional probabilities $P[X_{n+1} = j \mid X_n = i]$ actually depend on n, since the probability of moving from a low flow to an average flow, in particular, is not the same during the whole year. Indeed, this probability is assuredly smaller during the winter and higher in the spring. Therefore, we should, theoretically, use a *non*stationary Markov chain. In practice, we can use a stationary Markov chain, but for a shorter time period, such as the one covering only the thawing period in the spring.

In this book, we will consider only stationary Markov chains. Anyhow, the general case is not used much in practice. Moreover, in the absence of an indication to the contrary, we take for granted in the formulas and definitions that follow that the state space of the chain is the set $\{0, 1, \dots\}$.

Definition 3.2.2. *The* **one-step transition probability matrix P** *of a Markov chain is given by*

$$
\mathbf{P} = \begin{array}{c} \\ 0 \\ 1 \\ 2 \\ \vdots \end{array}
\begin{array}{cccc}
0 & 1 & 2 & \cdots
\end{array}
\left[\begin{array}{cccc}
p_{0,0} & p_{0,1} & p_{0,2} & \cdots \\
p_{1,0} & p_{1,1} & p_{1,2} & \cdots \\
p_{2,0} & p_{2,1} & p_{2,2} & \cdots \\
\vdots & \vdots & \vdots & \vdots
\end{array} \right] \tag{3.11}
$$

Remarks. i) We have indicated the possible states of the Markov chain to the left of and above the matrix, in order to facilitate the comprehension of this *transition matrix.* The state to the left is the one in which the process is at time n, and the state above that in which the process will be at time $n + 1$.

ii) Since the $p_{i,j}$'s are (conditional) probabilities, we have

$$p_{i,j} \geq 0 \quad \forall\, i, j \tag{3.12}$$

Moreover, because the process must be in *one and only one* state at time $n+1$, we may write that

$$\sum_{j=0}^{\infty} p_{i,j} = 1 \quad \forall\, i \tag{3.13}$$

A matrix that possesses these two properties is said to be *stochastic*. The sum $\sum_{i=0}^{\infty} p_{i,j}$, for its part, may take any nonnegative value. If we also have

$$\sum_{i=0}^{\infty} p_{i,j} = 1 \quad \forall\, j \tag{3.14}$$

the matrix $\mathbf{P}$ is called *doubly stochastic*. We will see, in Subsection 3.2.3, that for such a matrix the *limiting probabilities*, in the case when the state space is *finite*, are obtained without our having to do any calculations.

We now wish to generalize the transition matrix $\mathbf{P}$ by considering the case when the process moves from state i to state j in n steps. We then introduce the following notation.

Notation. The probability of moving from state i to state j in n steps (or transitions) is denoted by

$$p_{i,j}^{(n)} := P[X_{m+n} = j \mid X_m = i], \quad \text{for } m, n, i, j \geq 0 \tag{3.15}$$

From the $p_{i,j}^{(n)}$'s we can construct the matrix $\mathbf{P}^{(n)}$ of the transition probabilities in n steps. This matrix and $\mathbf{P}$ have the same dimensions. Moreover, we find that we can obtain $\mathbf{P}^{(n)}$ by raising the transition matrix $\mathbf{P}$ to the power n.

Proposition 3.2.1 (Chapman–Kolmogorov equations). *We have*

$$p_{i,j}^{(m+n)} = \sum_{k=0}^{\infty} p_{i,k}^{(m)} p_{k,j}^{(n)} \quad \text{for } m, n, i, j \geq 0 \tag{3.16}$$

Proof. Since we consider only stationary Markov chains, we can write, using the total probability rule, that

$$p_{i,j}^{(m+n)} \equiv P[X_{m+n} = j \mid X_0 = i] \tag{3.17}$$

$$= \sum_{k=0}^{\infty} P[X_{m+n} = j, X_m = k \mid X_0 = i] \tag{3.18}$$

We then deduce from the formula (see p. 74)

$$P[A \cap B \mid C] = P[A \mid B \cap C]P[B \mid C] \tag{3.19}$$

that

$$p_{i,j}^{(m+n)} = \sum_{k=0}^{\infty} P[X_{m+n} = j \mid X_m = k, X_0 = i]P[X_m = k \mid X_0 = i]$$

$$= \sum_{k=0}^{\infty} p_{k,j}^{(n)} p_{i,k}^{(m)} = \sum_{k=0}^{\infty} p_{i,k}^{(m)} p_{k,j}^{(n)} \quad \Box \tag{3.20}$$

In matricial form, the various equations (3.16) are written as follows:

$$\mathbf{P}^{(m+n)} = \mathbf{P}^{(m)}\mathbf{P}^{(n)} \tag{3.21}$$

which implies that

$$\mathbf{P}^{(n)} = \underbrace{\mathbf{P}^{(1)}\mathbf{P}^{(1)} \cdots \mathbf{P}^{(1)}}_{n\times} \tag{3.22}$$

Since $\mathbf{P}^{(1)} = \mathbf{P}$, we indeed have

$$\mathbf{P}^{(n)} = \mathbf{P}^n \tag{3.23}$$

Example 3.2.1. Suppose that the Markov chain $\{X_n, n = 0, 1, \ldots\}$, whose state space is the set $\{0, 1, 2\}$, has the (one-step) transition matrix

$$
\mathbf{P} = \begin{array}{c} \\ 0 \\ 1 \\ 2 \end{array}
\begin{array}{ccc} 0 & 1 & 2 \end{array}
\left[\begin{array}{ccc} 1/3 & 1/3 & 1/3 \\ 0 & 1/2 & 1/2 \\ 1 & 0 & 0 \end{array}\right]
$$

We have

$$\mathbf{P}^{(2)} = \mathbf{P}^2 = \begin{bmatrix} 4/9 & 5/18 & 5/18 \\ 1/2 & 1/4 & 1/4 \\ 1/3 & 1/3 & 1/3 \end{bmatrix}$$

Thus, $p_{0,0}^{(2)} = 4/9$. Note that to obtain this result, it is sufficient to know the first row and the first column of the matrix $\mathbf{P}$. Similarly, if we are looking for $p_{0,0}^{(5)}$, we can use the matrix $\mathbf{P}^{(2)}$ to calculate the first row of $\mathbf{P}^{(4)}$. We obtain

$$\mathbf{P}^{(4)} = \mathbf{P}^4 = \begin{bmatrix} 139/324 & 185/648 & 185/648 \\ - & - & - \\ - & - & - \end{bmatrix}$$

Next, we have

$$p_{0,0}^{(5)} = \frac{139}{324} \times \frac{1}{3} + \frac{185}{648} \times 0 + \frac{185}{648} \times 1 = \frac{833}{1944} \simeq 0.4285$$

So, in general, to obtain only the element $p_{i,j}^{(n)}$ of the matrix $\mathbf{P}^{(n)}$, it suffices to calculate the ith row of the matrix $\mathbf{P}^{(n-1)}$ and to multiply it by the jth column of $\mathbf{P}$.

In Subsection 3.2.3, we will be interested in finding (if it exists) the limit of the matrix $\mathbf{P}^{(n)}$ when n tends to infinity. To do so, we will give a theorem that enables us to calculate the *limiting probabilities* of the Markov chain. By making use of a mathematical software package, we can also multiply the matrix $\mathbf{P}$ by itself a sufficient number of times to see to what matrix it converges. Here, we find that if n is sufficiently large, then

$$\mathbf{P}^{(n)} = \mathbf{P}^n \simeq \begin{bmatrix} 0.4286 & 0.2857 & 0.2857 \\ 0.4286 & 0.2857 & 0.2857 \\ 0.4286 & 0.2857 & 0.2857 \end{bmatrix}$$

Observe that the three rows of the matrix above are identical, from which we deduce that whatever the state in which the process is at the initial time, there is a probability of approximately 0.4286 that it will be in state 0 after a very large number of transitions (this probability was equal to $\simeq 0.4285$ after only five transitions, from state 0). By using the theorem of Subsection 3.2.3, we can show that the *exact* probability that the process is in state 0 when it is *in equilibrium* is equal to 3/7.

Sometimes, rather than being interested in the transition probabilities in n steps of the Markov chain $\{X_n, n = 0, 1, \ldots\}$, we want to calculate the probability that the chain will move to state j, from $X_0 = i$, *for the first time* at time n.

Notation. The probability of moving to state j, from the initial state i, *for the first time* at the nth transition is denoted by

$$\rho_{i,j}^{(n)} := P[X_n = j, X_{n-1} \neq j, \ldots, X_1 \neq j \mid X_0 = i] \quad \text{for } n \geq 1 \text{ and } i, j \geq 0 \tag{3.24}$$

Remarks. i) When $i = j$, $\rho_{i,j}^{(n)}$ is the probability of first *return* to the initial state i after exactly n transitions.

ii) We have the following relation between $p_{i,j}^{(n)}$ and $\rho_{i,j}^{(n)}$:

$$p_{i,j}^{(n)} = \sum_{k=1}^{n} \rho_{i,j}^{(k)} p_{j,j}^{(n-k)} \tag{3.25}$$

When the $p_{i,j}^{(n)}$'s are known, we can use the preceding formula recursively to obtain the $\rho_{i,j}^{(k)}$'s, for $k = 1, 2, \ldots, n$.

Example 3.2.2. In some cases, it is easy to calculate directly the probabilities of first return in n transitions. For example, let

$$\mathbf{P} = \begin{array}{c} 0 \\ 1 \end{array} \begin{bmatrix} \begin{array}{cc} 0 & 1 \end{array} \\ \begin{array}{cc} 1/3 & 2/3 \\ 1 & 0 \end{array} \end{bmatrix}$$

be the matrix of transition probabilities in one step of a Markov chain whose possible values are 0 and 1. We find that

$$\rho_{0,1}^{(n)} = P[X_n = 1, X_{n-1} = 0, \ldots, X_1 = 0 \mid X_0 = 0] = (1/3)^{n-1}(2/3) = \frac{2}{3^n}$$

for $n = 1, 2, \ldots$. Similarly,

$$\rho_{1,0}^{(1)} = 1 \quad \text{and} \quad \rho_{1,0}^{(n)} = 0 \quad \text{for } n = 2, 3, \ldots$$

Next, we have

$$\rho_{0,0}^{(1)} = 1/3, \quad \rho_{0,0}^{(2)} = (2/3) \times 1 = 2/3, \quad \text{and} \quad \rho_{0,0}^{(n)} = 0 \quad \text{for } n = 3, 4, \ldots$$

Finally, we calculate $\rho_{1,1}^{(1)} = 0$, $\rho_{1,1}^{(2)} = 2/3$, and

$$\rho_{1,1}^{(n)} = 1 \times (1/3)^{n-2}(2/3) = \frac{2}{3^{n-1}} \quad \text{for } n = 3, 4 \ldots$$

Note that, in this example, we have $\sum_{n=1}^{\infty} \rho_{1,0}^{(n)} = 1$, $\sum_{n=1}^{\infty} \rho_{0,0}^{(n)} = 1$,

$$\sum_{n=1}^{\infty} \rho_{0,1}^{(n)} = \sum_{n=1}^{\infty} \frac{2}{3^n} = \frac{2}{3} \frac{1}{1 - \frac{1}{3}} = 1$$

and

$$\sum_{n=1}^{\infty} \rho_{1,1}^{(n)} = 0 + \frac{2}{3} + \sum_{n=3}^{\infty} \frac{2}{3^{n-1}} = \frac{2}{3} + \frac{2}{9} \frac{1}{1 - \frac{1}{3}} = 1$$

These results hold true because, here, whatever the initial state, the process is certain to eventually visit the other state.

Example 3.2.3. In Example 3.2.1, we calculated $\mathbf{P}^{(2)}$. We also find that the matrix $\mathbf{P}^{(3)}$ is given by

$$\mathbf{P}^{(3)} = \begin{array}{c} 0 \\ 1 \\ 2 \end{array} \begin{bmatrix} \begin{array}{ccc} 0 & 1 & 2 \end{array} \\ \begin{array}{ccc} 23/54 & 31/108 & 31/108 \\ 5/12 & 7/24 & 7/24 \\ 4/9 & 5/18 & 5/18 \end{array} \end{bmatrix}$$

From this, we deduce from Eq. (3.25) that, for instance,

$$\frac{1}{3} = \rho_{0,1}^{(1)} = \rho_{0,1}^{(1)} \rho_{1,1}^{(1-1)} = \rho_{0,1}^{(1)} \times 1 \implies \rho_{0,1}^{(1)} = \frac{1}{3}$$

and

$$\frac{5}{18} = p_{0,1}^{(2)} = \rho_{0,1}^{(1)} p_{1,1}^{(2-1)} + \rho_{0,1}^{(2)} p_{1,1}^{(2-2)}$$

$$= \frac{1}{3} \times \frac{1}{2} + \rho_{0,1}^{(2)} \times 1 \implies \rho_{0,1}^{(2)} = \frac{1}{9}$$

and

$$\frac{31}{108} = p_{0,1}^{(3)} = \rho_{0,1}^{(1)} p_{1,1}^{(3-1)} + \rho_{0,1}^{(2)} p_{1,1}^{(3-2)} + \rho_{0,1}^{(3)} p_{1,1}^{(3-3)}$$

$$= \frac{1}{3} \times \frac{1}{4} + \frac{1}{9} \times \frac{1}{2} + \rho_{0,1}^{(3)} \times 1 \implies \rho_{0,1}^{(3)} = \frac{4}{27}$$

and so on.

As in the preceding example, we can, at least for relatively small values of n, try to calculate directly $\rho_{0,1}^{(n)}$. Indeed, we have

$$\rho_{0,1}^{(1)} = p_{0,1}^{(1)} = \frac{1}{3}, \quad \rho_{0,1}^{(2)} = p_{0,0}^{(1)} \times p_{0,1}^{(1)} = \frac{1}{9}$$

and

$$\rho_{0,1}^{(3)} = p_{0,0}^{(1)} \times p_{0,0}^{(1)} \times p_{0,1}^{(1)} + p_{0,2}^{(1)} \times p_{2,0}^{(1)} \times p_{0,1}^{(1)} = \frac{1}{27} + \frac{1}{9} = \frac{4}{27}$$

However, in general, the technique that consists in summing up the probabilities of all the paths leading from i to j in exactly n transitions rapidly becomes of little use.

Until now, we have considered only *conditional* probabilities that the process $\{X_n, n = 0, 1, \dots\}$ finds itself in a state j at a time instant n, *given* that $X_0 = i$. To be able to calculate the *marginal* probability $P[X_n = j]$, we need to know the probability that $X_0 = i$, for every possible state i.

Definition 3.2.3. *The* **initial distribution** *of a Markov chain is the set* $\{a_i, i = 0, 1, \dots\}$, *where* a_i *is defined by*

$$a_i = P[X_0 = i] \tag{3.26}$$

Remarks. i) In many applications, the initial position is *deterministic*. For example, we often suppose that $X_0 = 0$. In this case, the initial distribution of the Markov chain becomes $a_0 = 1$ and $a_i = 0$, for $i = 1, 2, \dots$. In general, we have $\sum_{i=0}^{\infty} a_i = 1$.

ii) The marginal probability $a_j^{(n)} := P[X_n = j]$, for $n = 1, 2, \dots$, is obtained by conditioning on the initial state:

$$a_j^{(n)} \equiv P[X_n = j] = \sum_{i=0}^{\infty} P[X_n = j \mid X_0 = i] P[X_0 = i] = \sum_{i=0}^{\infty} p_{i,j}^{(n)} a_i \tag{3.27}$$

When $a_k = 1$ for some k, we simply have $a_j^{(n)} = p_{k,j}^{(n)}$.

Example 3.2.4. Suppose that $a_i = 1/3$, for $i = 0, 1$, and 2, in Example 3.2.1. Then, we may write that

$$a_0^{(2)} = \frac{1}{3}\left(\frac{4}{9} + \frac{1}{2} + \frac{1}{3}\right) = \frac{23}{54}$$

and

$$a_1^{(2)} = a_2^{(2)} = \frac{1}{3}\left(\frac{5}{18} + \frac{1}{4} + \frac{1}{3}\right) = \frac{31}{108}$$

It follows that

$$E[X_2] = 0 \times \frac{23}{54} + 1 \times \frac{31}{108} + 2 \times \frac{31}{108} = \frac{31}{36}$$

and

$$E[X_2^2] = 0 + 1^2 \times \frac{31}{108} + 2^2 \times \frac{31}{108} = \frac{155}{108}$$

so that

$$V[X_2] = \frac{155}{108} - \left(\frac{31}{36}\right)^2 = \frac{899}{1296}$$

Particular cases

1) A (classic) *random walk* is a Markov chain whose state space is the set $\{\ldots, -2, -1, 0, 1, 2, \ldots\}$ of all integers, and for which

$$p_{i,i+1} = p = 1 - p_{i,i-1} \quad \text{for } i = 0, \pm 1, \pm 2, \ldots \tag{3.28}$$

for some $p \in (0, 1)$. Its one-step transition matrix is therefore (with $q := 1 - p$):

$$\mathbf{P} = \begin{array}{c} \\ i \\ i+1 \\ \\ \end{array} \begin{bmatrix} \ddots & \ddots & \ddots & \\ q & 0 & p & \\ & q & 0 & p \\ & & \ddots & \ddots & \ddots \end{bmatrix} \begin{array}{c} {\scriptstyle i-1 \ \ i \ \ i+1 \ i+2} \end{array} \tag{3.29}$$

Thus, in the case of a classic random walk, the process can make a transition from a given state i only to one or the other of its two immediate neighbors: $i - 1$ or $i + 1$. Moreover, the length of the displacement is always the same, i.e., *one* unit. If the state space is *finite*, say the set $\{0, 1, \ldots, N\}$, then these properties can be checked for each *interior* state: $1, 2, \ldots, N - 1$. When the process reaches state 0 or state N, there are many possibilities. For example, if

$$p_{0,0} = p_{N,N} = 1 \tag{3.30}$$

we say that the states 0 and N are *absorbing*. If $p_{0,1} = 1$, the state 0 is said to be *reflecting*.

Remarks. i) If $p_{1,2} = p$, $p_{1,1} = cq$, and $p_{1,0} = (1 - c)q$, where $0 \leq c \leq 1$, the boundary at 0, when $p_{0,0} = 1$, is said to be *elastic*.

ii) We can also say that the process has a reflecting boundary at the point $1/2$ if the state space is $\{1, 2, \ldots\}$, $p_{1,1} = q$, and $p_{1,2} = p$.

2) We can easily construct a Markov chain from a set $Y_0, Y_1, \ldots$ of i.i.d. random variables, whose possible values are integers, by proceeding as follows: let

$$X_n := \sum_{k=0}^{n} Y_k \quad \text{for } n = 0, 1, \ldots \tag{3.31}$$

Then, $\{X_n, n = 0, 1, \ldots\}$ is a Markov chain and, if

$$\alpha_k := P[Y_n = k] \quad \forall \, k \tag{3.32}$$

we have

$$p_{i,j} := P[X_{n+1} = j \mid X_n = i] = P\left[\sum_{k=0}^{n+1} Y_k = j \,\middle|\, \sum_{k=0}^{n} Y_k = i\right] \tag{3.33}$$

$$= P\left[Y_{n+1} = j - i \,\middle|\, \sum_{k=0}^{n} Y_k = i\right] \stackrel{\text{ind.}}{=} P[Y_{n+1} = j - i] = \alpha_{j-i}$$

We say that the chain, in addition to being time-homogeneous, is also *homogeneous with respect to the state variable*, because the probability $p_{i,j}$ does not depend explicitly on i and j, but only on the difference $j - i$.

3.2.2 Properties

Generally, our first task in the study of a particular Markov chain is to determine whether, from an arbitrary state i, it is possible that the process *eventually* reaches any state j or, rather, some states are *inaccessible* from state i.

Definition 3.2.4. *We say that the state j is **accessible** from i if there exists an $n \geq 0$ such that $p_{i,j}^{(n)} > 0$. We write $i \to j$.*

Remark. Since we include $n = 0$ in the definition and since $p_{i,i}^{(0)} = 1 > 0$, any state is accessible from itself.

Definition 3.2.5. *If i is accessible from j, and j is accessible from i, we say that the states i and j **communicate**, or that they are **communicating**. We write $i \leftrightarrow j$.*

Remark. Any state communicates with itself. Moreover, the notion of *communication* is *commutative*: $i \leftrightarrow j \Rightarrow j \leftrightarrow i$. Finally, since (by the Chapman–Kolmogorov equations)

$$p_{i,k}^{(m+n)} \geq p_{i,j}^{(m)} p_{j,k}^{(n)} \quad \text{for all } m \text{ and } n \geq 0 \qquad (3.34)$$

we can assert that this notion is also *transitive*: if $i \leftrightarrow j$ and $j \leftrightarrow k$, then $i \leftrightarrow k$. Actually, it is sufficient to notice that if it is possible to move from state i to state j in m steps, and from j to k in n steps, then it is possible (by the Markov property) to go from i to k in $m + n$ steps.

Definition 3.2.6. *If $i \leftrightarrow j$, we say that the states i and j are in the same* class.

Remark. The transitivity property (see above) implies that two arbitrary classes are either identical or disjoint. Indeed, suppose that the state space is the set $\{0, 1, 2\}$ and that the chain has two classes: $\{0, 1\}$ and $\{1, 2\}$. Since $0 \leftrightarrow 1$ and $1 \leftrightarrow 2$, we have that $0 \leftrightarrow 2$, which contradicts the fact that the states 0 and 2 are not in the same class. Consequently, we can decompose the state space into disjoint classes.

Definition 3.2.7. *Let C be a subset of the state space of a Markov chain. We say that C is a* closed *set if, from any $i \in C$, the process always remains in C:*

$$P[X_{n+1} \in C \mid X_n = i \in C] = 1 \quad \text{for all } i \in C \qquad (3.35)$$

Definition 3.2.8. *A Markov chain is said to be* **irreducible** *if all the states communicate, that is, if the state space contains no closed set apart from the set of all states.*

The following result is easy to show.

Proposition 3.2.2. *If $i \to j$ or $j \to i$ for all pairs of states i and j of the Markov chain $\{X_n, n = 0, 1, \dots\}$, then the chain is irreducible.*

Remark. We can also give the following irreducibility criterion: if there exists a path, whose probability is *strictly* positive, which starts from any state i and returns to i after having visited at least once all other states of the chain, then the chain is irreducible. We can say that there exists a *cycle* with strictly positive probability.

Example 3.2.5. The Markov chain whose one-step transition matrix $\mathbf{P}$ is given in Example 3.2.1 is irreducible, because we may write that

$$p_{0,1} \times p_{1,2} \times p_{2,0} = \frac{1}{3} \times \frac{1}{2} \times 1 = \frac{1}{6} > 0$$

Note that here the cycle $0 \to 1 \to 2 \to 0$ is the shortest possible. However, we also have, for example,

$$p_{0,2} \times p_{2,0} \times p_{0,1} \times p_{1,2} \times p_{2,0} > 0$$

Remark. If a Markov chain (with at least two states) has an absorbing state, then it cannot be irreducible, since any absorbing state constitutes a closed set by itself.

Definition 3.2.9. *The state i is said to be* **recurrent** *if*

$$f_{i,i} := P\left[\bigcup_{n=1}^{\infty} \{X_n = i\} \,\middle|\, X_0 = i \right] = 1 \tag{3.36}$$

If $f_{i,i} < 1$, we say that i is a **transient** *state.*

Remarks. i) The quantity $f_{i,i}$ is the probability of an *eventual return* of the process to the initial state i. It is a particular case of

$$f_{i,j} := P\left[\bigcup_{n=1}^{\infty} \{X_n = j\} \,\middle|\, X_0 = i \right] \tag{3.37}$$

which denotes the probability that, starting from state i, the process will *eventually* visit state j. We may write that

$$f_{i,j} = \sum_{n=1}^{\infty} p_{i,j}^{(n)} \tag{3.38}$$

ii) It can be shown that the state space S_{X_n} of a Markov chain can be decomposed in a unique way as follows:

$$S_{X_n} = D \cup C_1 \cup C_2 \cup C_3 \ldots \tag{3.39}$$

where D is the set of transient states of the chain and the C_k's, $k = 1, 2, \ldots$, are closed and irreducible sets (that is, the states of each of these sets communicate) of recurrent states.

Proposition 3.2.3. *Let N_i be the number of times that state i will be visited, given that $X_0 = i$. The state i is recurrent if and only if $E[N_i] = \infty$.*

Proof. We have

$$P[N_i = n] = f_{i,i}^{n-1}(1 - f_{i,i}) \quad \text{for } n = 1, 2, \ldots \tag{3.40}$$

That is, $N_i \sim \text{Geom}(p := 1 - f_{i,i})$. Since $E[N_i] = 1/p$, we indeed have

$$E[N_i] = \frac{1}{1 - f_{i,i}} = \infty \quad \Longleftrightarrow \quad f_{i,i} = 1 \qquad \square \qquad (3.41)$$

Remark. Since the set $T = \{0, 1, \dots\}$ is countably infinite, that is, time continues forever, if the probability of revisiting the initial state is equal to 1, then this state will be visited an infinite number of times (because the process starts anew every time the initial state is visited), so that $P[N_i = \infty] = 1$, which directly implies that $E[N_i] = \infty$. Conversely, if i is transient, then we have

$$P[N_i = \infty] = \lim_{n \to \infty} P[N_i \geq n] = \lim_{n \to \infty} f_{i,i}^{n-1} = 0 \qquad (3.42)$$

That is, a transient state will be visited only a *finite* number of times. Consequently, if the state space of a Markov chain is *finite*, then at least one state must be recurrent (otherwise no states would be visited after a finite time).

Proposition 3.2.4. *The state i is recurrent if and only if $\sum_{n=0}^{\infty} p_{i,i}^{(n)} = \infty$.*

Proof. Let $I_{\{X_n = i\}}$ be the *indicator variable* of the event $\{X_n = i\}$. That is,

$$I_{\{X_n = i\}} := \begin{cases} 1 \text{ if } X_n = i \\ 0 \text{ if } X_n \neq i \end{cases} \qquad (3.43)$$

We only have to use the preceding proposition and notice that, given that $X_0 = i$, the random variable $I_{\{X_n = i\}}$ has a Bernoulli distribution with parameter $p := P[X_n = i \mid X_0 = i]$, so that

$$E[N_i] = E\left[\sum_{n=0}^{\infty} I_{\{X_n = i\}} \,\middle|\, X_0 = i\right] = \sum_{n=0}^{\infty} E[I_{\{X_n = i\}} \mid X_0 = i]$$

$$= \sum_{n=0}^{\infty} P[X_n = i \mid X_0 = i] = \sum_{n=0}^{\infty} p_{i,i}^{(n)} \qquad \square \qquad (3.44)$$

Remark. It can also be shown that if i is transient, then

$$\sum_{n=1}^{\infty} p_{k,i}^{(n)} < \infty \quad \text{for all states } k$$

The next result is actually a corollary of the preceding proposition.

Proposition 3.2.5. *Recurrence is a* class property: *if states i and j communicate, then they are either both recurrent or both transient.*

Proof. It is sufficient to show that if i is recurrent, then j too is recurrent (the other result being simply the *contrapositive* of this assertion).

By definition of communicating states, there exist integers k and l such that $p_{i,j}^{(k)} > 0$ and $p_{j,i}^{(l)} > 0$. Now, we have

$$p_{j,j}^{(l+n+k)} \geq p_{j,i}^{(l)} p_{i,i}^{(n)} p_{i,j}^{(k)} \tag{3.45}$$

Then

$$\sum_{n=0}^{\infty} p_{j,j}^{(l+n+k)} \geq \underbrace{p_{j,i}^{(l)} p_{i,j}^{(k)}}_{> 0} \underbrace{\sum_{n=0}^{\infty} p_{i,i}^{(n)}}_{= \infty} = \infty \tag{3.46}$$

Thus, j is recurrent. $\square$

Corollary 3.2.1. *All states of a* finite *and* irreducible *Markov chain are recurrent.*

Proof. This follows indeed from the preceding proposition and from the fact that a *finite* Markov chain must have at least one recurrent state (see p. 88). $\square$

Notation. Let i be a recurrent state. We denote by μ_i the average number of transitions needed by the process, starting from state i, to return to i for the first time. That is,

$$\mu_i := \sum_{n=1}^{\infty} n \rho_{i,i}^{(n)} \tag{3.47}$$

Definition 3.2.10. *Let i be a recurrent state. We say that i is*

$$\begin{cases} \text{positive recurrent } if \ \mu_i < \infty \\ \quad \text{null recurrent } if \ \mu_i = \infty \end{cases}$$

Remarks. i) It can be shown (see Prop. 3.2.4) that i is a null recurrent state if and only if $\sum_{n=0}^{\infty} p_{i,i}^{(n)} = \infty$, but $\lim_{n \to \infty} p_{i,i}^{(n)} = 0$. We then also have $\lim_{n \to \infty} p_{k,i}^{(n)} = 0$, for all states k.

ii) It can be shown as well that two recurrent states that are in the same class are both of the same type: either both positive recurrent or both null recurrent. Thus, the type of recurrence is also a class property.

iii) Finally, it is easy to accept the fact that any recurrent state of a *finite* Markov chain is *positive* recurrent.

Example 3.2.6. Because the Markov chain considered in Example 3.2.1 is finite and irreducible, we can at once assert that the three states—0, 1, and 2—are positive recurrent. However, the chain whose one-step transition matrix is given by

$$
\mathbf{P_1} = \begin{array}{c} \\ 0 \\ 1 \\ 2 \end{array}\overset{\begin{array}{ccc} 0 & 1 & 2 \end{array}}{\begin{bmatrix} 1/3 & 1/3 & 1/3 \\ 0 & 1/2 & 1/2 \\ 0 & 1 & 0 \end{bmatrix}}
$$

is not irreducible, because state 0 is not accessible from either 1 or 2. Since there is a probability of 2/3 that the process will leave state 0 on its very first transition and will never return to that state, we have

$$
f_{0,0} \leq 1 - 2/3 = 1/3 < 1 \quad (\text{actually, } f_{0,0} = 1/3)
$$

which implies that state 0 is *transient*. Next,

$$
p_{1,2} \times p_{2,1} = \frac{1}{2} > 0
$$

so that states 1 and 2 are in the same class. Because the chain must have at least one recurrent state, we conclude that 1 and 2 are *recurrent* states. We may write that $S_{X_n} = D \cup C_1$, where $D = \{0\}$ and $C_1 = \{1, 2\}$.

When

$$
\mathbf{P_2} = \begin{array}{c} \\ 0 \\ 1 \\ 2 \end{array}\overset{\begin{array}{ccc} 0 & 1 & 2 \end{array}}{\begin{bmatrix} 1/3 & 1/3 & 1/3 \\ 0 & 1/2 & 1/2 \\ 0 & 0 & 1 \end{bmatrix}}
$$

we find that each state constitutes a class. The classes $\{0\}$ and $\{1\}$ are transient, whereas $\{2\}$ is recurrent. These results are easy to check by using Proposition 3.2.4. We have

$$
\sum_{n=0}^{\infty} p_{0,0}^{(n)} = \sum_{n=0}^{\infty} \frac{1}{3^n} = \frac{1}{1 - \frac{1}{3}} = \frac{3}{2} < \infty
$$

$$
\sum_{n=0}^{\infty} p_{1,1}^{(n)} = \sum_{n=0}^{\infty} \frac{1}{2^n} = \frac{1}{1 - \frac{1}{2}} = 2 < \infty
$$

and

$$
\sum_{n=0}^{\infty} p_{2,2}^{(n)} = \sum_{n=0}^{\infty} 1 = \infty
$$

As we already mentioned, any absorbing state (like state 2 here) trivially constitutes a recurrent class. Moreover, the state space being finite, state 2 is positive recurrent. Finally, we have that $S_{X_n} = D \cup C_1$, where $D = \{0, 1\}$ and $C_1 = \{2\}$.

Definition 3.2.11. *A state i is said to be* **periodic** *with period d if $p_{i,i}^{(n)} = 0$ for any n that is not divisible by d, where d is the* largest *integer with this property. If $d = 1$, the state is called* **aperiodic.**

Remarks. i) If $p_{i,i}^{(n)} = 0$ for all $n > 0$, we consider i as an aperiodic state. That is, the chain may start from state i, but it leaves i on its first transition and never returns to i. For example, state 0 in the transition matrix

$$\mathbf{P} = \begin{matrix} & \begin{matrix} 0 & 1 & 2 \end{matrix} \\ \begin{matrix} 0 \\ 1 \\ 2 \end{matrix} & \begin{bmatrix} 0 & 1/2 & 1/2 \\ 0 & 1/2 & 1/2 \\ 0 & 0 & 1 \end{bmatrix} \end{matrix}$$

is such that $p_{0,0}^{(n)} = 0$ for all $n > 0$.

ii) As in the case of the other characteristics of the states of a Markov chain, it can be shown that periodicity is a class property: if i and j communicate and if i is periodic with period d, then j too is periodic with period d.

iii) If $p_{i,i}^{(1)} > 0$, then i is evidently aperiodic. Consequently, for an irreducible Markov chain, if there is at least one positive element on the diagonal (from $p_{0,0}$) of the transition matrix $\mathbf{P}$, then the chain is aperiodic, by the preceding remark.

iv) If $p_{i,i}^{(2)} > 0$ and $p_{i,i}^{(3)} > 0$, then i is aperiodic, because for any $n \in \{2, 3, \dots\}$ there exist integers k_1 and $k_2 \in \{0, 1, \dots\}$ such that $n = 2k_1 + 3k_2$.

v) If $d = 4$ for an arbitrary state i, then $d = 2$ too satisfies the definition of periodicity. Indeed, we then have

$$p_{i,i}^{(2n+1)} = 0 \quad \text{for } n = 0, 1, \dots$$

so that we can assert that $p_{i,i}^{(n)} = 0$ for any n that is not divisible by 2. Thus, we must really take the *largest* integer with the property in question.

Example 3.2.7. If

$$\mathbf{P} = \begin{matrix} & \begin{matrix} 0 & 1 \end{matrix} \\ \begin{matrix} 0 \\ 1 \end{matrix} & \begin{bmatrix} 0 & 1 \\ 1 & 0 \end{bmatrix} \end{matrix}$$

then the Markov chain is periodic with period $d = 2$, because it is irreducible and

$$p_{0,0}^{(2n+1)} = 0 \quad \text{and} \quad p_{0,0}^{(2n)} = 1 \quad \text{for } n = 0, 1, \dots$$

Note that d is not equal to 4, in particular because 2 is not divisible by 4 and $p_{0,0}^{(2)} = 1 > 0$.

Example 3.2.8. Let

$$\mathbf{P} = \begin{array}{c} \\ 0 \\ 1 \\ 2 \end{array} \begin{array}{ccc} 0 & 1 & 2 \\ \left[\begin{array}{ccc} 0 & 1/2 & 1/2 \\ 1/2 & 0 & 1/2 \\ 1/2 & 1/2 & 0 \end{array} \right] \end{array}$$

The Markov chain is aperiodic. Indeed, it is irreducible (since $p_{0,1} \times p_{1,2} \times p_{2,0} = 1/8 > 0$) and

$$p_{0,0}^{(2)} (= 1/2) > 0 \quad \text{and} \quad p_{0,0}^{(3)} (= 1/4) > 0$$

Example 3.2.9. Let

$$\mathbf{P} = \begin{array}{c} \\ 0 \\ 1 \\ 2 \\ 3 \\ 4 \end{array} \begin{array}{ccccc} 0 & 1 & 2 & 3 & 4 \\ \left[\begin{array}{ccccc} 0 & 1/2 & 1/2 & 0 & 0 \\ 0 & 0 & 0 & 1/3 & 2/3 \\ 0 & 0 & 0 & 2/3 & 1/3 \\ 1 & 0 & 0 & 0 & 0 \\ 1 & 0 & 0 & 0 & 0 \end{array} \right] \end{array}$$

be the one-step transition matrix of a Markov chain whose state space is the set $\{0, 1, 2, 3, 4\}$. We have

$$p_{0,1} \times p_{1,3} \times p_{3,0} \times p_{0,2} \times p_{2,4} \times p_{4,0} > 0$$

Therefore, the chain is irreducible. Moreover, we find that

$$p_{0,0}^{(n)} = \begin{cases} 1 \text{ if } n = 0, 3, 6, 9, \ldots \\ 0 \text{ otherwise} \end{cases}$$

Thus, we may conclude that the chain is periodic with period $d = 3$.

Example 3.2.10. In the case of a *classic* random walk, defined on all the integers, all the states communicate, which follows directly from the fact that the process cannot jump over a neighboring state and that it is *unconstrained*, that is, there are no boundaries (absorbing or else). The chain being irreducible, the states are either all recurrent or all transient. We consider state 0. Since the process moves exactly one unit to the right or to the left at each transition, it is clear that if it starts from 0, then it cannot be at 0 after an uneven number of transitions. That is,

$$p_{0,0}^{(2n+1)} = 0 \quad \text{for } n = 0, 1, \ldots$$

Next, for the process to be back to the initial state 0 after $2n$ transitions, there must have been exactly n transitions to the right (and thus n to the left). Since the transitions are independent and the probability that the process moves to the right is always equal to p, we may write that, for $n = 1, 2, \ldots$,

$$p_{0,0}^{(2n)} = P[\text{B}(2n, p) = n] = \binom{2n}{n} p^n (1-p)^n = \frac{(2n)!}{n!n!} p^n (1-p)^n$$

To determine whether state 0 (and therefore the chain) is recurrent, we consider the sum

$$\sum_{n=1}^{\infty} p_{0,0}^{(n)} = \sum_{n=1}^{\infty} p_{0,0}^{(2n)}$$

However, we do not need to know the exact value of this sum, but rather we need only know whether it converges or not. Consequently, we can use *Stirling's[2] formula*:

$$n! \sim n^{n+\frac{1}{2}} e^{-n} \sqrt{2\pi} \tag{3.48}$$

(that is, the ratio of both terms tends to 1 when n tends to infinity) to determine the behavior of the infinite sum. We find that

$$p_{0,0}^{(2n)} \sim \frac{[4p(1-p)]^n}{\sqrt{\pi n}}$$

which implies that

$$\sum_{n=1}^{\infty} p_{0,0}^{(n)} < \infty \quad \Longleftrightarrow \quad \sum_{n=1}^{\infty} \frac{[4p(1-p)]^n}{\sqrt{\pi n}} < \infty$$

Now, if $p = 1/2$, the stochastic process is called a *symmetric* random walk and the sum becomes

$$\sum_{n=1}^{\infty} \frac{1}{\sqrt{\pi n}} = \infty$$

because

$$\sum_{n=1}^{\infty} \frac{1}{\sqrt{\pi n}} \geq \frac{1}{\sqrt{\pi}} \sum_{n=1}^{\infty} \frac{1}{n} = \infty$$

When $p \neq 1/2$, we have

$$\sum_{n=1}^{\infty} \frac{[4p(1-p)]^n}{\sqrt{\pi n}} \leq \sum_{n=1}^{\infty} \frac{[4p(1-p)]^n}{\sqrt{\pi}} < \infty$$

Thus, the classic random walk is recurrent if $p = 1/2$ and is transient if $p \neq 1/2$.

Remark. It can be shown that the two-dimensional symmetric random walk too is recurrent. However, those of dimension $k \geq 3$ are transient. This follows from the fact that the probability of returning to the origin in $2n$ transitions is bounded by $c/n^{k/2}$, where c is a constant, and

$$\sum_{n=1}^{\infty} \frac{c}{n^{k/2}} < \infty \quad \text{if } k \geq 3 \tag{3.49}$$

---

[2] James Stirling, 1692–1770, who was born and died in Scotland, was a mathematician who worked especially in the field of differential calculus, particularly on the *gamma function*.

3.2.3 Limiting probabilities

In this section, we will present a theorem that enables us to calculate, if it exists, the limit $\lim_{n \to} p_{i,j}^{(n)}$ by solving a system of linear equations, rather than by trying to obtain directly $\lim_{n \to \infty} \mathbf{P}^{(n)}$, which is generally difficult. The theorem in question is valid when the chain is irreducible and *ergodic*, as defined below.

Definition 3.2.12. *Positive recurrent and aperiodic states are called* **ergodic** *states.*

Theorem 3.2.1. *In the case of an irreducible and ergodic Markov chain, the limit*

$$\pi_j := \lim_{n \to \infty} p_{i,j}^{(n)} \tag{3.50}$$

exists and is independent of i. Moreover, we have

$$\pi_j = \frac{1}{\mu_j} > 0 \quad \text{for all } j \in \{0, 1, \dots\} \tag{3.51}$$

where μ_j is defined in Eq. (3.47). To obtain the π_j's, we can solve the system

$$\pi = \pi \mathbf{P} \tag{3.52}$$

$$\sum_{j=0}^{\infty} \pi_j = 1 \tag{3.53}$$

where $\pi := (\pi_0, \pi_1, \dots)$. It can be shown that the preceding system possesses a unique *positive solution.*

Remarks. i) For a given state j, Eq. (3.52) becomes

$$\pi_j = \sum_{i=0}^{\infty} \pi_i \, p_{i,j} \tag{3.54}$$

Note that, if the state space is finite and comprises k states, then there are k equations like the following one:

$$\pi_j = \sum_{i=0}^{k-1} \pi_i \, p_{i,j} \tag{3.55}$$

With the condition $\sum_{j=0}^{k-1} \pi_j = 1$, there are thus $k+1$ equations and k unknowns. In practice, we drop one of the k equations given by (3.55) and make use of the condition $\sum_{j=0}^{k-1} \pi_j = 1$ to obtain a unique solution. Theoretically, we should make sure that the π_j's obtained satisfy the equation that was dropped.

ii) A solution $\{\pi_j, j = 0, 1, \ldots\}$ of the system (3.52), (3.53) is called a *stationary distribution* of the Markov chain. This terminology stems from the fact that if we set

$$P[X_0 = j] = \pi_j \quad \forall j \tag{3.56}$$

then, proceeding by *mathematical induction*, we may write that

$$P[X_{n+1} = j] = \sum_{i=0}^{\infty} P[X_{n+1} = j \mid X_n = i] P[X_n = i]$$

$$= \sum_{i=0}^{\infty} p_{i,j} \pi_i = \pi_j \tag{3.57}$$

where the last equality follows from Eq. (3.54). Thus, we find that

$$P[X_n = j] = \pi_j \quad \forall\, n, j \in \{0, 1, 2, \ldots\} \tag{3.58}$$

iii) If the chain is *not* irreducible, then there can be many stationary distributions. For example, let

$$\mathbf{P} = \begin{array}{c} \\ 0 \\ 1 \end{array} \!\! \begin{array}{cc} 0 & 1 \\ \begin{bmatrix} 1 & 0 \\ 0 & 1 \end{bmatrix} \end{array}$$

Because states 0 and 1 are absorbing, they are positive recurrent and aperiodic. However, the chain has two classes: $\{0\}$ and $\{1\}$. The system that must be solved to obtain the π_j's is

$$\pi_0 = \pi_0 \tag{3.59}$$
$$\pi_1 = \pi_1 \tag{3.60}$$
$$\pi_0 + \pi_1 = 1 \tag{3.61}$$

We see that $\pi_0 = c$, $\pi_1 = 1 - c$ is a valid solution for any $c \in [0, 1]$.

iv) The theorem asserts that the π_j's exist and can be obtained by solving the system (3.52), (3.53). If we assume that the π_j's exist, then it is easy to show that they must satisfy Eq. (3.52). Indeed, since $\lim_{n\to\infty} P[X_{n+1} = j] = \lim_{n\to\infty} P[X_n = j]$, we have

$$P[X_{n+1} = j] = \sum_{i=0}^{\infty} P[X_{n+1} = j \mid X_n = i] P[X_n = i]$$

$$\implies \lim_{n\to\infty} P[X_{n+1} = j] = \lim_{n\to\infty} \sum_{i=0}^{\infty} P[X_{n+1} = j \mid X_n = i] P[X_n = i]$$

$$\implies \pi_j = \sum_{i=0}^{\infty} p_{i,j} \pi_i = \sum_{i=0}^{\infty} \pi_i p_{i,j} \tag{3.62}$$

where we assumed that we can interchange the limit and the summation.

v) In addition to being the limiting probabilities, the π_j's also represent the *proportion of time* that the process spends in state j, over a long period of time. Actually, if the chain is positive recurrent but *periodic*, then this is the only interpretation of the π_j's. Note that, on average, the process spends one time unit in state j for μ_j time units, from which we deduce that $\pi_j = 1/\mu_j$, as stated in the theorem.

vi) When the Markov chain can be decomposed into subchains, we can apply the theorem to each of these subchains. For example, suppose that the transition matrix of the chain is

$$\mathbf{P} = \begin{array}{c} \\ 0 \\ 1 \\ 2 \end{array} \begin{array}{ccc} 0 & 1 & 2 \\ \left[\begin{array}{ccc} 1 & 0 & 0 \\ 0 & 1/2 & 1/2 \\ 0 & 1/2 & 1/2 \end{array} \right] \end{array}$$

In this case, the limit $\lim_{n\to\infty} p_{i,j}^{(n)}$ exists but is *not* independent of i. We easily find that

$$\lim_{n\to\infty} p_{i,0}^{(n)} = \begin{cases} 1 \text{ if } i = 0 \\ 0 \text{ if } i \neq 0 \end{cases} \tag{3.63}$$

and (by symmetry)

$$\lim_{n\to\infty} p_{i,1}^{(n)} = \lim_{n\to\infty} p_{i,2}^{(n)} = \begin{cases} 1/2 \text{ if } i = 1 \text{ or } 2 \\ 0 \quad \text{ if } i = 0 \end{cases} \tag{3.64}$$

However, we cannot write that $\pi_0 = 1$ and $\pi_1 = \pi_2 = 1/2$, since the π_j's do not exist in this example. At any rate, we see that the sum of the π_j's would then be equal to 2, and not to 1, as required.

Example 3.2.11. The Markov chain $\{X_n, n = 0, 1, \dots\}$ considered in Example 3.2.1 is irreducible and positive recurrent (see Example 3.2.6). Since the probability $p_{0,0}$ (or $p_{1,1}$) is strictly positive, the chain is aperiodic, and thus ergodic, so that Theorem 3.2.1 applies. We must solve the system

$$(\pi_0, \pi_1, \pi_2) = (\pi_0, \pi_1, \pi_2) \begin{bmatrix} 1/3 & 1/3 & 1/3 \\ 0 & 1/2 & 1/2 \\ 1 & 0 & 0 \end{bmatrix}$$

that is,

$$\pi_0 \overset{(1)}{=} \frac{1}{3}\pi_0 + \pi_2$$

$$\pi_1 \overset{(2)}{=} \frac{1}{3}\pi_0 + \frac{1}{2}\pi_1$$

$$\pi_2 \overset{(3)}{=} \frac{1}{3}\pi_0 + \frac{1}{2}\pi_1$$

subject to the condition $\pi_0 + \pi_1 + \pi_2 \overset{(*)}{=} 1$. Often, we try to express every limiting probability in terms of π_0, and then we make use of the condition (*) to evaluate π_0. Here, Eq. (1) implies that $\pi_2 = \frac{2}{3}\pi_0$, and Eq. (2) enables us to assert that $\pi_1 = \frac{2}{3}\pi_0$ too [so that $\pi_1 = \pi_2$, which actually follows directly from Eqs. (2) and (3)]. Substituting into Eq. (*), we obtain:

$$\pi_0 + \frac{2}{3}\pi_0 + \frac{2}{3}\pi_0 = 1 \quad \Rightarrow \quad \pi_0 = \frac{3}{7} \quad \Rightarrow \quad \pi_1 = \pi_2 = \frac{2}{7}$$

We can check that these values of the π_j's are also a solution of Eq. (3), which was not used.

Note finally that the π_j's correspond to the limits of the $p_{i,j}^{(n)}$'s computed in Example 3.2.1 by finding the approximate value of the matrix $\mathbf{P}^{(n)}$ when n is large.

When the state space S_{X_n} is finite, before trying to solve the system (3.52), (3.53), it is recommended to check whether the chain is doubly stochastic (see p. 79), as can be seen in the following proposition.

Proposition 3.2.6. *In the case of an irreducible and aperiodic Markov chain whose state space is the finite set $\{0, 1, \ldots, k\}$, if the chain is doubly stochastic, then the limiting probabilities exist and are given by*

$$\pi_j = \frac{1}{k+1} \quad for \ j = 0, 1, \ldots, k \tag{3.65}$$

Proof. Since the number of states is finite, the chain is positive recurrent. Moreover, as the chain is aperiodic (by assumption), it follows that it is ergodic. Thus, we can assert that the π_j's exist and are the unique positive solution of

$$\pi_j = \sum_{i=0}^{k} \pi_i\, p_{i,j} \quad \forall\, j \in \{0, 1, \ldots, k\} \tag{3.66}$$

$$\sum_{j=0}^{k} \pi_j = 1 \tag{3.67}$$

Now, if we set $\pi_j \equiv 1/(k+1) > 0$, we obtain

$$\sum_{j=0}^{k} \frac{1}{k+1} = 1 \tag{3.68}$$

and

$$\sum_{i=0}^{k} \frac{1}{k+1} p_{i,j} = \frac{1}{k+1} \sum_{i=0}^{k} p_{i,j} = \frac{1}{k+1} \times 1 = \frac{1}{k+1} \tag{3.69}$$

(because the chain is doubly stochastic). $\square$

Example 3.2.12. The matrix

$$
\begin{array}{c}
\quad\quad 0 \quad 1 \quad 2 \\
\mathbf{P} = \begin{array}{c} 0 \\ 1 \\ 2 \end{array}\begin{bmatrix} 1/3 & 1/3 & 1/3 \\ 0 & 1/2 & 1/2 \\ 2/3 & 1/6 & 1/6 \end{bmatrix}
\end{array}
$$

is irreducible and aperiodic (because $p_{0,0} > 0$, in particular). Since it is doubly stochastic, we can conclude that $\pi_j = 1/3$ for each of the three possible states. We can, of course, check these results by applying Theorem 3.2.1.

Example 3.2.13. The one-step transition matrix

$$
\begin{array}{c}
\quad\quad 0 \quad 1 \quad 2 \\
\mathbf{P} = \begin{array}{c} 0 \\ 1 \\ 2 \end{array}\begin{bmatrix} 1 & 0 & 0 \\ 0 & 1/2 & 1/2 \\ 0 & 1/2 & 1/2 \end{bmatrix}
\end{array}
$$

is doubly stochastic. However, the chain is not irreducible. Consequently, we cannot write that $\pi_j \equiv 1/3$. Actually, we calculated the limits $\lim_{n \to \infty} p_{i,j}^{(n)}$ in remark vi) on p. 96.

When the state space $\{0, 1, \dots\}$ is infinite, the calculation of the limiting probabilities is generally difficult. However, when the transition matrix is such that $p_{0,1} > 0$, $p_{j,j-1} \times p_{j,j+1} > 0 \ \forall \ j > 0$, and

$$
p_{i,j} = 0 \quad \text{if } | j - i | > 1 \tag{3.70}
$$

we can give a general formula for the π_j's. Indeed, we have

$$
\pi_0 = \pi_0 p_{0,0} + \pi_1 p_{1,0} \tag{3.71}
$$

and

$$
\pi_j = \pi_{j-1} p_{j-1,j} + \pi_j p_{j,j} + \pi_{j+1} p_{j+1,j} \tag{3.72}
$$

for $j = 1, 2, \dots$. We find that

$$
\pi_j = \frac{p_{0,1} \times p_{1,2} \times \cdots \times p_{j-1,j}}{p_{1,0} \times p_{2,1} \times \cdots \times p_{j,j-1}} \pi_0 \quad \text{for } j = 1, 2, \dots \tag{3.73}
$$

Then a *necessary* and *sufficient* condition for the existence of the π_j's is that

$$
\sum_{k=1}^{\infty} \frac{p_{0,1} \times p_{1,2} \times \cdots \times p_{k-1,k}}{p_{1,0} \times p_{2,1} \times \cdots \times p_{k,k-1}} < \infty \tag{3.74}
$$

If the sum above diverges, then we cannot have that $\sum_{j=0}^{\infty} \pi_j = 1$.

Note that this type of Markov chain includes the random walks on the set $\{0, 1, \ldots\}$, for which $p_{0,1} = p = 1 - p_{0,0}$ and $p_{i,i+1} = p = 1 - p_{i,i-1}$, for $i = 1, 2, \ldots$. We have

$$\sum_{k=1}^{\infty} \frac{p_{0,1} \times p_{1,2} \times \cdots \times p_{k-1,k}}{p_{1,0} \times p_{2,1} \times \cdots \times p_{k,k-1}} = \sum_{k=1}^{\infty} \frac{p^k}{(1-p)^k} \tag{3.75}$$

and this sum converges if and only if $p < 1/2$. Thus, we can write (with $q := 1 - p$) that

$$\pi_0 \sum_{k=0}^{\infty} (p/q)^k = 1 \quad \Longleftrightarrow \quad \pi_0 = 1 - \frac{p}{q} \tag{3.76}$$

and then

$$\pi_j = \left(\frac{p}{q}\right)^j \pi_0 = \left(\frac{p}{q}\right)^j \left(1 - \frac{p}{q}\right) \tag{3.77}$$

for $j = 1, 2, \ldots$.

Actually, we could have used Theorem 3.2.1 to obtain these results. Indeed, consider the transition matrix

$$\mathbf{P} = \begin{array}{c} \\ 0 \\ 1 \\ 2 \\ \vdots \\ \vdots \\ \vdots \\ k-1 \\ k \end{array} \begin{array}{cccccccc} 0 & 1 & 2 & 3 & \ldots & \ldots & k-2 & k-1 & k \end{array} \left[\begin{array}{ccccccccc} q & p & & & & & & & \\ q & 0 & p & & & & & & \\ & q & 0 & p & & & & & \\ & & & \ddots & \ddots & \ddots & & & \\ & & & & \ddots & \ddots & \ddots & & \\ & & & & & \ddots & \ddots & \ddots & \\ & & & & & & \ddots & \ddots & \ddots \\ & & & & & & q & 0 & p \\ & & & & & & & q & p \end{array} \right] \tag{3.78}$$

As the Markov chain is irreducible and ergodic, we can try to solve the system

$$\begin{aligned} \pi_0 &= q\,\pi_0 + q\,\pi_1 \\ \pi_j &= p\,\pi_{j-1} + q\,\pi_{j+1} \quad \text{for } j = 1, \ldots, k-1 \\ \pi_k &= p\,\pi_{k-1} + p\,\pi_k \end{aligned}$$

We find that

$$\pi_j = \left(\frac{p}{q}\right)^j \pi_0 \quad \text{for } j = 1, \ldots, k \tag{3.79}$$

Then, the condition $\sum_{j=0}^{k} \pi_j = 1$ enables us to write

$$\pi_0 = \left(1 - \frac{p}{q}\right) \left[1 - \left(\frac{p}{q}\right)^{k+1}\right]^{-1} \tag{3.80}$$

Taking the limit as k tends to infinity, we obtain (in the case when $p < q \Leftrightarrow p < 1/2$)

$$\pi_0 \to \left(1 - \frac{p}{q}\right) \quad \text{and} \quad \pi_j \to \left(\frac{p}{q}\right)^j \left(1 - \frac{p}{q}\right) \tag{3.81}$$

for $j = 1, 2, \ldots$, which effectively corresponds to the formulas (3.76) and (3.77).

3.2.4 Absorption problems

We already mentioned (see p. 87) that the state space of a Markov chain can be decomposed into the set D of transient states of the chain and the union of closed and irreducible sets C_k of recurrent states. We are interested in the problem of determining the probability that, starting from an element of D, the process will remain indefinitely in D or instead will enter one of the sets C_k, from where it cannot escape.

Notation. Let $i \in D$ and let C be a recurrent class. We set

$$r_i^{(n)}(C) = P[X_n \in C \mid X_0 = i] = \sum_{j \in C} p_{i,j}^{(n)} \quad \text{for } n = 1, 2, \ldots \tag{3.82}$$

and

$$r_i(C) = \lim_{n \to \infty} r_i^{(n)}(C) = P\left[\bigcup_{n=1}^{\infty} \{X_n \in C\} \middle| X_0 = i\right] \tag{3.83}$$

We have the following result.

Theorem 3.2.2. *The probability $r_i(C)$ is the smallest nonnegative solution of the system*

$$r_i(C) = \sum_{j \in D} p_{i,j} \, r_j(C) + \sum_{j \in C} p_{i,j} \quad \text{for all } i \in D \tag{3.84}$$

Moreover, if D is a finite *set, then the solution is* unique.

Remarks. i) The system (3.84) is a system of nonhomogeneous linear equations. When D is finite, we can try to solve this system by using results from linear algebra.

ii) Actually, it is not necessary that the set C be a class. It is sufficient that C be a *closed* set of recurrent states.

iii) We can imagine that all the states of the recurrent class C constitute a single absorbing state, since, once the process has entered this class, it cannot leave it again.

A classic example of an absorption problem is known as the *gambler's ruin problem* and is described as follows: a player, at each play of a game, wins one unit (for example, one dollar) with probability p and loses one unit with probability $q := 1 - p$. Assume that he initially possesses i units and that he plays independent repetitions of the game until his fortune reaches k units or he goes broke.

Let X_n be the fortune of the player at time n (that is, after n plays). Then $\{X_n, n = 0, 1, \ldots\}$ is a Markov chain whose state space is the set $\{0, 1, \ldots, k\}$. As states 0 and k are *absorbing*, we have that $p_{0,0} = p_{k,k} = 1$. For all the other states $i = 1, \ldots, k - 1$, we may write that

$$p_{i,i+1} = p = 1 - p_{i,i-1} \tag{3.85}$$

This is a random walk on the set $\{0, 1, \ldots, k\}$, with absorbing boundaries at 0 and k. The chain thus has three classes: $\{0\}$, $\{k\}$, and $\{1, 2, \ldots, k-1\}$. The first two are recurrent, because 0 and k are absorbing, whereas the third one is transient. Indeed, we have, in particular, that

$$P[X_1 = 0 \mid X_0 = 1] = q > 0 \quad \Longrightarrow \quad f_{1,1} < 1 \tag{3.86}$$

from which we can conclude that the player's fortune will reach 0 or k units after a *finite* number of repetitions.

Let $r_i(\{0\})$, for $i = 0, 1, \ldots, k$, be the probability that the player will be ruined, given that his initial fortune is equal to i units. That is, we write that C is the class $\{0\}$ in Eq. (3.83). We will first consider the case when $p = 1/2$. Note that, in this case, we have

$$E[X_1 \mid X_0 = i] = (i - 1) \times \frac{1}{2} + (i + 1) \times \frac{1}{2} = i \quad \text{for } i = 1, \ldots, k - 1 \tag{3.87}$$

We also have

$$E[X_1 \mid X_0 = 0] = 0 \quad \text{and} \quad E[X_1 \mid X_0 = k] = k \tag{3.88}$$

That is, in general,

$$E[X_{n+1} \mid X_n = i] = i \quad \text{for any } i \tag{3.89}$$

This type of Markov chain is a *martingale* and is very important for the applications, notably in financial mathematics.

Definition 3.2.13. *A Markov chain for which*

$$E[X_{n+1} \mid X_n = i] \equiv \sum_{j \in S_{X_n}} j\, p_{i,j} = i \quad \text{for all } i \in S_{X_n} \tag{3.90}$$

is called a **martingale.**

Remark. We can rewrite Eq. (3.90) as follows:

$$E[X_{n+1} \mid X_n] = X_n \tag{3.91}$$

Now, proceeding by *induction*, we may write that

$$i = E[X_n \mid X_0 = i] = \sum_{j=0}^{k} j\, p_{i,j}^{(n)} \tag{3.92}$$

Indeed, if we make the *induction assumption* that $E[X_{n-1} \mid X_0 = i] = i$, then we have

$$E[X_n \mid X_0 = i] = \sum_{j=0}^{k} j\, p_{i,j}^{(n)} = \sum_{j=0}^{k} \sum_{l=0}^{k} j\, p_{i,l}\, p_{l,j}^{(n-1)}$$

$$= \sum_{l=0}^{k} p_{i,l} \sum_{j=0}^{k} j\, p_{l,j}^{(n-1)} = \sum_{l=0}^{k} p_{i,l} E[X_{n-1} \mid X_0 = l]$$

$$= \sum_{l=0}^{k} p_{i,l}\, l = E[X_1 \mid X_0 = i] = i \tag{3.93}$$

Now, we have

$$\lim_{n \to \infty} p_{i,j}^{(n)} = 0 \quad \text{for all transient states } j \tag{3.94}$$

(otherwise the sum $\sum_{n=1}^{\infty} p_{i,j}^{(n)}$ would diverge, contradicting the remark on p. 88). It follows, taking the limit as n tends to infinity in Eq. (3.92), that

$$i = \lim_{n \to \infty} \left\{ 0\, p_{i,0}^{(n)} + \sum_{j=1}^{k-1} j\, p_{i,j}^{(n)} + k\, p_{i,k}^{(n)} \right\} = k \lim_{n \to \infty} p_{i,k}^{(n)} \tag{3.95}$$

That is,

$$\lim_{n \to \infty} p_{i,k}^{(n)} = \frac{i}{k} \tag{3.96}$$

from which we deduce that

$$\lim_{n \to \infty} p_{i,0}^{(n)} = 1 - \frac{i}{k} \quad \Longrightarrow \quad r_i(\{0\}) = 1 - \frac{i}{k} \tag{3.97}$$

Thus, the more ambitious the player is, the greater is the risk that he will be ruined.

In the general case where $p \in (0,1)$, the probability $r_i(\{0\})$, which will be denoted by r_i to simplify the formulas, is such that $r_0 = 1$, $r_k = 0$ and [see Eq. (3.84)]

$$r_1 = p\,r_2 + q \tag{3.98}$$
$$r_i = q\,r_{i-1} + p\,r_{i+1} \quad \text{for } i = 2,\dots,k-2 \tag{3.99}$$

and

$$r_{k-1} = q\,r_{k-2} \tag{3.100}$$

Rewriting Eq. (3.99) as follows:

$$(p+q)\,r_i = q\,r_{i-1} + p\,r_{i+1} \tag{3.101}$$

we obtain

$$r_{i+1} - r_i = \frac{q}{p}(r_i - r_{i-1}) \quad \text{for } i = 2,\dots,k-2 \tag{3.102}$$

which implies that

$$r_{i+1} - r_i = \left(\frac{q}{p}\right)^{i-1}(r_2 - r_1) = \left(\frac{q}{p}\right)^{i}(r_1 - 1) \tag{3.103}$$

where the last equality follows from Eq. (3.98).

Next, since $r_k = 0$, Eq. (3.100) can be rewritten as follows:

$$r_{k-1} = q\,r_{k-2} + p\,r_k \tag{3.104}$$

We can then state that Eq. (3.103) is valid for $i = 0,\dots,k-1$. Adding the equations for each of these values of i, from 0 to $j-1$, we find that

$$r_j - r_0 = (r_1 - 1)\sum_{i=0}^{j-1}(q/p)^i \tag{3.105}$$

Since $r_0 = 1$, we obtain

$$r_j = \begin{cases} 1 + \dfrac{1 - (q/p)^j}{1 - (q/p)}(r_1 - 1) & \text{if } p \neq q \\[2mm] 1 + j(r_1 - 1) & \text{if } p = q \end{cases} \tag{3.106}$$

for $j = 0,1,\dots,k$. Finally, the fact that $r_k = 0$ enables us to obtain an explicit expression for $r_1 - 1$ from the preceding formula, from which we find that

$$r_j = \begin{cases} 1 - \dfrac{1 - (q/p)^j}{1 - (q/p)^k} & \text{if } p \neq q \\ \\ 1 - \dfrac{j}{k} & \text{if } p = q \end{cases} \tag{3.107}$$

Note that for $p = q = 1/2$, we retrieve the formula (3.97).

We can calculate the limit of the probability r_j in the case when k tends to infinity. It is easy to check that

$$\lim_{k \to \infty} r_j = \begin{cases} (q/p)^j & \text{if } p > 1/2 \\ 1 & \text{if } p \leq 1/2 \end{cases} \tag{3.108}$$

Thus, if $p > 1/2$, there exists a strictly positive probability that the process will spend an infinite time in the set of transient states.

Another example for which the Markov chain may spend an infinite time in the set D of transient states of the chain is the following.

Example 3.2.14. Let $p_{0,0} = 1$ and

$$p_{j,0} = \alpha_j \, (> 0) \, = 1 - p_{j,j+1} \quad \text{for } j = 1, 2, \ldots$$

be the one-step transition probabilities of a Markov chain whose state space is the set $\{0, 1, \ldots\}$. This chain has two classes: $\{0\}$ (recurrent) and $\{1, 2, \ldots\}$ (transient). We calculate directly

$$r_j(\{0\}) = 1 - \prod_{i=0}^{\infty} (1 - \alpha_{j+i})$$

It can be shown that $r_j(\{0\}) < 1$ if and only if the sum $\sum_{i=0}^{\infty} \alpha_{j+i}$ converges. For example, if $\alpha_i = (1/2)^i$, for all $i \geq 1$, then we have

$$\sum_{i=0}^{\infty} \alpha_{j+i} = \sum_{i=0}^{\infty} (1/2)^{j+i} = (1/2)^{j-1} < \infty$$

3.2.5 Branching processes

In 19th-century England, some people got interested in the possibility that certain family names (particularly names of aristocratic families) would disappear, for lack of male descendants. Galton[3] formulated the problem mathematically in 1873, and he and Watson[4] published a paper on this subject in

---

[3] Francis Galton, 1822–1911, was born and died in England. He was a cousin of Charles Darwin. After having studied mathematics, he became an explorer and anthropologist.

[4] The Reverend Henry William Watson, 1827–1903, was born and died in England. He was a mathematician who wrote many books on various subjects. He became a priest in 1858.

1874. Because Bienaymé[5] had worked on this type of problem previously, the corresponding stochastic processes are sometimes called *(branching) processes of Bienaymé–Galton–Watson*. More simply, the term *branching processes* is also used.

Definition 3.2.14. *Let $\{Z_{n,j}, n = 0, 1, \ldots ; j = 1, 2, \ldots\}$ be a set of i.i.d. random variables whose possible values are nonnegative integers. That is, $S_{Z_{n,j}} \subset \{0, 1, \ldots\}$. A* **branching process** *is a Markov chain $\{X_n, n = 0, 1, \ldots\}$ defined by*

$$X_n = \begin{cases} \sum_{j=1}^{X_{n-1}} Z_{n-1,j} & \text{if } X_{n-1} > 0 \\ 0 & \text{if } X_{n-1} = 0 \end{cases} \tag{3.109}$$

for $n = 1, 2, \ldots$.

Remarks. i) In the case of the application to the problem of the disappearance of family names, we can interpret the random variables X_n and $Z_{n-1,j}$ as follows: X_0 is the number of members of the initial generation, that is, the number of *ancestors* of the population. Often, we assume that $X_0 = 1$, so that we are interested in a *lineage*. $Z_{n-1,j}$ denotes the number of descendants of the jth member of the $(n-1)$st generation.

ii) Let

$$p_i := P[Z_{n-1,j} = i] \quad \text{for all } n \text{ and } j \tag{3.110}$$

To avoid trivial cases, we assume that p_i is *strictly* smaller than 1, for all $i = 0, 1, \ldots$. We also assume that $p_0 > 0$; otherwise, the problem of the disappearance of family names would not exist.

The state space S_{X_n} of the Markov chain $\{X_n, n = 0, 1, \ldots\}$ is the set $\{0, 1, \ldots\}$. As state 0 is absorbing, we can decompose S_{X_n} into two sets:

$$S_{X_n} = D \cup \{0\} \tag{3.111}$$

where $D = \{1, 2, \ldots\}$ is the set of transient states. Indeed, since we assumed that $p_0 > 0$, we may write that

$$P[X_n \neq i \ \forall \ n \in \{1, 2, \ldots\} \mid X_0 = i] \geq p_{i,0} \stackrel{\text{ind.}}{=} p_0^i > 0 \tag{3.112}$$

Thus, $f_{i,i} < 1$, and all the states $i = 1, 2, \ldots$ are effectively transient. Now, given that a transient state is visited only a *finite* number of times, we can

---

[5] Irénée-Jules Bienaymé, 1796–1878, was born and died in France. He studied at the École Polytechnique de Paris. In 1848, he was named professor of probability at the Sorbonne. A friend of Chebyshev, he translated Chebyshev's works from Russian into French.

assert that the process cannot remain indefinitely in the set $\{1, 2, ..., k\}$, for any finite k. Thus, we conclude that the population will disappear *or* that its size will tend to infinity.

Suppose that $X_0 = 1$. Let's now calculate the average number μ_n of individuals in the nth generation, for $n = 1, 2, \ldots$. Note that $\mu_1 \equiv E[X_1]$ is the average number of descendants of an individual, in general. We have

$$\mu_n \equiv E[X_n] = \sum_{j=0}^{\infty} E[X_n \mid X_{n-1} = j] P[X_{n-1} = j]$$

$$= \sum_{j=0}^{\infty} j\mu_1 P[X_{n-1} = j] = \mu_1 E[X_{n-1}] \qquad (3.113)$$

Using this result recurrently, we obtain

$$\mu_n = \mu_1 E[X_{n-1}] = \mu_1^2 E[X_{n-2}] = \ldots = \mu_1^n E[X_0] = \mu_1^n \qquad (3.114)$$

Remark. Let $\sigma_1^2 := V[X_1]$. When $\mu_1 = 1$, from Eq. (1.96), which enables us to express the variance of X_n in terms of $E[X_n \mid X_{n-1}]$ and of $V[X_n \mid X_{n-1}]$, we may write that

$$
\begin{aligned}
V[X_n] &= E[V[X_n \mid X_{n-1}]] + V[E[X_n \mid X_{n-1}]] \\
&\overset{\text{i.i.d.}}{=} E[X_{n-1}\sigma_1^2] + V[X_{n-1} \times 1] \\
&= \sigma_1^2 \times 1 + V[X_{n-1}] = 2\sigma_1^2 + V[X_{n-2}] \\
&= \ldots = n\sigma_1^2 + V[X_0] = n\sigma_1^2 \qquad (3.115)
\end{aligned}
$$

since $V[X_0] = 0$ (X_0 being a constant). When $\mu_1 \neq 1$, we find that

$$V[X_n] = \sigma_1^2 \mu_1^{n-1} \left(\frac{\mu_1^n - 1}{\mu_1 - 1} \right) \qquad (3.116)$$

We wish to determine the probability of *eventual extinction* of the population, namely,

$$q_{0,i} := \lim_{n \to \infty} P[X_n = 0 \mid X_0 = i] \qquad (3.117)$$

By independence, we may write that

$$q_{0,i} = q_{0,1}^i \qquad (3.118)$$

Consequently, it suffices to calculate $q_{0,1}$, which will be denoted simply by q_0. We have

$$P[X_n = 0 \mid X_0 = 1] = 1 - P[X_n \geq 1 \mid X_0 = 1] = 1 - \sum_{k=1}^{\infty} P[X_n = k \mid X_0 = 1]$$

$$\geq 1 - \sum_{k=1}^{\infty} k\, P[X_n = k \mid X_0 = 1] = 1 - E[X_n \mid X_0 = 1]$$

$$= 1 - \mu_1^n \tag{3.119}$$

[by Eq. (3.114)]. It follows that, if $\mu_1 \in (0, 1)$, then

$$q_0 \geq \lim_{n \to \infty} 1 - \mu_1^n = 1 \quad \Longrightarrow \quad q_0 = 1 \tag{3.120}$$

which is a rather obvious result, since if each individual has less than one descendant, on average, we indeed expect the population to disappear.

When $\mu_1 \geq 1$, Eq. (3.119) implies only that $q_0 \geq 0$ (if $\mu_1 = 1$) or that $q_0 \geq -\infty$ (if $\mu_1 > 1$). However, the following theorem can be proved.

Theorem 3.2.3. *The probability q_0 of eventual extinction of the population is equal to 1 if $\mu_1 \leq 1$, while $q_0 < 1$ if $\mu_1 > 1$.*

Remarks. i) We deduce from the theorem that a *necessary* condition for the probability q_0 to be smaller than 1 is that p_j must be greater than 0 for at least one $j \geq 2$. Indeed, if $p_0 = p > 0$ and $p_1 = 1 - p$, then we directly have $\mu_1 = 1 - p < 1$.

ii) When $p_0 = 0$ and $p_1 = 1$, we have that $\mu_1 = 1$. According to the theorem, we should have $q_0 = 1$. Yet, if $p_1 = 1$, it is obvious that the size of the population will always remain equal to X_0, and then $q_0 = 0$. However, the theorem applies only when $p_0 > 0$.

Let F be the event defined by

$$F = \bigcup_{n=1}^{\infty} \{X_n = 0\} \tag{3.121}$$

so that $q_0 = P[F \mid X_0 = 1]$. To obtain the value of q_0, we can solve the following equation:

$$q_0 = \sum_{j=0}^{\infty} P[F \mid X_1 = j]\, p_j = \sum_{j=0}^{\infty} q_0^j\, p_j \tag{3.122}$$

The equation above possesses many solutions. It can be shown that when $\mu_1 > 1$, q_0 is the *smallest positive solution* of the equation.

Remark. Note that $q_0 = 1$ is always a solution of Eq. (3.122).

Example 3.2.15. Suppose that $p_0 = 1/3$ and $p_j = 2/9$, for $j = 1, 2, 3$. First, we calculate

$$\mu_1 = 0 + \frac{2}{9}(1 + 2 + 3) = \frac{4}{3} > 1$$

Thus, we may assert that $q_0 < 1$. Eq. (3.122) becomes

$$q_0 = \frac{1}{3} + \frac{2}{9}(q_0 + q_0^2 + q_0^3) \quad\Longleftrightarrow\quad q_0^3 + q_0^2 - \frac{7}{2}q_0 + \frac{3}{2} = 0$$

Since $q_0 = 1$ is a solution of this equation, we find that

$$(q_0 - 1)\left(q_0^2 + 2q_0 - \frac{3}{2}\right) = 0$$

The three solutions of the equation are 1 and $-1 \pm \sqrt{5/2}$. Therefore, we conclude that $q_0 = -1 + \sqrt{5/2} \simeq 0.5811$.

Remark. If $p_0 = 1/2$ and $p_j = 1/6$, for $j = 1, 2, 3$, then we have

$$\mu_1 = 0 + \frac{1}{6}(1 + 2 + 3) = 1$$

Consequently, Theorem 3.2.3 implies that $q_0 = 1$. Eq. (3.122) is now

$$q_0 = \frac{1}{2} + \frac{1}{6}(q_0 + q_0^2 + q_0^3) \quad\Longleftrightarrow\quad (q_0 - 1)^2(q_0 + 3) = 0$$

so that the three solutions of the equation are 1 (double root) and -3, which confirms the fact that $q_0 = 1$.

Similarly, if $p_0 = p_1 = 1/2$, then we find that

$$q_0 = \frac{1}{2}(1 + q_0) \quad\Longleftrightarrow\quad q_0 = 1$$

in accordance with Theorem 3.2.3, since $\mu_1 = 1/2 < 1$.

Assume again that $X_0 = 1$. Let

$$T_n := \sum_{k=0}^{n} X_k = 1 + \sum_{k=1}^{n} X_k \tag{3.123}$$

That is, T_n designates the total number of descendants of the population's ancestor, in addition to the ancestor himself. It can be shown that

$$\sum_{j=1}^{\infty} P\left[\lim_{n\to\infty} T_n = j\right] = q_0 \tag{3.124}$$

Thus, if $q_0 < 1$, then $T_\infty := \lim_{n\to\infty} T_n$ is a random variable called *defective*, which takes the value ∞ with probability $1 - q_0$. Moreover, if $q_0 < 1$, the mathematical expectation of T_∞ is evidently infinite. When $q_0 = 1$, we have the following result.

Proposition 3.2.7. *The mathematical expectation of the random variable T_∞ is given by*

$$E[T_\infty] = \frac{1}{1 - \mu_1} \quad \text{if } \mu_1 \leq 1 \tag{3.125}$$

Remark. We observe that when $\mu_1 = 1$, we have that $P[T_\infty < \infty] = 1$, but $E[T_\infty] = \infty$.

Example 3.2.16. If $p_0 = p \in (0,1)$ and $p_1 = 1 - p$, then we know that $q_0 = 1$, because $\mu_1 = 1 - p < 1$. In this case, we may write that

$$X_k = \begin{cases} 1 \text{ with probability } (1-p)^k \\ 0 \text{ with probability } 1 - (1-p)^k \end{cases}$$

Thus, we have

$$E[T_n] = 1 + \sum_{k=1}^{n}(1-p)^k \quad \stackrel{n \to \infty}{\longrightarrow} \quad \frac{1}{1-(1-p)} = \frac{1}{p}$$

which is indeed equal to $1/(1 - \mu_1)$.

3.3 Continuous-time Markov chains

3.3.1 Exponential and gamma distributions

In the case of discrete-time Markov chains, we said nothing about the time the processes spend in state i before making a transition to some state j. As in Example 2.1.1 on random walks, we may assume that this time is deterministic and is equal to one *unit*. On the other hand, an essential characteristic of continuous-time Markov chains is that the time that the processes spend in a given state has an exponential distribution, so that this time is random. Moreover, we will show that the sum of independent exponential random variables (having the same parameter) is a variable having a gamma distribution. We already mentioned the exponential and gamma distributions in Chapter 1. In the present section, we give the main properties of these two distributions.

Exponential distribution

Definition 3.3.1. (Reminder) *If the probability density function of the continuous random variable X, whose set of possible values is the interval $[0, \infty)$, is of the form*

$$f_X(x) = \begin{cases} \lambda e^{-\lambda x} & \text{if } x \geq 0 \\ 0 & \text{if } x < 0 \end{cases} \tag{3.126}$$

*we say that X has an **exponential distribution** with parameter $\lambda > 0$ and we write $X \sim Exp(\lambda)$.*

Remarks. i) Using the Heaviside function $u(x)$ (see p. 11), we may write that

$$f_X(x) = \lambda e^{-\lambda x}u(x) \quad (\forall x \in \mathbb{R}) \tag{3.127}$$

Note that we have the following relation between the functions $u(x)$ and $\delta(x)$ (see p. 63):

$$u(x) = \int_{-\infty}^{x} \delta(t)\, dt \tag{3.128}$$

ii) For some authors, a random variable X having an exponential distribution with parameter λ possesses the density function

$$f_X(x) = \frac{1}{\lambda} e^{-x/\lambda} u(x) \tag{3.129}$$

The advantage of this choice is that we then have $E[X] = \lambda$, while for us $E[X] = 1/\lambda$, as will be shown further on.

The distribution function of X is given by

$$F_X(x) = \begin{cases} 0 & \text{if } x < 0 \\ \int_0^x \lambda e^{-\lambda t}\, dt = 1 - e^{-\lambda x} & \text{if } x \geq 0 \end{cases} \tag{3.130}$$

Note that we have the following very simple formula:

$$P[X > x] = e^{-\lambda x} \quad \text{for } x \geq 0 \tag{3.131}$$

Some people write $\bar{F}(x)$ for the probability $P[X > x]$.

The main reason for which the exponential distribution is used so much, in particular in *reliability* and in the *theory of queues*, is the fact that it possesses the *memoryless property*, as we show below.

Proposition 3.3.1. (Memoryless property) *Suppose that $X \sim Exp(\lambda)$. We have*

$$P[X > s+t \mid X > t] = P[X > s] \quad \forall\, s, t \geq 0 \tag{3.132}$$

Proof. By the formula (3.131), we may write that

$$P[X > s+t \mid X > t] = \frac{P[X > s+t, X > t]}{P[X > t]} = \frac{P[X > s+t]}{P[X > t]}$$

$$= \frac{e^{-\lambda(s+t)}}{e^{-\lambda t}} = e^{-\lambda s} = P[X > s] \quad \square \tag{3.133}$$

Remark. Actually, the exponential random variables are the *only* r.v.s having this property for all nonnegative s and t. The geometric distribution possesses the memoryless property, but only for integer (and positive) values of s and t.

We can easily calculate the moment-generating function of the r.v. $X \sim Exp(\lambda)$:

$$M_X(t) \equiv E[e^{tX}] = \int_0^\infty e^{tx} \lambda e^{-\lambda x} \, dx = \frac{\lambda}{\lambda - t} \quad \text{if } t < \lambda \qquad (3.134)$$

Then

$$M_X'(t) = \frac{\lambda}{(\lambda - t)^2} \quad \text{and} \quad M_X''(t) = \frac{2\lambda}{(\lambda - t)^3} \qquad (3.135)$$

which implies that

$$E[X] = M_X'(0) = \frac{1}{\lambda}, \quad E[X^2] = M_X''(0) = \frac{2}{\lambda^2}, \quad \text{and} \quad V[X] = \frac{1}{\lambda^2} \quad (3.136)$$

Remark. Note that the mean and the standard deviation of X are equal. Consequently, if we seek a model for some data, the exponential distribution should only be considered if the mean and the standard deviation of the observations are approximately equal. Otherwise, we must transform the data, for example, by subtracting or by adding a constant to the *raw data*.

Example 3.3.1. Suppose that the *lifetime* X of a car has an exponential distribution with parameter λ. What is the probability that a car having already reached its expected lifetime will function (in all) more than twice its expected lifetime?

Solution. We seek

$$P\left[X > \frac{2}{\lambda} \,\middle|\, X > \frac{1}{\lambda}\right] = P\left[X > \frac{1}{\lambda}\right] = e^{-1} \simeq 0.3679$$

Remark. The assumption that the lifetime of a car has an exponential distribution is certainly not entirely realistic, since cars *age*. However, this assumption may be acceptable for a time period during which the *(major) failure rate* of cars is more or less constant, for example, during the first three years of use. Incidentally, most car manufacturers offer a three-year warranty.

As the following proposition shows, the geometric distribution may be considered as the discrete version of the exponential distribution.

Proposition 3.3.2. *Let $X \sim Exp(\lambda)$ and $Y := int(X) + 1$, where "int" designates the integer part. We have*

$$P[Y = k] = (e^{-\lambda})^{k-1}(1 - e^{-\lambda}) \quad \text{for } k = 1, 2, \ldots \qquad (3.137)$$

That is, $Y \sim Geom(p := 1 - e^{-\lambda})$.

Proof. First, since $int(X) \in \{0, 1, \ldots\}$, we indeed have that $S_Y = \{1, 2, \ldots\}$. We calculate

$$P[Y = k] = P[int(X) = k - 1] = P[k - 1 \le X < k] \qquad (3.138)$$

$$= \int_{k-1}^{k} \lambda e^{-\lambda x} \, dx = - \left. e^{-\lambda x} \right|_{k-1}^{k} = e^{-\lambda(k-1)}(1 - e^{-\lambda}) \quad \square$$

Remark. If we define the geometric distribution by

$$P[Y = k] = q^k p \quad \text{for } k = 0, 1, 2, \ldots \tag{3.139}$$

(as many authors do), then we simply have that $Y := \text{int}(X) \sim \text{Geom}(p = 1 - e^{-\lambda})$.

We can also consider the exponential distribution on the entire real line.

Definition 3.3.2. *Let X be a continuous random variable whose density function is given by*

$$f_X(x) = \frac{\lambda}{2} e^{-\lambda|x|} \quad \text{for } x \in \mathbb{R} \tag{3.140}$$

*where λ is a positive parameter. We say that X has a **double exponential** distribution or a **Laplace**[6] distribution.*

Remark. We find that the mean value of X is equal to zero and that $V[X] = 2/\lambda^2$. The fact that $E[X] = 0$ follows from the symmetry of the function f_X about the origin (and from the existence of this mathematical expectation).

Proposition 3.3.3. *If $X \sim Exp(\lambda)$, then for all $x_0 \geq 0$, we have*

$$E[X \mid X > x_0] = x_0 + E[X] \quad \text{and} \quad V[X \mid X > x_0] = V[X] \tag{3.141}$$

Proof. The formula $P[X > x_0] = e^{-\lambda x_0}$ implies that

$$f_X(x \mid X > x_0) = \lambda e^{-\lambda(x-x_0)} \quad \text{for } x > x_0 \tag{3.142}$$

from which we have

$$E[X \mid X > x_0] = \int_{x_0}^{\infty} x \, \lambda e^{-\lambda(x-x_0)} \, dx \overset{y=x-x_0}{=} \int_{0}^{\infty} (y + x_0)\lambda e^{-\lambda y} \, dy$$
$$= E[X] + x_0 P[X \in [0, \infty)] = E[X] + x_0 \tag{3.143}$$

Next, we calculate

$$E[X^2 \mid X > x_0] = \int_{x_0}^{\infty} x^2 \, \lambda e^{-\lambda(x-x_0)} \, dx \overset{y=x-x_0}{=} \int_{0}^{\infty} (y + x_0)^2 \lambda e^{-\lambda y} \, dy$$
$$= E[X^2] + 2x_0 E[X] + x_0^2 P[X \in [0, \infty)]$$
$$= E[X^2] + 2x_0 E[X] + x_0^2 \tag{3.144}$$

---

[6] See p. 19.

Then we obtain

$$V[X \mid X > x_0] = E[X^2 \mid X > x_0] - (E[X \mid X > x_0])^2$$
$$= E[X^2] - (E[X])^2 = V[X] \quad \square \qquad (3.145)$$

Remark. The preceding proposition actually follows directly from the memoryless property of the exponential distribution.

A result that will be used many times in this book is given in the following proposition.

Proposition 3.3.4. *Let $X_1 \sim Exp(\lambda_1)$ and $X_2 \sim Exp(\lambda_2)$ be two independent random variables. We have*

$$P[X_1 < X_2] = \frac{\lambda_1}{\lambda_1 + \lambda_2} \qquad (3.146)$$

Proof. By conditioning on the possible values of X_1, we obtain

$$P[X_2 > X_1] = \int_0^\infty P[X_2 > X_1 \mid X_1 = x] f_{X_1}(x) \, dx \qquad (3.147)$$

$$\overset{\text{ind.}}{=} \int_0^\infty P[X_2 > x] \lambda_1 e^{-\lambda_1 x} \, dx = \int_0^\infty e^{-\lambda_2 x} \lambda_1 e^{-\lambda_1 x} \, dx$$

$$= \int_0^\infty \lambda_1 e^{-(\lambda_1 + \lambda_2)x} \, dx = \frac{\lambda_1}{\lambda_1 + \lambda_2} \quad \square \qquad (3.148)$$

Remarks. i) We have that $P[X_1 < X_2] = 1/2$ if $\lambda_1 = \lambda_2$, which had to be the case, by symmetry and continuity.

ii) The proposition may be rewritten as follows:

$$P[X_1 = \min\{X_1, X_2\}] = \frac{\lambda_1}{\lambda_1 + \lambda_2} \qquad (3.149)$$

Moreover, we can generalize the result: let $X_1, \ldots, X_n$ be independent random variables, where $X_k \sim Exp(\lambda_k)$, for all k. We have

$$P[X_1 = \min\{X_1, \ldots, X_n\}] = \frac{\lambda_1}{\lambda_1 + \ldots + \lambda_n} \qquad (3.150)$$

To prove this formula, we can make use of the following proposition.

Proposition 3.3.5. *Let $X_1 \sim Exp(\lambda_1)$, $\ldots$, $X_n \sim Exp(\lambda_n)$ be independent random variables, and let $Y := \min\{X_1, \ldots, X_n\}$. The r.v. Y has an exponential distribution with parameter $\lambda := \lambda_1 + \ldots + \lambda_n$.*

Proof. Since $\min\{X_1, X_2, X_3\} = \min\{X_1, \min\{X_2, X_3\}\}$, it suffices to prove the result for $n = 2$. We have, for $y \geq 0$,

$$P[Y > y] = P[X_1 > y, X_2 > y] \overset{\text{ind.}}{=} P[X_1 > y]P[X_2 > y]$$
$$= e^{-\lambda_1 y}e^{-\lambda_2 y} = e^{-(\lambda_1+\lambda_2)y}$$
$$\implies \quad f_Y(y) = (\lambda_1 + \lambda_2)e^{-(\lambda_1+\lambda_2)y} u(y) \quad \square \qquad (3.151)$$

Example 3.3.2. Suppose that X_1, X_2, and X_3 are independent random variables, all of which have an $\text{Exp}(\lambda)$ distribution. Calculate the probability $P[X_1 < (X_2 + X_3)/2]$.

Solution. We have $P[X_1 < (X_2 + X_3)/2] = P[Y < X_2 + X_3]$, where $Y := 2X_1$. Moreover, for $y \geq 0$,

$$P[Y \leq y] = P[X_1 \leq y/2] = 1 - e^{-\lambda y/2} \quad \implies \quad Y \sim \text{Exp}(\lambda/2)$$

Then, by the memoryless property of the exponential distribution,

$$P[2X_1 < X_2 + X_3] = 1 - P[Y \geq X_2 + X_3]$$
$$= 1 - P[Y \geq X_2 + X_3 \mid Y \geq X_2]P[Y \geq X_2]$$
$$= 1 - P[Y \geq X_3]P[Y \geq X_2] \overset{\text{i.d.}}{=} 1 - \left(\frac{\lambda}{\lambda + \frac{\lambda}{2}}\right)^2$$
$$= 1 - (2/3)^2 = 5/9$$

Finally, we would like to have a two-dimensional version of the exponential distribution. We can, of course, define

$$f_{X_1,X_2}(x_1, x_2) = \lambda_1\lambda_2 e^{-(\lambda_1 x_1 + \lambda_2 x_2)} \quad \text{for } x_1 \geq 0,\ x_2 \geq 0 \qquad (3.152)$$

However, the random variables X_1 and X_2 are then independent. To obtain a nontrivial generalization of the exponential distribution to the two-dimensional case, we can write

$$P[X_1 > x_1, X_2 > x_2] = \exp\{-\lambda_1 x_1 - \lambda_2 x_2 - \lambda_{12} \max\{x_1, x_2\}\} \qquad (3.153)$$

for $x_1 \geq 0$, $x_2 \geq 0$, where λ_{12} is a positive constant. We indeed find that $X_i \sim \text{Exp}(\lambda_i)$, for $i = 1, 2$.

Another possibility is the random vector whose joint density function is (see Ref. [5])

$$f_{X_1,X_2}(x_1, x_2) = \frac{\lambda_1\lambda_2}{1-\rho} \exp\left[-\frac{\lambda_1 x_1 + \lambda_2 x_2}{1-\rho}\right] I_0\left[\frac{2(\rho\lambda_1\lambda_2 x_1 x_2)^{1/2}}{1-\rho}\right] \qquad (3.154)$$

for $x_1 \geq 0$, $x_2 \geq 0$, where $\rho \in [0, 1)$ is the correlation coefficient of X_1 and X_2, and $I_0(\cdot)$ is a *modified Bessel[7] function of the first kind* (of order 0) defined by (see p. 375 of Ref. [1])

---

[7] Friedrich Wilhelm Bessel, 1784–1846, was born in Germany and died in Königsberg, in Prussia (now Kaliningrad, in Russia). He was an astronomer and mathematician. The mathematical functions that he introduced in 1817 are important in applied mathematics, in physics, and in engineering.

$$I_0(z) = 1 + \frac{\frac{1}{4}z^2}{(1!)^2} + \frac{(\frac{1}{4}z^2)^2}{(2!)^2} + \frac{(\frac{1}{4}z^2)^3}{(3!)^2} + \cdots \qquad (3.155)$$

Here, too, we find that $X_1 \sim \text{Exp}(\lambda_1)$ and $X_2 \sim \text{Exp}(\lambda_2)$.

Gamma distribution

Definition 3.3.3 (Reminder). *We say that the continuous and nonnegative random variable X has a* **gamma distribution** *with parameters $\alpha > 0$ and $\lambda > 0$, and we write that $X \sim G(\alpha, \lambda)$, if*

$$f_X(x) = \frac{(\lambda x)^{\alpha-1}\lambda e^{-\lambda x}}{\Gamma(\alpha)} u(x) \qquad (3.156)$$

where $\Gamma(\cdot)$ is the **gamma function,** *defined (for $\alpha > 0$) by*

$$\Gamma(\alpha) = \int_0^\infty t^{\alpha-1}e^{-t}\, dt \qquad (3.157)$$

Remarks. i) The function $\Gamma(\alpha)$ is strictly positive for any positive α. For $\alpha > 1$, we have

$$\Gamma(\alpha) = -t^{\alpha-1}e^{-t}\big|_0^\infty + (\alpha-1)\int_0^\infty t^{\alpha-2}e^{-t}\, dt$$
$$\stackrel{\alpha \geq 1}{=} 0 + (\alpha-1)\Gamma(\alpha-1) = (\alpha-1)\Gamma(\alpha-1) \qquad (3.158)$$

Then, since

$$\Gamma(1) = \int_0^\infty e^{-t}\, dt = 1 \qquad (3.159)$$

we have

$$\Gamma(n) = (n-1)\Gamma(n-1) = (n-1)(n-2)\Gamma(n-2)$$
$$= \cdots = (n-1)(n-2)\cdots 1 \cdot \Gamma(1) = (n-1)! \qquad (3.160)$$

Thus, the gamma function generalizes the factorial function. We also have

$$\Gamma(1/2) = \int_0^\infty t^{-1/2}e^{-t}\, dt \stackrel{s=\sqrt{2t}}{=} \int_0^\infty \sqrt{2}e^{-s^2/2}\, ds \qquad (3.161)$$
$$= 2\sqrt{\pi}\int_0^\infty \frac{1}{\sqrt{2\pi}}e^{-s^2/2}\, ds = 2\sqrt{\pi}P[N(0,1) \geq 0] = \sqrt{\pi}$$

ii) Contrary to the exponential distribution, whose density function always has the same form, the shape of the density function f_X changes with each value of the parameter α, which makes it a very useful model for the applications. We say that the parameter α is a *shape parameter*, while λ is a *scale* parameter (see

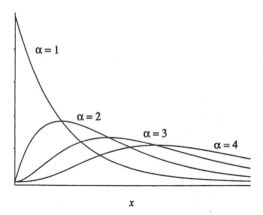

Fig. 3.1. Examples of probability density functions of $G(\alpha, \lambda = 1)$ random variables.

Fig. 3.1). In reality, the shape of the function f_X varies mostly when α is small. When α increases, the gamma distribution tends to a Gaussian distribution, which follows from the fact that if $X \sim G(n, \lambda)$, then the random variable X can be represented as the sum of n independent r.v.s $X_k \sim \text{Exp}(\lambda)$, for $k = 1, \ldots, n$ (see Prop. 3.3.6).

iii) If $X \sim G(\alpha = 1, \lambda)$, we have

$$f_X(x) = \lambda e^{-\lambda x} u(x) \tag{3.162}$$

Thus, the gamma distribution generalizes the exponential distribution, since $G(\alpha = 1, \lambda) \equiv \text{Exp}(\lambda)$.

iv) The parameter α may take any real positive value. When $\alpha = n \in \mathbb{N}$, the gamma distribution is also named the *Erlang*[8] distribution. Moreover, we have that $G(\alpha = n/2, \lambda = 1/2) \equiv \chi_n^2$. That is, the *chi-square distribution with* n *degrees of freedom*, which is very important in statistics, is a particular case of the gamma distribution, too.

The moment-generating function of $X \sim G(\alpha, \lambda)$ is

$$M_X(t) = \int_0^\infty e^{tx} \frac{\lambda e^{-\lambda x}(\lambda x)^{\alpha-1}}{\Gamma(\alpha)} \, dx = \frac{\lambda^\alpha}{\Gamma(\alpha)} \int_0^\infty e^{(t-\lambda)x} x^{\alpha-1} \, dx$$

$$\overset{y=(\lambda-t)x}{=} \frac{\lambda^\alpha}{\Gamma(\alpha)(\lambda-t)^\alpha} \int_0^\infty e^{-y} y^{\alpha-1} \, dy = \frac{\lambda^\alpha}{\Gamma(\alpha)(\lambda-t)^\alpha} \Gamma(\alpha)$$

---

[8] Agner Krarup Erlang, 1878–1929, was born and died in Denmark. He was first educated by his father, who was a schoolmaster. He studied mathematics and natural sciences at the University of Copenhagen and taught in schools for several years. After meeting the chief engineer for the Copenhagen telephone company, he joined this company in 1908. He then started to apply his knowledge of probability theory to the resolution of problems related to telephone calls. He was also interested in mathematical tables.

$$= \left(\frac{\lambda}{\lambda - t}\right)^\alpha \quad \text{for } t < \lambda \tag{3.163}$$

We then calculate

$$E[X] = M_X'(0) = \frac{\alpha}{\lambda} \quad \text{and} \quad E[X^2] = M_X''(0) = \frac{\alpha(\alpha + 1)}{\lambda^2} \tag{3.164}$$

so that

$$V[X] = \frac{\alpha(\alpha + 1)}{\lambda^2} - \left(\frac{\alpha}{\lambda}\right)^2 = \frac{\alpha}{\lambda^2} \tag{3.165}$$

We can transform a $G(\alpha, \lambda)$ random variable X into an r.v. having a $G(\alpha, 1)$ distribution, sometimes called the *standard gamma distribution*, by setting $Y = \lambda X$. Indeed, we then have

$$f_Y(y) = f_X(y/\lambda) \left|\frac{d(y/\lambda)}{dy}\right| = \frac{y^{\alpha-1}e^{-y}}{\Gamma(\alpha)} u(y) \tag{3.166}$$

The distribution function of Y can be expressed as follows:

$$F_Y(y) = \frac{\gamma(\alpha, y)}{\Gamma(\alpha)} \quad \text{for } y \geq 0 \tag{3.167}$$

where $\gamma(\alpha, y)$ is the *incomplete gamma function*, defined by

$$\gamma(\alpha, y) = \int_0^y t^{\alpha-1}e^{-t}\, dt \tag{3.168}$$

We have the following formula (see p. 272 of Ref. [1]):

$$\gamma(\alpha, y) = \alpha^{-1}y^\alpha e^{-y} M(1, 1 + \alpha, y) \tag{3.169}$$

where $M(\cdot, \cdot, \cdot)$ is a *confluent hypergeometric function*, defined by (see p. 504 of Ref. [1])

$$M(a, b, z) = 1 + \frac{a}{b}z + \frac{a(a + 1)}{b(b + 1)}\frac{z^2}{2!} + \frac{a(a + 1)(a + 2)}{b(b + 1)(b + 2)}\frac{z^3}{3!} + \dots \tag{3.170}$$

When $\alpha = n \in \mathbb{N}$, the function $\gamma(\alpha, y)$ becomes

$$\gamma(n, y) = \Gamma(n)\left[1 - e^{-y}\sum_{k=0}^{n-1}\frac{y^k}{k!}\right] \tag{3.171}$$

which implies that

$$F_Y(y) = 1 - \sum_{k=0}^{n-1} e^{-y}\frac{y^k}{k!} \quad \text{for } y \geq 0 \tag{3.172}$$

This formula can be rewritten as follows:

$$P[Y \leq y] = 1 - P[W \leq n-1] = P[W \geq n], \quad \text{where } W \sim \text{Poi}(y) \quad (3.173)$$

Remarks. i) The formula (3.173) can be obtained by doing the integral

$$\int_0^y t^{n-1} e^{-t}\, dt \quad (\equiv \gamma(n, y)) \tag{3.174}$$

by parts (repeatedly).

ii) We will see in Section 5.1 that, in the case of a *Poisson process* (with *rate* $\lambda = 1$), the random variable Y represents the time needed for n events to occur, whereas W is the number of events that occur in the interval $[0, y]$. In other words, the relation between Y and W is expressed as follows: the nth event of the Poisson process occurs at the latest at time y if and only if there are at least n events in the interval $[0, y]$.

Example 3.3.3. Suppose that the duration T (in hours) of a major power failure is a random variable having a gamma distribution with parameters $\alpha = 2$ and $\lambda = 1/2$ (so that the average duration is equal to four hours). What is the probability that an arbitrary (major) power failure lasts more than six hours?

Solution. First, we have

$$P[T > 6] = P[X > 3], \quad \text{where } X \sim G(2, 1)$$

Then, by using the formula (3.173), we can write that

$$P[X > 3] = P[\text{Poi}(3) \leq 1] = e^{-3}(1 + 3) = 4\,e^{-3} \simeq 0.1991$$

Remarks. i) It is important not to forget that T (or X) is a *continuous* random variable, while W is *discrete*.

ii) When the parameter α is small, as in this example, we can simply integrate by parts:

$$P[X > 3] = \int_3^\infty \frac{x^{2-1} e^{-x}}{\Gamma(2)}\, dx = \int_3^\infty x\, e^{-x}\, dx$$

$$= -x\, e^{-x}\Big|_3^\infty + \int_3^\infty e^{-x}\, dx = 3\,e^{-3} + e^{-3} = 4\,e^{-3}$$

We have seen that the exponential distribution is a particular case of the gamma distribution. We will now show that the sum of independent exponential random variables, with the same parameter, has a gamma distribution.

Proposition 3.3.6. *Let* $X_1, \ldots, X_n$ *be independent random variables. If* X_i *has an exponential distribution with parameter* λ, *for all* i, *then*

$$\sum_{i=1}^{n} X_i \sim G(\alpha = n, \lambda) \tag{3.175}$$

Proof. Let $S := \sum_{i=1}^{n} X_i$. We have

$$M_S(t) \equiv E[e^{tS}] = E\left[\exp\left\{t \sum_{i=1}^{n} X_i\right\}\right] \stackrel{\text{ind.}}{=} \prod_{i=1}^{n} M_{X_i}(t)$$

$$= \prod_{i=1}^{n} \frac{\lambda}{\lambda - t} = \left(\frac{\lambda}{\lambda - t}\right)^n \quad \text{for } t < \lambda \tag{3.176}$$

Since $[\lambda/(\lambda - t)]^n$ is the moment-generating function of an r.v. having a $G(\alpha = n, \lambda)$ distribution, the result is then obtained by *uniqueness*. Indeed, only the $G(\alpha = n, \lambda)$ distribution has this moment-generating function. $\square$

The exponential and gamma distributions are models that are widely used in reliability. Another continuous random variable, which also generalizes the exponential distribution and which is very commonly used in reliability and in many other applications, is the *Weibull* [9] distribution.

Definition 3.3.4. *Let* X *be a continuous random variable whose probability density function is given by*

$$f_X(x) = \frac{\beta}{\delta}\left(\frac{x - \gamma}{\delta}\right)^{\beta - 1} \exp\left[-\left(\frac{x - \gamma}{\delta}\right)^{\beta}\right] u(x - \gamma) \tag{3.177}$$

We say that X *has a* **Weibull distribution** *with parameters* $\beta > 0$, $\gamma \in \mathbb{R}$, *and* $\delta > 0$.

Remark. The exponential distribution is the particular case where $\beta = 1$, $\gamma = 0$, and $\delta = 1/\lambda$. Like the gamma distribution, this distribution has a shape parameter, namely β. The parameter γ is a *position* parameter, while δ is a *scale* parameter. When $\gamma = 0$ and $\delta = 1$, we have

$$f_X(x) = \beta x^{\beta - 1} e^{-x^{\beta}} u(x) \tag{3.178}$$

This variable is called the *standard Weibull distribution*. Its distribution function is

$$F_X(x) = 1 - e^{-x^{\beta}} \quad \text{for } x \geq 0 \tag{3.179}$$

---

[9] E. H. Wallodi Weibull, 1887–1979, was born in Sweden and died in France. In addition to his scientific papers on the distribution that bears his name, he is the author of numerous papers on strength of materials, fatigue, and reliability.

Because the exponential, gamma, and Weibull distributions are very important in reliability, we will end this section with the definition of the *failure rate* of a device, or a system, etc.

Definition 3.3.5. *The* **failure rate** *(or* **hazard rate***) of a device, whose lifetime X is a continuous and nonnegative random variable, is defined by*

$$r_X(t) = \frac{f_X(t)}{1 - F_X(t)} \tag{3.180}$$

Remarks. i) We have that $r_X(t) \simeq P[X \in (t, t + dt] \mid X > t]/dt$. That is, the failure rate $r_X(t)$, multiplied by dt, is approximately equal to the probability that a device being t time unit(s) old will fail in the interval $(t, t + dt]$, given that it is functioning at time t.

ii) To each function $r_X(t)$ there corresponds *one and only one* distribution function $F_X(t)$. Indeed, we have

$$r_X(t) := \frac{f_X(t)}{1 - F_X(t)} = \frac{\frac{d}{dt}F_X(t)}{1 - F_X(t)} \tag{3.181}$$

from which we can write (since $F_X(0) = 0$) that

$$\int_0^t r_X(s)\, ds = \int_0^t \frac{\frac{d}{ds}F_X(s)}{1 - F_X(s)}\, ds = -\ln[1 - F_X(s)]\big|_0^t = -\ln[1 - F_X(t)]$$

$$\implies \quad F_X(t) = 1 - \exp\left\{ -\int_0^t r_X(s)\, ds \right\} \tag{3.182}$$

Particular cases

1) If $X \sim \text{Exp}(\lambda)$, then

$$r_X(t) = \frac{\lambda e^{-\lambda t}}{e^{-\lambda t}} = \lambda \quad \forall\, t \geq 0 \tag{3.183}$$

Note that this result is a consequence of the memoryless property of the exponential distribution. We say that the parameter λ is the *rate* of the exponential distribution.

2) In the case of the standard gamma distribution, we may write that

$$r_X(t) = \frac{t^{\alpha-1}e^{-t}}{\Gamma(\alpha) - \gamma(\alpha, t)} \tag{3.184}$$

3) Finally, if X has a standard Weibull distribution, we deduce from Eqs. (3.178) and (3.179) that

$$r_X(t) = \beta t^{\beta-1} \qquad (3.185)$$

In practice, the function $r_X(t)$ has more or less the shape of a *bathtub*. That is, at the beginning the failure rate decreases, then this rate is rather constant, and finally the failure rate increases (which is equally true for the death rate of humans). To obtain this kind of curve, we can consider a random variable that is a linear combination of three Weibull distributions:

$$X := c_1 X_1 + c_2 X_2 + c_3 X_3 \qquad (3.186)$$

with $c_i > 0$, for all i, and $c_1 + c_2 + c_3 = 1$ (called a *mixed Weibull distribution*), where X_1 has a β parameter smaller than 1, X_2 a β parameter equal to 1, and X_3 a β parameter greater than 1.

3.3.2 Continuous-time Markov chains

Let $\{X(t), t \geq 0\}$ be a continuous-time and discrete-state *Markovian* [see Eq. (3.1)] stochastic process, and let τ_i be the time that the process spends in state i before making a transition to some other state. We may write that

$$P[\tau_i > s + t \mid \tau_i > t] = P[\tau_i > s] \quad \forall\, s, t \geq 0 \qquad (3.187)$$

Indeed, since the process is Markovian, the time it has spent in a given state does not influence the future. Consequently, whatever the time that the process has already spent in i, it is as likely that it will remain there during at least s additional time units than if it had just entered this state. Eq. (3.187) means that the continuous random variable τ_i possesses the memoryless property. As we already mentioned (see p. 110), only the exponential distribution possesses this property. We can therefore conclude that τ_i has an exponential distribution, with parameter denoted by ν_i, which, in general, depends on state i.

Moreover, the Markov property also implies that the next state visited, j, is *independent* of τ_i. Thus, when the process leaves state i, it enters state j ($\neq i$) with probability $p_{i,j}$ (by definition), where

$$p_{i,i} = 0 \quad \forall\, i \quad \text{and} \quad \sum_{j=0}^{\infty} p_{i,j} = 1 \quad \forall\, i \qquad (3.188)$$

The $p_{i,j}$'s are the one-step transition probabilities of the *embedded* (or *associated*) discrete-time Markov chain. Note, however, that, contrary to the transition matrices in the preceding section, all the terms on the main (decreasing from top left to bottom right) diagonal of the matrix are necessarily equal to zero, by definition of the $p_{i,j}$'s in the present case.

The process $\{X(t), t \geq 0\}$ is called a *continuous-time Markov chain*, which is now defined formally.

Definition 3.3.6. *Let* $\{X(t), t \geq 0\}$ *be a continuous-time stochastic process whose state space is* $\mathbb{N}^0 = \{0, 1, \dots\}$. *We say that* $\{X(t), t \geq 0\}$ *is a* **continuous-time Markov chain** *if*

$$P[X(t+s) = j \mid X(s) = i, X(r) = x_r, 0 \leq r < s]$$
$$= P[X(t+s) = j \mid X(s) = i] = p_{i,j}(t) \tag{3.189}$$

$\forall\, s, t \geq 0$ *and* $\forall\, i, j, x_r \in \mathbb{N}^0$.

Remarks. i) As in the case of the discrete-time Markov chains, we assume that the chains considered have *stationary* or *time-homogeneous* transition probabilities. We could treat the general case and denote the conditional probability $P[X(t) = j \mid X(s) = i]$ by $p_{i,j}(s, t)$, where $t \geq s$, but since the most important processes for the applications are indeed such that $p_{i,j}(s, t) = p_{i,j}(t - s)$, it will not be necessary.

ii) Continuous-time Markov chains are also known as *Markov jump processes*.

iii) The function $p_{i,j}(t)$ is called the *transition function* of the continuous-time Markov chain.

iv) The probabilities $p_{i,j}(t)$ correspond to the $p_{i,j}^{(n)}$'s in discrete-time Markov chains. If there exist a $t \geq 0$ for which $p_{i,j}(t) > 0$ and a $t^* \geq 0$ for which $p_{j,i}(t^*) > 0$, we say that states i and j *communicate*. The chain is *irreducible* if all states communicate.

v) We may write that

$$\sum_{j=0}^{\infty} p_{i,j}(t) = 1 \quad \forall\, i \tag{3.190}$$

since the process must be in some state at time $t + s$, regardless of the state it was in at time s.

vi) For the sake of simplicity, we will assume in the sequel that the state space of the Markov chain $\{X(t), t \geq 0\}$ is, save indication to the contrary, the set $\{0, 1, \dots\}$. However, as in the discrete case, the state space can actually be a set $S_{X(t)} \subset \{0, 1, \dots\}$, or, more generally, a finite or countably infinite set of real numbers.

Example 3.3.4. Since all the elements of the main diagonal of the matrix **P** in Example 3.2.8 are equal to zero, we can consider this matrix as the transition matrix of the embedded discrete-time Markov chain of a continuous-time Markov chain. The fact that $p_{i,j} = 1/2$, for all $i \neq j$, does not mean that all the random variables τ_i have the same parameter ν_i, for $i = 0, 1, 2$.

The following proposition is the equivalent, in the continuous case, of Proposition 3.2.1.

Proposition 3.3.7 (Chapman–Kolmogorov equations (2)). *For all s, t nonnegative, we have*

$$p_{i,k}(t+s) = \sum_{j=0}^{\infty} p_{i,j}(t)p_{j,k}(s) = \sum_{j=0}^{\infty} p_{i,j}(s)p_{j,k}(t) \qquad (3.191)$$

Proof. The proof is similar to that of Proposition 3.2.1. □

Example 3.3.5. In general, it is not easy to calculate the functions $p_{i,j}(t)$ explicitly. T heorems 3.3.1 and 3.3.2 provide differential equations whose solutions are these $p_{i,j}(t)$'s. In the case of a *Poisson process* (see Section 5.1), we find directly that

$$p_{i,j}(t) = \begin{cases} e^{-\lambda t}\dfrac{(\lambda t)^{j-i}}{(j-i)!} & \text{for } j-i \geq 0 \\[2mm] 0 & \text{for } j-i < 0 \end{cases}$$

where λ is the *rate* of the process. We indeed have, for $k \geq i$,

$$\sum_{j=0}^{\infty} p_{i,j}(t)p_{j,k}(s) = \sum_{j=i}^{k} p_{i,j}(t)p_{j,k}(s) = \sum_{j=i}^{k} e^{-\lambda t}\frac{(\lambda t)^{j-i}}{(j-i)!}e^{-\lambda s}\frac{(\lambda s)^{k-j}}{(k-j)!}$$

$$= e^{-\lambda(t+s)}\lambda^{k-i}\sum_{j=i}^{k}\frac{t^{j-i}s^{k-j}}{(j-i)!(k-j)!} = e^{-\lambda(t+s)}\lambda^{k-i}\sum_{j=0}^{k-i}\frac{t^{j}s^{k-i-j}}{j!(k-i-j)!}$$

$$= e^{-\lambda(t+s)}\lambda^{k-i}\sum_{j=0}^{k-i}\binom{k-i}{j}\frac{t^{j}s^{k-i-j}}{(k-i)!} \overset{(*)}{=} e^{-\lambda(t+s)}\lambda^{k-i}\frac{(t+s)^{k-i}}{(k-i)!}$$

$$= e^{-\lambda(t+s)}\frac{[\lambda(t+s)]^{k-i}}{(k-i)!} = p_{i,k}(t+s)$$

where Eq. (*) is obtained by *Newton's*[10] *binomial theorem.*

Notation. We denote by $p_j(t)$ the (marginal) probability that the process $\{X(t), t \geq 0\}$ will be in state j at time t:

$$p_j(t) := P[X(t) = j] \qquad (3.192)$$

If $a_i := P[X(0) = i]$, for $i = 0, 1, \ldots$, then we may write that

$$p_j(t) = \sum_{i=0}^{\infty} a_i\, p_{i,j}(t) \qquad (3.193)$$

When $P[X(0) = k] = 1$ for some $k \in \{0, 1, \ldots\}$, we simply have that $p_j(t) = p_{k,j}(t)$.

---

[10] Sir Isaac Newton, 1643–1727, was born and died in England. Newton was a scholar who is famous for his contributions to the fields of mechanics, optics, and astronomy. He is one of the inventors of differential calculus. He also wrote theological books.

3.3.3 Calculation of the transition function $p_{i,j}(t)$

First, if state i is *absorbing*, which means that the parameter ν_i of the random variable τ_i is equal to 0, we may write that

$$p_{i,j}(t) = \delta_{ij} := \begin{cases} 1 \text{ if } i = j \\ 0 \text{ if } i \neq j \end{cases} \tag{3.194}$$

for all $t \geq 0$. In the case of nonabsorbing states, we will obtain two systems of differential equations, which, when we can solve them, give us the value of $p_{i,j}(t)$, for all states $i, j \in \{0, 1, \dots\}$.

Remark. It can be shown that the $p_{i,j}(t)$'s are continuous functions of t, for every pair (i, j).

Definition 3.3.7. *The quantities*

$$\nu_{i,j} := \nu_i \, p_{i,j} \quad \forall \, i \neq j \in \{0, 1, \dots\} \tag{3.195}$$

are the **infinitesimal parameters** *or* **instantaneous transition rates** *of the continuous-time Markov chain* $\{X(t), t \geq 0\}$.

Remark. We have

$$\sum_{j \neq i} \nu_{i,j} = \nu_i \sum_{j \neq i} p_{i,j} = \nu_i \quad (\text{because } p_{i,i} = 0) \tag{3.196}$$

We set

$$\nu_{i,i} = -\nu_i \tag{3.197}$$

It follows that

$$\sum_{j=0}^{\infty} \nu_{i,j} = 0 \tag{3.198}$$

Definition 3.3.8. *The matrix*

$$\mathbf{G} = \begin{array}{c} 0 \\ 1 \\ 2 \\ \vdots \end{array} \left[\begin{array}{cccc} \nu_{0,0} & \nu_{0,1} & \nu_{0,2} & \cdots \\ \nu_{1,0} & \nu_{1,1} & \nu_{1,2} & \cdots \\ \nu_{2,0} & \nu_{2,1} & \nu_{2,2} & \cdots \\ \vdots & \vdots & \vdots & \vdots \end{array} \right] \tag{3.199}$$

is known as the **generating matrix** *of the continuous-time Markov chain* $\{X(t), t \geq 0\}$.

Remarks. i) If we know the quantities $\nu_{i,j}$, for all $i \neq j$, then we can calculate the rates ν_i and the probabilities $p_{i,j}$ from Eq. (3.195). Moreover, we will show in this section that the $p_{i,j}(t)$'s depend only on the ν_i's and the $p_{i,j}$'s, from which one derives the name of *generator* of the Markov chain for the set of all $\nu_{i,j}$'s.

ii) The matrix $\mathbf{G}$ corresponds to the transition matrix $\mathbf{P}$ for a discrete-time Markov chain. However, note that the $\nu_{i,j}$'s are *not* probabilities and that the sum of the elements of each row of $\mathbf{G}$ is equal to 0 rather than to 1.

Notation. If the function $g(x)$ is such that

$$\lim_{x \to 0} \frac{g(x)}{x} = 0 \qquad (3.200)$$

then we write that $g(x) = o(x)$.

A function $g(x)$ that "is" $o(x)$ must therefore tend to 0 more rapidly than the identity function $f(x) = x$ when $x \to 0$. Thus, $g_1(x) := x^2$ is $o(x)$, while $g_2(x) := \sqrt{x}$ is not $o(x)$. Moreover, if $g_i(x) = o(x)$, for $i = 1, \dots, n$, then

$$\sum_{i=1}^{n} c_i g_i(x) = o(x) \quad \forall \, c_i \in \mathbb{R} \qquad (3.201)$$

Proposition 3.3.8. *The probability that a continuous-time Markov chain, $\{X(t), t \geq 0\}$, makes two or more transitions in an interval of length δ is $o(\delta)$.*

Proof. We know that the time τ_i that the process spends in state i has an exponential distribution with parameter ν_i, for $i = 0, 1, \dots$. Suppose first that $\nu_i = \nu \; \forall \, i$ and, without loss of generality, that $X(0) = 0$. Let N be the number of transitions of the Markov chain in the interval $[0, \delta]$. We have

$$P[N \geq 2] = 1 - P[N = 0] - P[N = 1]$$

$$= 1 - P[\tau_0 > \delta] - \sum_{k=1}^{\infty} \int_0^{\delta} f_{\tau_0}(u) \, p_{0,k} P[\tau_k > \delta - u] \, du$$

$$= 1 - e^{-\nu\delta} - \sum_{k=1}^{\infty} p_{0,k} \int_0^{\delta} \nu e^{-\nu u} \, e^{-\nu(\delta-u)} \, du$$

$$= 1 - e^{-\nu\delta} - \sum_{k=1}^{\infty} p_{0,k} \, e^{-\nu\delta} \nu\delta = 1 - e^{-\nu\delta}(1 + \nu\delta) \qquad (3.202)$$

Using the series expansion

$$e^x = 1 + x + \frac{x^2}{2!} + \dots = 1 + x + o(x) \qquad (3.203)$$

we may write that

$$P[N \geq 2] = 1 - e^{-\nu\delta}(1 + \nu\delta) = 1 - [1 - \nu\delta + o(\delta)](1 + \nu\delta) = o(\delta) \qquad (3.204)$$

Now, if the ν_i's are not all equal, it suffices to replace each ν_i by $\nu :=$ $\max\{\nu_0, \nu_1, \ldots\}$ to prove the result. Indeed, since $E[\tau_i] = 1/\nu_i$, the larger ν_i is, the shorter the average time that the process spends in state i is. But, since for the largest ν_i we can assert that $P[N \geq 2] = o(\delta)$, this must hold true when $\nu_i \leq \nu \; \forall \, i$. $\square$

Proposition 3.3.9. *We may write that*

$$\nu_{i,j} = \frac{d}{dt} p_{i,j}(t) \Big|_{t=0} \quad \forall \, i, j \in \{0, 1, \ldots\} \tag{3.205}$$

Proof. Suppose first that $i = j$. We know, by the preceding proposition, that the probability of two or more transitions in an interval of length δ is $o(\delta)$. It follows that

$$p_{i,i}(\delta) = P[\tau_i > \delta] + o(\delta) = e^{-\nu_i \delta} + o(\delta) = 1 - \nu_i \delta + o(\delta) \tag{3.206}$$

Since $p_{i,i}(0) = 1$, we may write that

$$p_{i,i}(\delta) - p_{i,i}(0) = -\nu_i \delta + o(\delta) \iff \frac{p_{i,i}(\delta) - p_{i,i}(0)}{\delta} = -\nu_i + \frac{o(\delta)}{\delta} \tag{3.207}$$

Taking the limit on both sides as δ decreases to 0, we obtain

$$p'_{i,i}(0) = -\nu_i = \nu_{i,i} \tag{3.208}$$

When $i \neq j$, we have that $p_{i,j}(0) = 0$. Then

$$p_{i,j}(\delta) = P[\tau_i < \delta] \, p_{i,j} + o(\delta) = (1 - e^{-\nu_i \delta}) \, p_{i,j} + o(\delta) = \nu_i \, \delta \, p_{i,j} + o(\delta) \tag{3.209}$$

so that

$$p'_{i,j}(0) = \lim_{\delta \downarrow 0} \frac{p_{i,j}(\delta) - p_{i,j}(0)}{\delta} = \nu_i \, p_{i,j} + \lim_{\delta \downarrow 0} \frac{o(\delta)}{\delta} = \nu_i \, p_{i,j} = \nu_{i,j} \quad \square \tag{3.210}$$

We now give the first system of differential equations that enables us to calculate the $p_{i,j}(t)$'s.

Theorem 3.3.1 (Kolmogorov backward equations). *For all states i, j $\in \mathbb{N}^0$, and for all $t \geq 0$, we have*

$$p'_{i,j}(t) = \sum_{k=0}^{\infty} \nu_{i,k} \, p_{k,j}(t) \tag{3.211}$$

Proof. We decompose the probability $p_{i,j}(t)$ into two incompatible cases:

$$p_{i,j}(t) = P[\tau_i \leq t, X(t) = j \mid X(0) = i] + P[\tau_i > t, X(t) = j \mid X(0) = i] \tag{3.212}$$

Since $P[\tau_i > t] = e^{-\nu_i t}$, we may write that

$$P[\tau_i > t, X(t) = j \mid X(0) = i] = e^{-\nu_i t} \delta_{ij} \qquad (3.213)$$

Next, we have

$$P[\tau_i \leq t, X(t) = j \mid X(0) = i] = \int_0^t \nu_i e^{-\nu_i s} \left(\sum_{k \neq i} p_{i,k} \, p_{k,j}(t - s) \right) ds \qquad (3.214)$$

so that

$$p_{i,j}(t) = e^{-\nu_i t} \delta_{ij} + \int_0^t \nu_i e^{-\nu_i s} \left(\sum_{k \neq i} p_{i,k} \, p_{k,j}(t - s) \right) ds \qquad (3.215)$$

Remark. This *integral equation* allows us to state that the $p_{i,j}(t)$'s are *continuous* functions of t, as we already mentioned above. It follows that the function that we integrate is a continuous function as well.

Finally, to obtain the Kolmogorov backward equations, it suffices to first rewrite Eq. (3.215) as follows:

$$p_{i,j}(t) = e^{-\nu_i t} \delta_{ij} + \nu_i e^{-\nu_i t} \int_0^t e^{\nu_i u} \left(\sum_{k \neq i} p_{i,k} \, p_{k,j}(u) \right) du \qquad (3.216)$$

and then to differentiate this last equation:

$$p_{i,j}'(t) = -\nu_i \left\{ e^{-\nu_i t} \delta_{ij} + \nu_i e^{-\nu_i t} \int_0^t e^{\nu_i u} \left(\sum_{k \neq i} p_{i,k} \, p_{k,j}(u) \right) du \right\}$$

$$+ \nu_i e^{-\nu_i t} e^{\nu_i t} \sum_{k \neq i} p_{i,k} \, p_{k,j}(t)$$

$$= -\nu_i \, p_{i,j}(t) + \nu_i \sum_{k \neq i} p_{i,k} \, p_{k,j}(t) = \nu_{i,i} p_{i,j}(t) + \sum_{k \neq i} \nu_{i,k} \, p_{k,j}(t)$$

$$= \sum_{k=0}^{\infty} \nu_{i,k} \, p_{k,j}(t) \quad \Box \qquad (3.217)$$

Remark. If we set $t = 0$ in the Kolmogorov backward equations, we obtain

$$p_{i,j}'(0) = \sum_{k=0}^{\infty} \nu_{i,k} \, p_{k,j}(0) = \nu_{i,j} \, p_{j,j}(0) = \nu_{i,j} \qquad (3.218)$$

which confirms the result in Proposition 3.3.9. Actually, from Proposition 3.3.9, we can show directly the validity of the Kolmogorov backward equations. Indeed, we deduce from the Chapman–Kolmogorov equations that

$$\frac{d}{ds}p_{i,j}(s+t) = \sum_{k=0}^{\infty} \frac{d}{ds}p_{i,k}(s)\,p_{k,j}(t) \quad \forall\, s,t \geq 0 \tag{3.219}$$

$$\implies \quad p'_{i,j}(t) = \sum_{k=0}^{\infty} p'_{i,k}(0)\,p_{k,j}(t) \quad \text{(with } s=0\text{)}$$

$$\implies \quad p'_{i,j}(t) = \sum_{k=0}^{\infty} \nu_{i,k}\,p_{k,j}(t)$$

where we assumed, in the first equation, that we can interchange the derivative and the summation.

The second system of differential equations that we can use to calculate the $p_{i,j}(t)$'s is valid when, as in the remark above, we can interchange the derivative and the summation when we differentiate the Chapman–Kolmogorov equations. Now, this interchange is allowed when the state space of the Markov chain is *finite* and, also, in particular, for the *birth and death processes*, which will be defined in the next subsection.

Theorem 3.3.2 (Kolmogorov forward equations). *Under the condition mentioned above, for all $i,j \in \mathbb{N}^0$, and for any $t \geq 0$, we have*

$$p'_{i,j}(t) = \sum_{k=0}^{\infty} p_{i,k}(t)\,\nu_{k,j} \tag{3.220}$$

Proof. We have, under the condition in question,

$$\frac{d}{ds}p_{i,j}(t+s) = \sum_{k=0}^{\infty} p_{i,k}(t)\,\frac{d}{ds}p_{k,j}(s) \quad \forall\, s,t \geq 0 \tag{3.221}$$

$$\stackrel{\text{if } s=0}{\implies} \quad p'_{i,j}(t) = \sum_{k=0}^{\infty} p_{i,k}(t)\,p'_{k,j}(0)$$

$$\implies \quad p'_{i,j}(t) = \sum_{k=0}^{\infty} p_{i,k}(t)\,\nu_{k,j}$$

by Proposition 3.3.9. $\square$

Remarks. i) Since there is a Kolmogorov equation for every pair (i,j), the two systems of differential equations comprise m^2 equations each, where m is the number of elements in the state space of the Markov chain $\{X(t), t \geq 0\}$ (which can be infinite). Thus, obtaining an explicit solution for the probabilities $p_{i,j}(t)$ from these systems of equations is generally very difficult. In the next subsection, we will calculate the $p_{i,j}(t)$'s when the state space has only two elements.

ii) In certain particular cases, for example, in the case of the Poisson process, we can determine the $p_{i,j}(t)$'s without having to solve the Kolmogorov differential equations.

Example 3.3.6. Suppose that the Markov chain $\{X(t), t \geq 0\}$ has state space $\{0, 1, \ldots\}$, and that the process, from state 0, can only move to state 1. That is, $p_{0,1} = 1$. We then have

$$\nu_{0,k} = \begin{cases} -\nu_0 & \text{if } k = 0 \\ \nu_0 & \text{if } k = 1 \\ 0 & \text{otherwise} \end{cases}$$

It follows that the Kolmogorov backward equation for the pair $(0, j)$ is given by

$$p'_{0,j}(t) = \nu_0[p_{1,j}(t) - p_{0,j}(t)]$$

If state 0 is absorbing instead, we have the following trivial result:

$$p_{0,0}(t) = 1 \quad \text{and} \quad p_{0,j}(t) = 0 \quad \text{for any } j \neq 0$$

Note that since the two functions are constant, the differential equation above, which becomes $p'_{0,j}(t) = 0$ (since $\nu_0 = 0$), is satisfied. Conversely, the solution of the differential equation $p'_{0,j}(t) = 0$ is $p_{0,j}(t) \equiv c$, and the fact that $p_{0,j}(0) = \delta_{0j}$, that is, the *initial condition*, enables us to determine the value of the constant c.

3.3.4 Particular processes

A continuous-time, two-state Markov chain

The first particular case that we consider is that for which the state space of the continuous-time Markov chain $\{X(t), t \geq 0\}$ is the set $\{0, 1\}$. We assume that these states are not absorbing, so that they communicate. Since the process, from state 0 (respectively, 1), can only move to state 1 (resp., 0), we have that $p_{0,1} = p_{1,0} = 1$. The generating matrix of the chain is given by

$$\mathbf{G} = \begin{matrix} 0 \\ 1 \end{matrix} \begin{bmatrix} -\nu_0 & \nu_0 \\ \nu_1 & -\nu_1 \end{bmatrix} \tag{3.222}$$

from which we deduce that the four Kolmogorov backward equations are, for any $t \geq 0$,

$$\left. \begin{array}{lll} p'_{0,0}(t) & \overset{(1)}{=} & -\nu_0\, p_{0,0}(t) + \nu_0\, p_{1,0}(t) \\ p'_{0,1}(t) & \overset{(2)}{=} & -\nu_0\, p_{0,1}(t) + \nu_0\, p_{1,1}(t) \\ p'_{1,0}(t) & \overset{(3)}{=} & \nu_1\, p_{0,0}(t) - \nu_1\, p_{1,0}(t) \\ p'_{1,1}(t) & \overset{(4)}{=} & \nu_1\, p_{0,1}(t) - \nu_1\, p_{1,1}(t) \end{array} \right\} \tag{3.223}$$

Moreover, we have the following initial conditions:

$$p_{i,j}(0) = \delta_{ij} \quad \text{for } i, j = 0, 1 \tag{3.224}$$

Since

$$p_{0,0}(t) + p_{0,1}(t) = p_{1,0}(t) + p_{1,1}(t) = 1 \tag{3.225}$$

it is sufficient to consider Eqs. (1) and (3) of the system (3.223). We deduce from these equations that

$$p'_{0,0}(t) + p'_{1,0}(t) = (\nu_1 - \nu_0)[p_{0,0}(t) - p_{1,0}(t)] \tag{3.226}$$

Thus, if $\nu_0 = \nu_1$, we have

$$p'_{0,0}(t) + p'_{1,0}(t) = 0 \quad \Longrightarrow \quad p_{0,0}(t) + p_{1,0}(t) \equiv c \tag{3.227}$$

a constant. The initial conditions (3.224) imply that $c = 1$. Eq. (1) can therefore be rewritten as follows:

$$p'_{0,0}(t) = -\nu_0[p_{0,0}(t) - 1 + p_{0,0}(t)] \quad \Longleftrightarrow \quad p'_{0,0}(t) + 2\nu_0\, p_{0,0}(t) = \nu_0 \tag{3.228}$$

Multiplying on both sides by $e^{2\nu_0 t}$, we may write that

$$\frac{d}{dt}\left(e^{2\nu_0 t}\, p_{0,0}(t)\right) = \nu_0\, e^{2\nu_0 t} \tag{3.229}$$

The general solution of this ordinary differential equation is

$$e^{2\nu_0 t}\, p_{0,0}(t) = \frac{1}{2} e^{2\nu_0 t} + c_0 \tag{3.230}$$

where c_0 is a constant. Making use of the condition $p_{0,0}(0) = 1$, we find that $c_0 = 1/2$, so that

$$p_{0,0}(t) = \frac{1}{2}\left(1 + e^{-2\nu_0 t}\right) \quad \text{for } t \geq 0 \tag{3.231}$$

From this function, we can calculate the other three functions $p_{i,j}(t)$.

Remark. The general solution of the ordinary differential equation (o.d.e.)

$$F'(x) + c\,F(x) = G(x) \tag{3.232}$$

where c is a constant, is (for $x \geq 0$)

$$F(x) = e^{-cx}\left(F(0) + \int_0^x e^{cy} G(y)\, dy\right) \tag{3.233}$$

When $\nu_0 \neq \nu_1$, subtracting Eq. (3) from Eq. (1) in the system (3.223), we find that

$$p'_{0,0}(t) - p'_{1,0}(t) = -(\nu_0 + \nu_1)[p_{0,0}(t) - p_{1,0}(t)] \tag{3.234}$$

Using the preceding remark with $F(t) := p_{0,0}(t) - p_{1,0}(t)$ and $G(t) \equiv 0$, we may write that the solution of this o.d.e. is

$$p_{0,0}(t) - p_{1,0}(t) = [p_{0,0}(0) - p_{1,0}(0)]e^{-(\nu_0+\nu_1)t} = e^{-(\nu_0+\nu_1)t} \qquad (3.235)$$

Substituting into Eq. (1), we obtain

$$p'_{0,0}(t) = -\nu_0 e^{-(\nu_0+\nu_1)t} \qquad (3.236)$$

from which we deduce that

$$\int_0^t p'_{0,0}(s)\, ds = -\int_0^t \nu_0 e^{-(\nu_0+\nu_1)s}\, ds$$
$$\Longleftrightarrow p_{0,0}(t) = p_{0,0}(0) - \frac{\nu_0}{\nu_0+\nu_1}\left(1 - e^{-(\nu_0+\nu_1)t}\right) \qquad (3.237)$$

That is,

$$p_{0,0}(t) = \frac{\nu_1}{\nu_0+\nu_1} + \frac{\nu_0}{\nu_0+\nu_1}e^{-(\nu_0+\nu_1)t} \quad \forall\, t \geq 0 \qquad (3.238)$$

Note that if $\nu_0 = \nu_1$ in this formula, then we retrieve the formula (3.231). Next, by using Eq. (3.235), we find that

$$p_{1,0}(t) = p_{0,0}(t) - e^{-(\nu_0+\nu_1)t} = \frac{\nu_1}{\nu_0+\nu_1} - \frac{\nu_1}{\nu_0+\nu_1}e^{-(\nu_0+\nu_1)t} \quad \forall\, t \geq 0$$
$$(3.239)$$

Finally, Eq. (3.225) implies that

$$p_{0,1}(t) = 1 - p_{0,0}(t) = \frac{\nu_0}{\nu_0+\nu_1} - \frac{\nu_0}{\nu_0+\nu_1}e^{-(\nu_0+\nu_1)t} \quad \forall\, t \geq 0 \qquad (3.240)$$

and

$$p_{1,1}(t) = 1 - p_{1,0}(t) = \frac{\nu_0}{\nu_0+\nu_1} + \frac{\nu_1}{\nu_0+\nu_1}e^{-(\nu_0+\nu_1)t} \quad \forall\, t \geq 0 \qquad (3.241)$$

As in the discrete case, an important problem (which will be treated in the next subsection) is that of determining, if it exists, the limiting probability $\lim_{t\to\infty} p_{i,j}(t)$. Here, we easily find that

$$\lim_{t\to\infty} p_{i,j}(t) = \begin{cases} \dfrac{\nu_1}{\nu_0+\nu_1} & \text{if } j = 0 \text{ and } i = 0,1 \\[2mm] \dfrac{\nu_0}{\nu_0+\nu_1} & \text{if } j = 1 \text{ and } i = 0,1 \end{cases} \qquad (3.242)$$

Note that the limit exists and does not depend on the initial state i.

Pure birth processes

Definition 3.3.9. *Let* $\{X(t), t \geq 0\}$ *be a continuous-time Markov chain whose state space is the set* $\{0, 1, \ldots\}$. *If*

$$p_{i,i+1} = 1 \quad for \ i = 0, 1, \ldots \tag{3.243}$$

the process is called a **pure birth process**.

Remarks. i) A pure birth process is thus a continuous-time Markov chain for which transitions can only be made from an arbitrary state to its right-hand neighbor. It is a particular case of the *birth and death processes*, which will be treated further on.

ii) We can also define a pure birth process by setting

$$\nu_{i,j} = \begin{cases} \nu_i \ \text{if} \ j = i+1 \\ 0 \ \text{otherwise} \end{cases} \tag{3.244}$$

for all $i \neq j \in \{0, 1, \ldots\}$.

The Kolmogorov forward equations for a pure birth process are the following:

$$p'_{i,i}(t) = -\nu_i \, p_{i,i}(t) \tag{3.245}$$

and

$$p'_{i,j}(t) = \nu_{j-1} \, p_{i,j-1}(t) - \nu_j \, p_{i,j}(t) \quad \text{if} \ j = i+1, i+2, \ldots \tag{3.246}$$

Proposition 3.3.10. *The function* $p_{i,j}(t)$ *for a pure birth process is given, for any* $t \geq 0$ *and for all* $i \in \{0, 1, \ldots\}$, *by* $p_{i,j}(t) = 0$ *if* $j < i$, *and*

$$p_{i,j}(t) = \begin{cases} e^{-\nu_i t} & \text{if} \ j = i \\ \nu_{j-1} e^{-\nu_j t} \int_0^t e^{\nu_j s} p_{i,j-1}(s) \, ds & \text{if} \ j > i \end{cases} \tag{3.247}$$

Proof. Since the process can move only to the right, we indeed have $p_{i,j}(t) = 0$ if $j < i$.

Next, let F be the following event: the process makes no transitions in the interval $[0, t]$. Since the process cannot move backward, we may write that

$$p_{i,i}(t) = P[F] = P[\tau_i > t] = e^{-\nu_i t} \tag{3.248}$$

Finally, by (3.233), the solution of Eq. (3.246) is

$$p_{i,j}(t) = e^{-\nu_j t} \left\{ p_{i,j}(0) + \nu_{j-1} \int_0^t e^{\nu_j s} p_{i,j-1}(s) \, ds \right\} \tag{3.249}$$

and the result follows from the fact that $p_{i,j}(0) = 0$, for $j = i+1, i+2, \ldots$. $\square$

From the formulas (3.247), we can calculate $p_{i,j}(t)$ recursively. For instance, we have

$$p_{i,i+1}(t) = \nu_i e^{-\nu_{i+1}t} \int_0^t e^{\nu_{i+1}s} p_{i,i}(s)\, ds$$

$$= \nu_i e^{-\nu_{i+1}t} \int_0^t e^{\nu_{i+1}s} e^{-\nu_i s}\, ds = \nu_i e^{-\nu_{i+1}t} \int_0^t e^{(\nu_{i+1}-\nu_i)s}\, ds$$

$$= \begin{cases} \nu_i\, t\, e^{-\nu_i t} & \text{if } \nu_i = \nu_{i+1} \\[2mm] \dfrac{\nu_i}{\nu_{i+1} - \nu_i} \left(e^{-\nu_i t} - e^{-\nu_{i+1}t}\right) & \text{if } \nu_i \neq \nu_{i+1} \end{cases} \tag{3.250}$$

Particular cases

1) **Poisson process.** This process is obtained by setting $\nu_i = \lambda$, for all $i \in \{0, 1, \ldots\}$. To obtain its transition function $p_{i,j}(t)$, we can use Eqs. (3.247) and (3.250), with $\nu_i = \nu_{i+1} = \lambda$. We have

$$p_{i,i+1}(t) = \lambda\, t\, e^{-\lambda t} \implies p_{i,i+2}(t) = \lambda e^{-\lambda t} \int_0^t e^{\lambda s} \lambda\, s\, e^{-\lambda s}\, ds = \frac{(\lambda t)^2}{2} e^{-\lambda t} \tag{3.251}$$

Proceeding recursively, we obtain the following general formula:

$$p_{i,i+k}(t) = \frac{(\lambda t)^k}{k!} e^{-\lambda t} \quad \text{for } t \geq 0 \text{ and } k = 0, 1, \ldots \tag{3.252}$$

That is,

$$p_{i,i+k}(t) = P[\mathrm{Poi}(\lambda t) = k] \tag{3.253}$$

The parameter λ is called the *rate* of the process.

We can also obtain the preceding formula without making use of the Kolmogorov equations, proceeding as follows: since $\nu_i \equiv \lambda$, the random variables τ_i all have an exponential distribution with parameter λ. Moreover, they are independent, by the Markov property. It follows, by Proposition 3.3.6, that

$$S := \tau_i + \ldots + \tau_{i+k-1} \sim G(k, \lambda) \quad \forall\, k \geq 1 \tag{3.254}$$

Then, we may write that

$$p_{i,i+k}(t) = \int_0^t f_S(s)\, P[\tau_{i+k} > t - s]\, ds = \int_0^t \frac{(\lambda s)^{k-1}}{(k-1)!} \lambda e^{-\lambda s}\, e^{-\lambda(t-s)}\, ds$$

$$= e^{-\lambda t} \frac{\lambda^k}{(k-1)!} \int_0^t s^{k-1}\, ds = e^{-\lambda t} \frac{(\lambda t)^k}{k!} \tag{3.255}$$

for $i, k \in \{0, 1, \dots\}$ (since $p_{i,i}(t) \stackrel{(3.248)}{=} e^{-\lambda t}$) and $\forall\, t \geq 0$.

2) **Yule[11] process.** This process, studied by Yule in the framework of the theory of evolution, can be interpreted as follows: we consider a population whose members cannot die and such that each member, independently of the others, gives birth to descendants according to a Poisson process with rate λ. It follows that

$$\nu_i = i\lambda \quad \forall\, i \geq 0 \tag{3.256}$$

Indeed, when $X(t) = i$, the time τ_i that the process spends in this state is given by

$$\tau_i = \min\{X_1, \dots, X_i\} \tag{3.257}$$

where the random variables $X_1, \dots, X_i$ are independent and all have an exponential distribution with parameter λ. By Proposition 3.3.5, we may write that $\tau_i \sim \mathrm{Exp}(\nu_i = i\lambda)$.

Remark. Note that if $X(0) = 0$, then the process remains in this state forever.

We can calculate the transition function of the process in various ways. We find that

$$p_{i,j}(t) = \binom{j-1}{j-i} e^{-i\lambda t}(1 - e^{-\lambda t})^{j-i} \quad \text{if } j \geq i \geq 1 \tag{3.258}$$

To check this result, we use the fact that Eq. (3.246) becomes

$$p_{i,j}'(t) = (j-1)\lambda\, p_{i,j-1}(t) - j\lambda\, p_{i,j}(t) \quad \text{for } j = i+1, i+2, \dots \tag{3.259}$$

Next, from the formula (3.258), we may write that

$$p_{i,j-1}(t) = \frac{j-i}{j-1}\left(\frac{1}{1 - e^{-\lambda t}}\right) p_{i,j}(t) \tag{3.260}$$

Moreover, we calculate [also from (3.258)]

$$p_{i,j}'(t) = -i\lambda\, p_{i,j}(t) + (j-1)\lambda e^{-\lambda t} p_{i,j-1}(t) \tag{3.261}$$

We have

$$-i\lambda\, p_{i,j}(t) + (j-1)\lambda e^{-\lambda t} p_{i,j-1}(t) = (j-1)\lambda\, p_{i,j-1}(t) - j\lambda\, p_{i,j}(t)$$
$$\iff \quad (j-1)(1 - e^{-\lambda t}) p_{i,j-1}(t) = (j-i) p_{i,j}(t) \tag{3.262}$$

---

[11] George Udny Yule, 1871–1951, was born in Scotland and died in England. He first obtained an engineering degree. Next, he became interested in the field of statistics and wrote important papers on regression and correlation theory. His book *Introduction to the Theory of Statistics*, published for the first time in 1911, was very successful.

which holds true by Eq. (3.260).

Finally, the function given in (3.258) does indeed satisfy the initial condition

$$p_{i,j}(0) = \delta_{ij} \qquad (3.263)$$

Remark. When $i = 1$, the formula (3.258) becomes

$$p_{1,j}(t) = e^{-\lambda t}(1 - e^{-\lambda t})^{j-1} \quad \text{if } j \geq 1 \qquad (3.264)$$

We can then write that

$$p_{1,j}(t) = P[\text{Geom}(p := e^{-\lambda t}) = j] \quad \text{if } j \geq 1 \qquad (3.265)$$

Now, when $X(0) = i > 1$, it is as if we added i independent random variables, $X_1, \ldots, X_i$, all of which have a geometric distribution with parameter $p = e^{-\lambda t}$. We can show that such a sum S has a *negative binomial* or *Pascal*[12] *distribution* with parameters i and p, whose probability mass function is given by

$$P[S = j] = \binom{j-1}{j-i} p^i(1-p)^{j-i} \quad \text{for } j = i, i+1, \ldots \qquad (3.266)$$

The formula for $p_{i,j}(t)$ thus follows directly from that for $p_{1,j}(t)$.

Birth and death processes

Definition 3.3.10. *If the instantaneous transition rates $\nu_{i,j}$ of the continuous-time Markov chain $\{X(t), t \geq 0\}$ are such that*

$$\nu_{i,j} = 0 \quad \text{if } |j - i| > 1 \qquad (3.267)$$

the process is said to be a **birth and death process.**

Definition 3.3.11. *The parameters*

$$\lambda_i := \nu_{i,i+1} \quad (\text{for } i \geq 0) \quad \text{and} \quad \mu_i := \nu_{i,i-1} \quad (\text{for } i \geq 1) \qquad (3.268)$$

are called, respectively, the **birth** *and* **death rates** *of the birth and death process.*

---

[12] Blaise Pascal, 1623–1662, was born and died in France. He is one of the founders of the theory of probability. He was also interested in geometry and in physics, in addition to publishing books on philosophy and on theology.

Remark. Other terms used to designate the parameters λ_i and μ_i are the following: *growth* or *arrival* rates, and *mortality* or *departure* rates, respectively.

From the instantaneous transition rates $\nu_{i,j}$, we can calculate the parameters ν_i of the random variables τ_i, as well as the probabilities $p_{i,j}$, for all states i and j in the set $\{0, 1, \ldots\}$. We have

$$\nu_0 = -\nu_{0,0} = \nu_{0,1} = \lambda_0 \tag{3.269}$$

and

$$\nu_i = -\nu_{i,i} = \nu_{i,i+1} + \nu_{i,i-1} = \lambda_i + \mu_i \quad \text{if } i \geq 1 \tag{3.270}$$

In the case when the state space is the *finite* set $\{0, 1, \ldots, m\}$, we have

$$\nu_m = -\nu_{m,m} = \nu_{m,m-1} = \mu_m \tag{3.271}$$

If $\lambda_i = \mu_i = 0$, then state i is absorbing. For all nonabsorbing states $i \in \{0, 1, \ldots\}$, we have that $p_{0,1} = 1$ and, using the formula (3.195),

$$p_{i,i+1} = \frac{\nu_{i,i+1}}{\nu_i} = \frac{\lambda_i}{\lambda_i + \mu_i} = 1 - p_{i,i-1} \quad \text{if } i \geq 1 \tag{3.272}$$

Remarks. i) A birth and death process is a continuous-time Markov chain for which transitions, from state i, can only be made to $i-1$ (if $i > 0$) or $i+1$. In a short interval, of length δ, the probability that the process moves from state i to $i+1$ (respectively, $i-1$) is equal to $\lambda_i \delta + o(\delta)$ (resp., $\mu_i \delta + o(\delta)$). Thus, $1 - (\lambda_i + \mu_i)\delta + o(\delta)$ is the probability that the process will still be (or will be back) in state i after δ unit(s) of time.

ii) In many applications, the state of the process at a given time instant is the number of individuals in the *system* at this time instant. A birth and death process may then be interpreted as follows: when $X(t) = i$, the waiting time until the next *arrival* is a random variable X_i having an $\text{Exp}(\lambda_i)$ distribution and which is independent of the waiting time Y_i, having an $\text{Exp}(\mu_i)$ distribution, until the next *departure*. We then have $\tau_0 = X_0 \sim \text{Exp}(\nu_0 = \lambda_0)$ and

$$\tau_i = \min\{X_i, Y_i\} \sim \text{Exp}(\nu_i = \lambda_i + \mu_i) \quad \text{for } i > 0 \tag{3.273}$$

Moreover, we indeed have $p_{0,1} = 1$ and

$$p_{i,i+1} = P[X_i < Y_i] = P[\text{Exp}(\lambda_i) < \text{Exp}(\mu_i)] \stackrel{\text{ind.}}{=} \frac{\lambda_i}{\lambda_i + \mu_i} \quad \text{if } i > 0 \tag{3.274}$$

Similarly,

$$p_{i,i-1} = \frac{\mu_i}{\lambda_i + \mu_i} \quad \text{if } i > 0 \tag{3.275}$$

iii) A *pure birth process* is the particular case where the death rates μ_i are all equal to zero. A process for which $\lambda_i = 0$ for all i is called a *pure death process*.

Particular cases

1) The continuous-time Markov chain whose state space is the set $\{0,1\}$, considered at the beginning of this subsection (see p. 129), is an example of a birth and death process, for which

$$\lambda_0 = \nu_0 \quad \text{and} \quad \mu_1 = \nu_1 \tag{3.276}$$

(all the other birth and death rates being equal to zero).

2) Suppose that we modify the Yule process as follows: after an exponential time with parameter λ, an individual either gives birth to a descendant, with probability p, or disappears from the population, with probability $1 - p$. Let $X(t)$ be the number of individuals in the population at time t. The process $\{X(t), t \geq 0\}$ is now a birth and death process whose birth and death rates are given by

$$\lambda_i = ip\lambda \quad \text{and} \quad \mu_i = i(1 - p)\lambda \quad \text{for } i \geq 0 \tag{3.277}$$

If we assume rather that, when an individual gives birth to a descendant, there is a probability p (respectively, $1 - p$) that this individual remains in (resp., disappears from) the population, then the process $\{X(t), t \geq 0\}$ is no longer a continuous-time Markov chain. Indeed, suppose that $X(0) = 1$. We may write that

$$P[\tau_1 \leq t] = \sum_{k=1}^{\infty} p(1 - p)^{k-1} P[X_1 + \ldots + X_k \leq t] \tag{3.278}$$

where the X_i's are independent random variables that all have an exponential distribution with parameter λ. This result follows from the fact that we perform independent trials for which the probability of success, that is, the case when the individual remains in the population, is equal to p. Since the random variable τ_1 does *not* have an exponential distribution, $\{X(t), t \geq 0\}$ is not a continuous-time Markov chain anymore. Actually, τ_1 is an infinite linear combination of independent gamma distributions.

3) **The queueing model** $M/M/s$. This process will be studied in detail in Section 6.3.1. We suppose that customers arrive at a system according to a Poisson process with rate λ. There are s servers, but the customers form a single queue. The *service times* are independent and have an $\text{Exp}(\mu)$ distribution. We have

$$\lambda_i = \lambda \ \forall \, i \geq 0, \quad \mu_i = i\mu \ \text{ if } 1 \leq i \leq s, \quad \text{and} \quad \mu_i = s\mu \ \text{ if } i > s \tag{3.279}$$

(since if all the servers are occupied, then the departure rate from the system is equal to $s\mu$, for any number of persons in the system).

Remark. M means that the waiting (and service) times are exponential, thus *Markovian.*

To end this subsection, we give the Kolmogorov backward and forward equations for a birth and death process. First, the backward equations are

$$p'_{0,j}(t) = -\lambda_0\, p_{0,j}(t) + \lambda_0\, p_{1,j}(t) \tag{3.280}$$

$$p'_{i,j}(t) \stackrel{\text{if } i>0}{=} \mu_i\, p_{i-1,j}(t) - (\lambda_i + \mu_i)\, p_{i,j}(t) + \lambda_i\, p_{i+1,j}(t) \tag{3.281}$$

In the case of the Kolmogorov forward equations, we find that

$$p'_{i,0}(t) = \mu_1\, p_{i,1}(t) - \lambda_0\, p_{i,0}(t) \tag{3.282}$$

$$p'_{i,j}(t) \stackrel{\text{if } j>0}{=} \lambda_{j-1}\, p_{i,j-1}(t) - (\lambda_j + \mu_j)\, p_{i,j}(t) + \mu_{j+1}\, p_{i,j+1}(t) \tag{3.283}$$

3.3.5 Limiting probabilities and balance equations

Since it is generally very difficult to solve explicitly the Kolmogorov equations to obtain the transition functions $p_{i,j}(t)$, we must often content ourselves with the computation of the *limiting probability* that the process will be in a given state when it is *in equilibrium*. To obtain these limiting probabilities, we can try to solve a system of linear equations called the *balance equations* of the process.

Definition 3.3.12. *Let* π_0, π_1, ... *be nonnegative real numbers such that*

$$\sum_{j=0}^{\infty} \pi_j = 1 \tag{3.284}$$

If the equation

$$\sum_{i=0}^{\infty} \pi_i\, p_{i,j}(t) = \pi_j \tag{3.285}$$

is satisfied for all $j \in \{0, 1, \ldots\}$ *and for all* $t \geq 0$, *then* $\pi := (\pi_0, \pi_1, \ldots)$ *is a* stationary distribution.

The expression *stationary distribution* is used because if we assume that

$$P[X(0) = j] = \pi_j \quad \text{for all } j \in \{0, 1, \ldots\} \tag{3.286}$$

then we have

$$P[X(t) = j] = \sum_{i=0}^{\infty} P[X(t) = j \mid X(0) = i] P[X(0) = i]$$

$$= \sum_{i=0}^{\infty} p_{i,j}(t)\, \pi_i \stackrel{(3.285)}{=} \pi_j \tag{3.287}$$

That is,

$$P[X(t) = j] = P[X(0) = j] = \pi_j \tag{3.288}$$

$\forall\, j \in \{0, 1, \dots\}$ and $\forall\, t \geq 0$.

As in the discrete case, we can define the notion of *recurrence* of a state.

Notation. We denote by $T_{i,i}$ the time elapsed between two consecutive visits to state i of the continuous-time Markov chain $\{X(t), t \geq 0\}$.

Definition 3.3.13. *We say that state i is* **recurrent** *if $P[T_{i,i} < \infty] = 1$ and* **transient** *if $P[T_{i,i} < \infty] < 1$. Moreover, let*

$$\mu_{i,i} := E[T_{i,i}] \tag{3.289}$$

The recurrent state i is said to be **positive** *(respectively,* **null***) recurrent if $\mu_{i,i} < \infty$ (resp., $= \infty$).*

We can prove the following theorem.

Theorem 3.3.3. *If the continuous-time Markov chain $\{X(t), t \geq 0\}$ is irreducible and positive recurrent, then it has a unique stationary distribution $\pi = (\pi_0, \pi_1, \dots)$, where π_j is the limiting probability*

$$\pi_j := \lim_{t \to \infty} p_{i,j}(t) \quad \text{for all } j \in \{0, 1, \dots\} \tag{3.290}$$

Remarks. i) Note that the π_j's do not depend on the initial state i. The quantity π_j is also the proportion of time that the Markov chain spends in state j, on the long run.

ii) If the Markov chain is transient or null recurrent, then it does not have a stationary distribution.

iii) Contrary to the discrete-time Markov chains, a continuous-time Markov chain *cannot* be periodic. This follows from the fact that the time the process spends in an arbitrary state is a random variable having an exponential distribution, which is continuous. Consequently, if the limiting probabilities exist, then we say that the Markov chain is *ergodic*.

iv) We find that

$$\pi_j = \frac{1}{\nu_j \mu_{j,j}} \quad \text{for all } j \in S_{X(t)} = \{0, 1, \dots\} \tag{3.291}$$

where the state space $S_{X(t)}$ may, of course, be finite, but must comprise more than one state. The interpretation of this result is the following: ν_j^{-1} is the average time that the process spends in state j on an arbitrary visit to this state, and $\mu_{j,j}$ is the average time between two consecutive visits to state j, so that $t/\mu_{j,j}$ is the average number of visits to state j over a period of length t, where t is large. Thus, over this long period of time, the proportion of time during which the chain is in state j is effectively given by the ratio $\nu_j^{-1}/\mu_{j,j}$.

v) If state j is absorbing, we consider it as *positive recurrent*. The theorem applies even in the case when the state space contains a single state, say 0. We then have, trivially, that $\pi_0 = 1$.

Now, if we differentiate both sides of Eq. (3.285) with respect to t, we obtain

$$\sum_{i=0}^{\infty} \pi_i\, p'_{i,j}(t) = 0 \quad \forall\, j \in \{0, 1, \dots\} \text{ and } \forall\, t \geq 0 \qquad (3.292)$$

With $t = 0$, we may write, by Proposition 3.3.9, that

$$\sum_{i=0}^{\infty} \pi_i\, \nu_{i,j} = 0 \quad \forall\, j \in \{0, 1, \dots\} \qquad (3.293)$$

Remarks. i) If the number of states is *finite*, we can indeed interchange the derivative and the summation. However, when the state space is infinite, this interchange, though allowed here, must be justified.

ii) It can be shown that Eq. (3.293) is satisfied *if and only if* Eq. (3.285) is also satisfied.

Since $\nu_{j,j} = -\nu_j$, we can rewrite Eq. (3.293) as follows:

$$\pi_j \nu_j = \sum_{i \neq j} \pi_i\, \nu_{i,j} \quad \forall\, j \in \{0, 1, \dots\} \qquad (3.294)$$

Definition 3.3.14. *The equations above are called the* **balance equations** *of the stochastic process* $\{X(t), t \geq 0\}$.

Remarks. i) We can interpret the balance equations as follows: the rate at which the process *leaves* state j must be equal to the rate at which it *enters* state j, for all j. Now, the rate at which transitions occur, on the long run, from state i to any other state $j \neq i$ of the Markov chain is given by $\pi_i \nu_{i,j}$, since π_i is the proportion of time that the process spends in state i during the period considered, and $\nu_{i,j}$ is the rate at which, *when it is in state i*, it enters j. Given that the process spends an exponential time with parameter ν_j in state j, $\pi_j \nu_j$ is the departure rate from j, and the sum of the terms $\pi_i \nu_{i,j}$ over all states $i \neq j$ is the arrival rate to j.

ii) The limiting probabilities π_j can be obtained by solving the system (3.294), under the condition $\sum_{j=0}^{\infty} \pi_j = 1$ [see (3.284)].

We will now obtain a general formula for the π_j's when $\{X(t), t \geq 0\}$ is a birth and death process. First, we give a proposition that tells us when the π_j's do exist.

Proposition 3.3.11. *Let $\{X(t), t \geq 0\}$ be an irreducible birth and death process, whose state space is the set $\{0, 1, \dots\}$. We set*

$$S_1 = \sum_{k=1}^{\infty} \frac{\lambda_0 \lambda_1 \cdots \lambda_{k-1}}{\mu_1 \mu_2 \cdots \mu_k} \quad and \quad S_2 = \sum_{k=1}^{\infty} \frac{\mu_1 \mu_2 \cdots \mu_k}{\lambda_1 \lambda_2 \cdots \lambda_k} \tag{3.295}$$

The Markov chain is

$$\left. \begin{array}{c} \textit{positive recurrent if } S_1 < \infty \\ \textit{null recurrent if } S_1 = S_2 = \infty \\ \textit{transient if } S_2 < \infty \end{array} \right\} \tag{3.296}$$

Remark. If the state space of the irreducible birth and death process is *finite*, then we necessarily have that $S_1 < \infty$, so that the Markov chain is positive recurrent.

With the help of this proposition, we can prove the following theorem.

Theorem 3.3.4. *For an irreducible and positive recurrent birth and death process $\{X(t), t \geq 0\}$, whose state space is the set $\{0, 1, \dots\}$, the limiting probabilities are given by*

$$\pi_j = \frac{\Pi_j}{\sum_{k=0}^{\infty} \Pi_k} \quad \textit{for } j = 0, 1, \dots \tag{3.297}$$

where

$$\Pi_0 := 1 \quad and \quad \Pi_j := \frac{\lambda_0 \lambda_1 \cdots \lambda_{j-1}}{\mu_1 \mu_2 \cdots \mu_j} \quad \textit{for } j \geq 1 \tag{3.298}$$

Proof. In the case of a birth and death process, the balance equations become

state j departure rate from j = arrival rate to j

0	$\lambda_0 \pi_0 = \mu_1 \pi_1$
1	$(\lambda_1 + \mu_1)\pi_1 = \mu_2 \pi_2 + \lambda_0 \pi_0$
$\vdots$	$\vdots \; \vdots \; \vdots$
$k \; (\geq 1)$	$(\lambda_k + \mu_k)\pi_k = \mu_{k+1}\pi_{k+1} + \lambda_{k-1}\pi_{k-1}$

Adding the equations for $j = 0$ and $j = 1$, then for $j = 1$ and $j = 2$, etc., we find that

$$\lambda_j \pi_j = \mu_{j+1} \pi_{j+1} \quad \forall\, j \geq 0 \tag{3.299}$$

which implies that

$$\pi_j = \frac{\lambda_{j-1} \cdots \lambda_1 \lambda_0}{\mu_j \cdots \mu_2 \mu_1} \pi_0 \tag{3.300}$$

Using the fact that $\sum_{j=0}^{\infty} \pi_j = 1$, we obtain

$$\pi_0 = \left[1 + \sum_{k=1}^{\infty} \frac{\lambda_0 \lambda_1 \cdots \lambda_{k-1}}{\mu_1 \mu_2 \cdots \mu_k} \right]^{-1} = \frac{\Pi_0}{\sum_{k=0}^{\infty} \Pi_k} \tag{3.301}$$

so that

$$\pi_j = \frac{\Pi_j}{\sum_{k=0}^{\infty} \Pi_k} \quad \text{for } j = 0, 1, \dots \tag{3.302}$$

Finally, it can be shown that $\pi = (\pi_0, \pi_1, \dots)$ is a stationary distribution of the Markov chain $\{X(t), t \geq 0\}$. By Theorem 3.3.3, this distribution is *unique*. Thus, the limiting probabilities are indeed given by the preceding equations.
□

Remark. We can rewrite the formula (3.297) as follows:

$$\pi_j = \frac{\Pi_j}{1 + S_1} \quad \text{for } j = 0, 1, \dots \tag{3.303}$$

where S_1 is defined in (3.295).

Example 3.3.7. In the case of the two-state, continuous-time Markov chain (see pp. 129 and 137), we have

$$S_1 = \frac{\lambda_0}{\mu_1} = \frac{\nu_0}{\nu_1}$$

from which we deduce that

$$\pi_0 = \frac{1}{1 + \frac{\nu_0}{\nu_1}} = \frac{\nu_1}{\nu_0 + \nu_1} = 1 - \pi_1$$

Note that these results correspond to those given by Eq. (3.242).

When $\nu_0 = \nu_1$, we have that $\pi_0 = \pi_1 = 1/2$, which had to be the case, by symmetry.

3.4 Exercises

Section 3.2

Question no. 1

We suppose that the probability that a certain machine functions without failure today is equal to

0.7 if the machine functioned without failure yesterday and the day before yesterday (state 0),

0.5 if the machine functioned without failure yesterday, but not the day before yesterday (state 1),

0.4 if the machine functioned without failure the day before yesterday, but not yesterday (state 2),

0.2 if the machine did not function without failure, neither yesterday nor the day before yesterday (state 3).

(a) Find the one-step transition probability matrix of the Markov chain associated with the functioning state of the machine.

(b) Calculate $p_{0,1}^{(2)}$, that is, the probability of moving from state 0 to state 1 in two steps.

(c) Calculate the average number of days without failure of the machine over the next two days, given that the Markov chain is presently in state 0.

Question no. 2

Let $\{X_n, n = 0, 1, \ldots\}$ be a Markov chain whose state space is the set $\{0, 1\}$ and whose one-step transition probability matrix $\mathbf{P}$ is given by

$$\mathbf{P} = \begin{bmatrix} 1/2 & 1/2 \\ p & 1-p \end{bmatrix}$$

where $0 \leq p \leq 1$.

(a) Suppose that $p = 1$ and that $X_0 = 0$. Calculate $E[X_2]$.

(b) Suppose that $p = 1/2$ and that $P[X_0 = 0] = P[X_0 = 1] = 1/2$. We define the continuous-time stochastic process $\{Y(t), t \geq 0\}$ by $Y(t) = tX_{[t]}$, for $t \geq 0$, where $[t]$ denotes the integer part of t.

　(i) Calculate $C_Y(t, t+1)$.

　(ii) Is the stochastic process $\{Y(t), t \geq 0\}$ wide-sense stationary? Justify.

　(iii) Calculate $\lim_{n \to \infty} P[X_n = 0]$.

Question no. 3

We consider a Markov chain $\{X_n, n = 0, 1, \ldots\}$ having states 0 and 1. On each step, the process moves from state 0 to state 1 with probability $p \in (0, 1)$, or from state 1 to state 0 with probability $1 - p$.

(a) Calculate $p_{1,1}^{(10)}$.

(b) Suppose that $X_0 = 0$. Calculate the autocorrelation function $R_X(1, 13)$.

Question no. 4

The one-step transition probability matrix $\mathbf{P}$ of a Markov chain whose state space is $\{0,1\}$ is given by

$$\mathbf{P} = \begin{bmatrix} 1/2 & 1/2 \\ 0 & 1 \end{bmatrix}$$

Calculate $E[X_2]$ if $P[X_0 = 0] = 1/3$.

Question no. 5

Let $Y_1, Y_2, \ldots$ be an infinite sequence of independent random variables, all having a Bernoulli distribution with parameter $p = 1/3$. We define $X_n = \sum_{k=1}^n Y_k$, for $n = 1, 2, \ldots$. Then (see p. 85) the stochastic process $\{X_n, n = 1, 2, \ldots\}$ is a Markov chain. Calculate $p_{0,2}^{(3)}$.

Question no. 6

Let $\{X_n, n = 0, 1, \ldots\}$ be a random walk for which

$$p_{i,i+1} = \frac{2}{3} \quad \text{and} \quad p_{i,i-1} = \frac{1}{3} \quad \text{for } i \in \{0, \pm 1, \pm 2, \ldots\}$$

Calculate $E[X_2 \mid X_0 = 0]$.

Question no. 7

Let $\{X_n, n = 0, 1, \ldots\}$ be a Markov chain whose state space is the set $\{0,1\}$ and whose one-step transition probability matrix is given by

$$\mathbf{P} = \begin{bmatrix} 0 & 1 \\ 1 & 0 \end{bmatrix}$$

(a) Calculate $C_X(t_1, t_2)$ at $t_1 = 0$ and $t_2 = 1$ if $P[X_0 = 0] = P[X_0 = 1] = 1/2$.

(b) Find $\lim_{n \to \infty} P[X_n = 0 \mid X_0 = 0]$.

Question no. 8

Let $X_1, X_2, \ldots$ be an infinite sequence of independent random variables, all having a Poisson distribution with parameter $\alpha = 1$. We define

$$Y_n = \sum_{k=1}^n X_k \quad \text{for } n = 1, 2, \ldots$$

Then, $\{Y_n, n = 1, 2, \ldots\}$ is a Markov chain (see p. 85). Calculate $p_{1,3}^{(4)}$, that is, the probability of moving from state 1 to state 3 in four steps.

Question no. 9

The flow of a certain river can be one of the following three states:

0: low flow
1: average flow
2: high flow

We suppose that the stochastic process $\{X_n, n = 0, 1, \ldots\}$, where X_n represents the state of the river flow on the nth day, is a Markov chain. Furthermore, we estimate that the probability that the flow moves from state i to state j in one day is given by the formula

$$p_{i,j} = \frac{1}{2} - |i - j|\theta_i$$

where $0 < \theta_i < 1$, for $i, j = 0, 1, 2$.

(a) Calculate the probability that the river flow moves from state 0 to state 1 in one day.

(b) What is the probability that the river flow moves from state 0 to state 2 in two days?

Question no. 10
A machine is made up of two components that operate independently. The lifetime T_i (in days) of component i has an exponential distribution with parameter λ_i, for $i = 1, 2$.

Suppose that the two components are placed in parallel and that $\lambda_1 = \lambda_2 = \ln 2$. When the machine breaks down, the two components are replaced by new ones at the beginning of the following day. Let X_n be the number of components that operate at the end of n days. Then the stochastic process $\{X_n, n = 0, 1, \ldots\}$ is a Markov chain. Calculate its one-step transition probability matrix.

Question no. 11
Let $\{X_n, n = 0, 1, \ldots\}$ be a Markov chain whose state space is the set $\{0, 1, 2, 3, 4\}$ and whose one-step transition probability matrix is

$$\mathbf{P} = \begin{bmatrix} 1 & 0 & 0 & 0 & 0 \\ 0.5 & 0.2 & 0.3 & 0 & 0 \\ 0 & 0 & 0 & 1 & 0 \\ 0 & 0 & 0 & 0 & 1 \\ 0 & 0 & 1 & 0 & 0 \end{bmatrix}$$

(a) Calculate the probability that the process will move from state 1 to state 2 in four steps.

(b) Suppose that $X_0 = 1$. Let N_1 be the number of times that state 1 will be visited, including the initial state. Calculate $E[N_1]$.

Question no. 12
A Markov chain $\{X_n, n = 0, 1, \ldots\}$ with state space $\{0, 1, 2, 3\}$ has the following one-step transition probability matrix:

$$\mathbf{P} = \begin{bmatrix} 1/2 & 1/2 & 0 & 0 \\ 1/2 & 1/4 & 1/4 & 0 \\ 0 & 0 & 1/4 & 3/4 \\ 0 & 0 & 0 & 1 \end{bmatrix}$$

Assuming that $X_0 = 1$, calculate the probability that state 0 will be visited before state 3.

Question no. 13

A Markov chain has the following one-step transition probabilities:

$$p_{0,0} = 1$$
$$p_{i,i} = p = 1 - p_{i,i-1} \quad \text{for } i = 1, 2, 3, \ldots .$$

Calculate the probability $\rho_{i,0}^{(n)}$ that the chain will move from state i to state 0 for the first time after exactly n transitions, for $i = 1, 2, \ldots$

Question no. 14

Let

$$\mathbf{P} = \begin{bmatrix} q & p & 0 & 0 & 0 \\ q & 0 & p & 0 & 0 \\ 0 & q & 0 & p & 0 \\ 0 & 0 & q & 0 & p \\ 0 & 0 & 0 & q & p \end{bmatrix}$$

where $p + q = 1$ and $0 < p < 1$, be the one-step transition probability matrix of a Markov chain whose state space is the set $\{0, 1, 2, 3, 4\}$.

(a) Is the chain periodic or aperiodic? Justify.

(b) Calculate, if they exist, the limiting probabilities π_i.

Question no. 15

We perform repeated trials that are *not* Bernoulli trials (see p. 51). We suppose that the probability p_n of a *success* on the nth trial is given by

$$p_n = \begin{cases} 1/2 & \text{for } n = 1, 2 \\ \dfrac{X_n + 1}{X_n + 2} & \text{for } n = 3, 4, \ldots \end{cases}$$

where X_n is the total number of successes obtained on the $(n - 2)$nd and $(n - 1)$st trials. Calculate $\lim_{n \to \infty} p_n$.

Question no. 16

In the gambler's ruin problem (see p. 101), let Y_i be the number of plays needed to end the game (with the player being ruined or having reached his objective of k units), given that his initial fortune is equal to i units, for $i = 0, 1, \ldots, k$. Show that

$$E[Y_i] = \begin{cases} (2p - 1)^{-1} \left\{ \dfrac{k \left[1 - (q/p)^i \right]}{1 - (q/p)^k} - i \right\} & \text{if } p \neq 1/2 \\ \\ i(k - i) & \text{if } p = 1/2 \end{cases}$$

Indication. We may write that $E[Y_0] = E[Y_k] = 0$ and (by conditioning on the result of the first play)

$$E[Y_i] = 1 + p\,E[Y_{i+1}] + q\,E[Y_{i-1}] \quad \text{for } i = 1,\dots,k-1$$

Question no. 17

We consider a branching process for which $p_i = (1-p)^i p$, for $i = 0,1,\dots$, where $0 < p < 1$. That is, $Y := Z+1$, where Z is the number of descendants of an arbitrary individual, has a geometric distribution with parameter p. Show that the probability q_0 of eventual extinction of the population is given by

$$q_0 = \begin{cases} p/(1-p) & \text{if } p < 1/2 \\ 1 & \text{if } p \geq 1/2 \end{cases}$$

Question no. 18

Let $\{X_n, n = 0,1,\dots\}$ be an irreducible and ergodic Markov chain. Suppose that the chain is in *equilibrium*, so that we can write that

$$P[X_n = i] = \pi_i \quad \text{for all states } i \text{ and for any } n$$

We can show that the process $\{X_k, k = \dots, n+1, n, \dots\}$, for which the time is reversed, is also a Markov chain, whose transition probabilities are given by

$$q_{i,j} := P[X_n = j \mid X_{n+1} = i] = p_{j,i}\frac{\pi_j}{\pi_i} \quad \text{for all states } i, j$$

We say that the chain $\{X_n, n = 0,1,\dots\}$ is *time-reversible* if $q_{i,j} = p_{i,j}$, for all i, j. Show that the Markov chain whose state space is the set $\{0,1,2,3\}$ and with one-step transition probability matrix

$$\mathbf{P} = \begin{bmatrix} 1/2 & 1/2 & 0 & 0 \\ 1/2 & 0 & 1/2 & 0 \\ 0 & 1/2 & 0 & 1/2 \\ 0 & 0 & 1/2 & 1/2 \end{bmatrix}$$

is time-reversible.

Question no. 19

In the urn model of P. and T. Ehrenfest for the movement of molecules in a gas (see p. 75), suppose that $X_0 = 1$. Let $m_n := E[X_n]$ be the average number of molecules in urn I after n shifts. Show that

$$m_n = \frac{N}{2} + \left(1 - \frac{2}{N}\right)^n \left(1 - \frac{N}{2}\right)$$

for $n = 1, 2, \dots$.

Indication. Show first that $m_{n+1} = 1 + \frac{(N-2)}{N}\,m_n$.

Question no. 20

A Markov chain has the following one-step transition probabilities: $p_{0,1} = 1$ and

$$p_{i,i+1} = \alpha_i \ (> 0) = 1 - p_{i,0} \quad \text{for } i = 1, 2, 3, \ldots$$

(a) Show that all states are recurrent if and only if

$$\lim_{n \to \infty} \prod_{k=1}^{n} \alpha_k = 0$$

(b) Show that if the chain is recurrent, then all states are positive recurrent if and only if

$$\sum_{n=1}^{\infty} \prod_{k=1}^{n} \alpha_k < \infty$$

Indication. If N is a random variable taking its values in the set $\mathbb{N}^0 :=$ $\{0, 1, \ldots\}$, then we have

$$E[N] = \sum_{n=1}^{\infty} P[N \geq n]$$

Use this result with the variable N denoting the number of transitions needed to return to state 0.

Question no. 21

The one-step transition probabilities of a Markov chain whose state space is $\{-3, -2, -1, 0, 1, 2, 3\}$ are given by

$$\begin{aligned}
p_{i,i+1} &= 2/3 \quad \text{for } i = -3, -2, \ldots, 2 \\
p_{i,i-1} &= 1/3 \quad \text{for } i = -2, -1, \ldots, 3 \\
p_{-3,3} &= 1/3 = 1 - p_{3,-3}
\end{aligned}$$

Remark. This chain can be considered as a particular case of a random walk defined on a circle.

(a) Show that the Markov chain is irreducible.

(b) Determine the period of the chain.

(c) Calculate the fraction of time that the process spends, over a long period, in state i, for $i = -3, \ldots, 3$.

Question no. 22

(a) Show that in a *symmetric* random walk $\{X_n, n = 0, 1, \ldots\}$, starting from $X_0 = 0$, the probability that state $a > 0$ will be visited before state $-b < 0$ is equal to $b/(a + b)$.

(b) In a random walk starting from $X_0 = 0$, what is the probability that there will be exactly one visit to state 0?

(c) In a random walk on $\{0, 1, 2, \ldots\}$, starting from $X_0 = 0$ and for which $p_{0,1} = 1$, what is the probability that there will be exactly $k \in \{1, 2 \ldots\}$ visit(s) to state 0?

Question no. 23

We consider a branching process $\{X_n, n = 0, 1, \ldots\}$ for which $X_0 = 1$ and

$$p_i = e^{-\lambda}\frac{\lambda^i}{i!} \quad \text{for } i = 0, 1, \ldots$$

That is, the number of descendants of an arbitrary individual has a Poisson distribution with parameter λ. Determine the probability q_0 of eventual extinction of the population if (a) $\lambda = \ln 2$ and (b) $\lambda = \ln 4$.

Question no. 24

A Markov chain with state space $\{-2, -1, 0, 1, 2\}$ has the following one-step transition probabilities:

$$p_{i,i+1} = 3/4 \quad \text{for } i = -2, -1, 0, 1$$
$$p_{i,i-1} = 1/4 \quad \text{for } i = -1, 0, 1, 2$$
$$p_{-2,2} = 1/4 = 1 - p_{2,-2}$$

(a) Determine the classes of the Markov chain. For each class, establish whether it is recurrent or transient and whether it is periodic or aperiodic.

(b) Do the limiting probabilities exist? If they do, compute them.

(c) Is the chain time-reversible (see p. 147)? Justify.

Question no. 25

We consider an irreducible Markov chain whose state space is the set $\{0, 1, 2, 3, 4\}$ and whose one-step transition probability matrix is given by

$$\mathbf{P} = \begin{bmatrix} 0 & 1/3 & 2/3 & 0 & 0 \\ 0 & 0 & 0 & 1/4 & 3/4 \\ 0 & 0 & 0 & 1/4 & 3/4 \\ 1 & 0 & 0 & 0 & 0 \\ 1 & 0 & 0 & 0 & 0 \end{bmatrix}$$

(a) What is the period of the chain?

(b) What is the fraction of time, π_j, that the process spends in state j on the long run, for $j = 0, 1, 2, 3, 4$?

Question no. 26

Let (X_n, Y_n) be the position of a particle that moves in the plane. We suppose that $\{X_n, n = 0, 1, \ldots\}$ and $\{Y_n, n = 0, 1, \ldots\}$ are two independent symmetric random walks such that $X_0 = Y_0 = 0$. We define $D_n = \sqrt{X_n^2 + Y_n^2}$. That is, D_n represents the distance of the particle from the origin after n transitions. Show that if n is large, then

$$P[D_n \leq d] \simeq 1 - e^{-d^2/(2n)} \quad \text{for } d \geq 0$$

Indication. Express X_n (and Y_n) in terms of a binomial distribution, and use the fact that if $Z_1, Z_2, \ldots, Z_k$ are independent random variables having a standard Gaussian distribution, then

$$Z_1^2 + Z_2^2 + \ldots + Z_k^2 \sim G\,(\alpha = k/2, \lambda = 1/2)$$

Question no. 27

Let $Y_1, Y_2, \ldots$ be an infinite sequence of independent random variables, all distributed as the discrete r.v. Y whose set of possible values is the set $\mathbf{Z} := \{0, \pm 1, \pm 2, \ldots\}$ of all integers. We know (see p. 85) that the stochastic process $\{X_n, n = 1, 2, \ldots\}$ defined by

$$X_n = Y_1 + Y_2 + \ldots + Y_n \quad \text{for } n = 1, 2, \ldots$$

is a Markov chain and that $p_{i,j} = \alpha_{j-i}$, where $\alpha_i := P[Y = i]$, for any i.

We suppose that $|E[Y]| < \infty$ and $V[Y] < \infty$. Show that the chain is transient if $E[Y] \neq 0$.

Question no. 28

We consider a Markov chain with transition probabilities given by

$$p_{k,0} = \frac{k+1}{k+2} \quad \text{and} \quad p_{k,k+1} = \frac{1}{k+2}$$

for $k = 0, 1, 2, \ldots$.

(a) Show that the chain is irreducible and ergodic.

Indication. See Question no. 20.

(b) Calculate the limiting probabilities π_k, for $k = 0, 1, 2, \ldots$.

Question no. 29

The one-step transition probabilities of an irreducible Markov chain are given by

$$p_{0,1} = 1, \quad p_{k,0} = \frac{1}{k+1}, \quad \text{and} \quad p_{k,k+1} = \frac{k}{k+1}$$

for $k = 1, 2, \ldots$.

(a) Calculate $p_{0,k}^{(k)}$, for $k = 1, 2, \ldots$.

(b) Calculate $p_{0,0}^{(k)}$, for $k = 2$ and 3. What is the period of the chain?

(c) Is the Markov chain transient, positive recurrent, or null recurrent? Justify.

Indication. See Question no. 20.

Question no. 30

We consider a Markov chain defined by the one-step transition probability matrix

$$\mathbf{P} = \begin{bmatrix} 1/2 & 1/4 & 1/4 \\ \alpha & 1-\alpha & 0 \\ 0 & \alpha & 1-\alpha \end{bmatrix}$$

where $0 \leq \alpha \leq 1$. The state space is the set $\{0, 1, 2\}$.

(a) For what values of the constant α is the Markov chain irreducible and ergodic? Justify.

(b) Calculate the limiting probabilities for the values of α found in part (a).

Question no. 31

   A system is made up of two identical components in *standby redundancy*. That is, only one component is active at a time, and the other is in standby. We assume that the lifetime of each component has an exponential distribution and that the probability that the active component will fail during a given day is equal to 0.1. The other component, if it is not also down, then relieves the failed one at the beginning of the following day. There is a single technician who repairs the components. Moreover, he only starts repairing a failed component at the beginning of his workday, and he needs two working days to complete a repair.

   Let X_n be the condition of the components at the end of the nth day. The process $\{X_n, n = 0, 1, 2, \ldots\}$ is a Markov chain having the following states:

   0: neither component is down
   1: a single component is down and it will be repaired one day
      from now
   2: a single component is down and it will be repaired two days
      from now
   3: both components are down (and a component will be repaired
      one day from now)

(a) What is the matrix $\mathbf{P}$ of one-step transition probabilities of the chain?

(b) For each class of the chain, determine whether it is transient or recurrent. Justify.

(c) Suppose that $X_0 = 0$. Calculate the probability that there will be a single component that is not down at the end of the second day.

Question no. 32
   Let

$$\mathbf{P} = \begin{bmatrix} 1/3 & 0 & 1/3 & 1/3 \\ 1 & 0 & 0 & 0 \\ 0 & 1 & 0 & 0 \\ 0 & 1 & 0 & 0 \end{bmatrix}$$

be the one-step transition probability matrix of an irreducible Markov chain whose state space is the set $\{0, 1, 2, 3\}$.

(a) What is the period of state 1? Justify.

(b) Calculate, if they exist, the quantities π_j, where π_j is the proportion of time that the process spends in state j, over a long period of time, for $j = 0, 1, 2, 3$.

(c) Suppose that the elements of the first row of the matrix $\mathbf{P}$ are the p_j's of a branching process. Calculate the probability that the population will die out if (i) $X_0 = 2$, (ii) $X_1 \leq 1$ (and $X_0 = 2$).

Question no. 33
    John plays independent repetitions of the following game: he tosses two fair dice simultaneously. If he gets a sum of 7 or 11 (respectively, 2, 3, or 12), he wins (resp., loses) $1. Otherwise, he neither wins nor loses anything. John has an initial fortune of i, where $i = 1$ or 2, and he will stop playing when either he goes broke or his fortune reaches $3. Let X_n be John's fortune after n repetitions, for $n = 0, 1, \ldots$. Then $\{X_n, n = 0, 1, \ldots\}$ is a Markov chain.

(a) Find the matrix $\mathbf{P}$ of one-step transition probabilities of the Markov chain.

(b) For each class of the chain, specify whether it is recurrent or transient. Justify.

(c) Calculate the mathematical expectation of X_1 if the initial distribution of the chain is $a_0 = 0$, $a_1 = 1/2$, $a_2 = 1/2$, and $a_3 = 0$.

Question no. 34
    A Markov chain whose state space is the set $\{0, 1, 2\}$ has the following one-step transition probability matrix:

$$\mathbf{P} = \begin{bmatrix} 0 & 1 & 0 \\ 1-p & 0 & p \\ 0 & 1 & 0 \end{bmatrix}$$

where $0 < p < 1$.

(a) Calculate $\mathbf{P}^{(n)}$, for $n \geq 2$.

(b) Find the period of every state of the Markov chain.

(c) (i) Calculate the proportion of time that the process spends, on the long run, in state 0.

    (ii) Is this proportion equal to the limit $\lim_{n \to \infty} p_{i,0}^{(n)}$? Justify.

Question no. 35
    A machine is made up of two components placed in parallel and that operate independently of each other. The lifetime (in months) of each component has an exponential distribution with mean equal to two months. When the machine breaks down, the two components are replaced at the beginning of the next month. Let X_n, $n = 0, 1, \ldots$, be the number of components functioning after n months.

(a) Justify why the stochastic process $\{X_n, n = 0, 1, \ldots\}$ is a Markov chain.

(b) Calculate the matrix $\mathbf{P}$ of one-step transition probabilities of the Markov chain.

(c) Identify each class of the chain as recurrent or transient. Justify.

(d) Calculate $V[X_1]$ if the two components are functioning at time $n = 0$.

Question no. 36
    A Markov chain whose state space is the set $\{0, 1, \ldots\}$ has the following one-step transition probabilities:

$$p_{i,0} = 2/3 \quad \text{and} \quad p_{i,k} = (1/4)^k \quad \text{for } i = 0, 1, \ldots \text{ and } k = 1, 2, \ldots$$

(a) Show that the limiting probabilities π_j exist, and calculate them.

(b) Suppose that the $p_{i,k}$'s are the p_k's, $k \geq 0$, of a branching process. Calculate the probability of eventual extinction of the population if $X_0 = 1$.

Question no. 37

Let $X_1, X_2, \ldots$ be an infinite sequence of independent random variables, all having a Bernoulli distribution with parameter $p \in (0, 1)$. We define

$$Y_n = \sum_{k=1}^{n} X_k \quad \text{for } n = 1, 2, \ldots$$

Then the stochastic process $\{Y_n, n = 1, 2, \ldots\}$ is a Markov chain (see p. 85).

(a) Calculate the one-step transition probability matrix $\mathbf{P}$ of the Markov chain.

(b) For every class of the chain, find out whether it is recurrent or transient. Justify.

(c) Calculate $V[Y_n \mid X_1 = 1]$.

(d) Let $T_1 := \min\{n \geq 1 : Y_n = 1\}$. What is the distribution of the random variable T_1?

Question no. 38

Let

$$\mathbf{P} = \begin{bmatrix} 0 & 1 & 0 & 0 \\ 1/2 & 0 & 1/2 & 0 \\ 0 & 1/2 & 0 & 1/2 \\ 0 & 0 & 1 & 0 \end{bmatrix}$$

be the one-step transition probability matrix of a Markov chain $\{X_n, n = 0, 1, \ldots\}$ whose state space is the set $\{0, 1, 2, 3\}$.

(a) Let π_j be the proportion of time that the process spends in state j, over a long period of time. Show that the π_j's exist and calculate them.

(b) Calculate the period of the Markov chain.

(c) Let $T_j := \min\{n \geq 0 : X_n = j\}$. Calculate $P[T_0 < T_3 \mid X_0 = 1]$. That is, calculate the probability that, from state 1, the process will visit state 0 before state 3.

Question no. 39

We define the Markov chain $\{Y_n, n = 1, 2, \ldots\}$ by

$$Y_n = \sum_{k=1}^{n} X_k \quad \text{for } n = 1, 2, \ldots$$

where $X_1, X_2, \ldots$ is an infinite sequence of independent random variables having a binomial distribution with parameters $n = 2$ and $p = 1/2$ (see p. 85).

(a) Calculate the transition matrix of the Markov chain.

(b) For every class of the chain, establish whether it is recurrent or transient. Justify.

(c) What is the distribution of the random variable Y_2?

(d) Let $T_1 := \min\{n \geq 1 : Y_n = 1\}$. Calculate
   (i) $P[T_1 = k]$, for $k = 1, 2, \ldots,$
   (ii) $P[T_1 < \infty]$.

Question no. 40

We consider the particular case of the gambler's ruin problem (see p. 101) for which $k = 4$.

(a) (i) Calculate the matrix $\mathbf{P}$ if the gambler, on any play, actually has a probability equal to

$$\begin{cases} 1/4 \text{ of winning the play} \\ 1/4 \text{ of losing the play} \\ 1/2 \text{ of neither winning nor losing the play} \end{cases}$$

(ii) Do the probabilities $\pi_j^* := \lim_{n \to \infty} P[X_n = j \mid X_0 = 2]$ exist? If they do, calculate these probabilities.

(b) Suppose that the gambler, on an arbitrary play, bets

$$\begin{cases} \$1 \text{ if his fortune is equal to } \$1 \text{ or } \$3 \\ \$2 \text{ if his fortune is equal to } \$2 \end{cases}$$

Calculate, under the same assumptions as in (a), the probability that the gambler will eventually be ruined if $X_0 = \$3$.

Question no. 41

We consider a population of constant size, N, composed of individuals of type A and of type B. We assume that before reproducing, and disappearing from the population, an arbitrary individual of type A (respectively, B) is mutated into an individual of type B (resp., A) with probability α (resp., β). Moreover, we assume that at the moment of reproduction, each individual in the population has a probability p of giving birth to an offspring of type A, where p is the mathematical expectation of the proportion of individuals of type A in the population after mutation. Finally, we assume that the individuals are independent from one another. Let X_n be the number of individuals of type A in the nth generation before mutation. Then $\{X_n, n = 0, 1, \ldots\}$ is a Markov chain.

(a) Calculate $p \ (= p_i)$ if $X_n = i$.

(b) Calculate $p_{i,j}$, for $i, j \in \{0, \ldots, N\}$.

(c) Suppose that $\alpha = 0$ and $\beta \in (0, 1)$. For each class of the Markov chain, determine whether it is recurrent or transient. Justify.

(d) Suppose that there are exactly $N_A > 0$ individuals of type A in the initial generation after mutation. Given that the first of these N_A individuals gave birth to an offspring of type A, calculate, with $p = N_A/N$, (i) the mean and (ii) the variance of X_1.

Question no. 42

Suppose that

$$\mathbf{P} = \begin{bmatrix} 1/4 & 1/2 & 1/4 \\ \alpha & 0 & 1-\alpha \\ \beta & 0 & 1-\beta \end{bmatrix}$$

is the one-step transition probability matrix of a Markov chain with state space $\{0, 1, 2\}$.

(a) For what values of α and β is the Markov chain irreducible?

(b) Suppose that $\alpha = \beta = 1/2$. Compute, if they exist, the limiting probabilities π_j, for $j = 0, 1, 2$.

(c) Suppose that $X_0 = 0$. Compute the probability that the process will (i) visit state 1 before state 2 and (ii) return to (or stay in) state 0 before visiting state 2.

Question no. 43

Let $\{X_n, n = 0, 1, \ldots\}$ be a Markov chain with state space $\{0, 2\}$ and with one-step transition probability matrix

$$\mathbf{P}_X = \begin{bmatrix} 1/2 & 1/2 \\ 1/2 & 1/2 \end{bmatrix}$$

and let $\{Y_n, n = 0, 1, \ldots\}$ be a Markov chain whose state space is $\{3, 4\}$ and whose matrix $\mathbf{P}$ is given by

$$\mathbf{P}_Y = \begin{bmatrix} 0 & 1 \\ 1 & 0 \end{bmatrix}$$

Assume that the random variables X_n and Y_n are independent, for all n. We define $Z_n = X_n + Y_n$. We can show that the stochastic process $\{Z_n, n = 0, 1, \ldots\}$ is a Markov chain whose state space is $\{3, 4, 5, 6\}$.

(a) Calculate the matrix $\mathbf{P}_Z$ of the Markov chain $\{Z_n, n = 0, 1, \ldots\}$.

(b) For every class of the chain $\{Z_n, n = 0, 1, \ldots\}$, determine whether it is recurrent or transient. Justify.

(c) Give the period of every class of the chain $\{Z_n, n = 0, 1, \ldots\}$.

(d) Calculate the matrix $\mathbf{P}_Z^{(n)}$ and deduce from it the value of the limit $\lim_{n \to \infty} P[Z_n = 3 \mid Z_0 = 3]$.

Question no. 44

Let

$$\mathbf{P} = \begin{bmatrix} 0 & 1/2 & 1/2 \\ 0 & \alpha & 1-\alpha \\ \beta & 1-\beta & 0 \end{bmatrix}$$

be the one-step transition probability matrix of a Markov chain $\{X_n, n = 0, 1, \ldots\}$ having state space $\{0, 1, 2\}$.

(a) For what values of α and β is the Markov chain irreducible?

(b) Suppose that $\alpha = 1/2$ and $\beta = 1/3$. Calculate, if they exist, the limiting probabilities π_j, for $j = 0, 1, 2$.

(c) Give the values of α and β for which the limiting probabilities π_j exist and are equal to $1/3$, for $j = 0, 1, 2$. Justify.

(d) Suppose that $\alpha = 1/2$, $\beta = 1$, and $X_0 = 1$. Let N_0 be the number of transitions needed for the process to visit state 0 for the first time. Calculate $E[N_0]$.

Question no. 45

A person buys stocks of a certain company at the price of 3 cents per share (what is known as a *penny stock*). The investor decides to sell her shares if their value decreases to 1 cent or becomes greater than or equal to 5 cents. Let X_n be the value of the shares (for the investor) after n days. We suppose that $\{X_n, n = 0, 1, \ldots\}$ is a Markov chain having as state space the set $\{0, 1, \ldots, 7\}$, and for which rows 3 to 5 (corresponding to states $2, 3, 4$) of the one-step transition probability matrix $\mathbf{P}$ are the following:

$$\begin{bmatrix} 1/8 & 1/4 & 1/4 & 1/4 & 1/8 & 0 & 0 & 0 \\ 1/12 & 1/7 & 1/4 & 1/21 & 1/4 & 1/7 & 1/12 & 0 \\ 0 & 1/12 & 1/7 & 1/4 & 1/21 & 1/4 & 1/7 & 1/12 \end{bmatrix}$$

(a) Give the other rows of the matrix $\mathbf{P}$.

Indication. Once the investor has sold her shares, or their price went to zero, their value (for the investor) does not change anymore.

(b) For every class of the chain $\{Z_n, n = 0, 1, \ldots\}$, establish whether it is recurrent or transient. Justify.

(c) Calculate $E[X_1 \mid \{X_1 = 3\} \cup \{X_1 = 4\}]$.

(d) What is the probability that, after exactly two days, the investor (i) sells his shares with a profit? (ii) sells her shares with a loss? (iii) loses all the money she invested?

Question no. 46

A Markov chain $\{X_n, n = 0, 1, \ldots\}$ with state space $\{0, 1, 2\}$ has the following one-step transition probability matrix:

$$\mathbf{P} = \begin{bmatrix} 1/2 & 0 & 1/2 \\ 0 & 1 & 0 \\ 1 & 0 & 0 \end{bmatrix}$$

Calculate $\lim_{n \to \infty} p_{i,j}^{(n)}$, for all $i, j \in \{0, 1, 2\}$.

Question no. 47

We consider the gambler's ruin problem (see p. 101). Suppose that $X_0 = \$1$ and that $p = 1/4$. However, if the player wins, he wins $2 and if he loses, he loses $1. His objective is to reach at least $4. Calculate the probability that he will achieve his objective.

Question no. 48

(a) Calculate the probability q_0 of eventual extinction of the population in a branching process for which $p_0 = 1/4$, $p_1 = 1/4$, and $p_2 = 1/2$.

(b) Suppose that the individuals can only give birth to twins, so that the probabilities in (a) become $p_0^* = 1/4$, $p_2^* = 1/4$, and $p_4^* = 1/2$. Can we assert that $q_0^* = q_0^2$? Justify.

Question no. 49

Let $\{X_n, n = 0, 1, \ldots\}$ be a symmetric random walk defined on the set $\{0, \pm 1, \pm 2, \ldots\}$ and such that $X_0 = 0$. We set

$$Y_n = X_n^2 \quad \text{for } n = 0, 1, \ldots$$

It can be shown that $\{Y_n, n = 0, 1, \ldots\}$ is a Markov chain whose state space is $\{0, 1, 4, 9, \ldots\}$.

(a) Calculate the one-step transition probability matrix of the chain $\{Y_n, n = 0, 1, \ldots\}$.

(b) Is the stochastic process $\{Y_n, n = 0, 1, \ldots\}$ a random walk? Justify.

(c) For each class of the chain $\{Y_n, n = 0, 1, \ldots\}$, (i) determine whether it is transient or recurrent and (ii) find its period.

Question no. 50

Let

$$P = \begin{bmatrix} 1/2 & 0 & 1/4 & 1/4 \\ 0 & 1 & 0 & 0 \\ 1 & 0 & 0 & 0 \\ 0 & 0 & 1/4 & 3/4 \end{bmatrix}$$

be the one-step transition probability matrix of a Markov chain $\{X_n, n = 0, 1, \ldots\}$ whose state space is $\{0, 1, 2, 3\}$. Calculate the limit $\lim_{n \to \infty} p_{i,j}^{(n)}$, for all $i, j \in \{0, 1, 2, 3\}$.

Question no. 51

Suppose that $X_0 = \$i$ in the gambler's ruin problem (see p. 101) and that $p \neq 1/2$. Suppose also that if the player loses, then someone lends him (only once) $1 and he starts to play again, independently from what occurred previously. However, the probability p becomes $p/2$. His objective is to reach k (without taking into account the dollar that someone may have lent him), where $k > i$. Calculate the probability that he will achieve his objective.

Question no. 52

(a) Calculate, assuming that $X_0 = 1$, the probability q_0 of eventual extinction of the population in a branching process for which $p_i = 1/4$, for $i = 0, 1, 2, 3$.

(b) Find another distribution of the p_i's for which the value of the probability q_0 is the same as that in (a).

Question no. 53

A machine is composed of two components. The lifetime T_i of component i has an exponential distribution with parameter λ_i, for $i = 1, 2$. When the machine breaks down, a technician replaces the failed component(s) at the beginning of the next time unit. Let X_n, $n = 0, 1, \ldots$, be the number of components that are not down after n time unit(s).

(a) Is the stochastic process $\{X_n, n = 0, 1, \ldots\}$ a Markov chain if the components are placed in series? If it is, justify and calculate the one-step transition probability matrix $\mathbf{P}$ of the chain. If it's not, justify.

(b) Suppose that the components are placed in parallel and operate independently from each other, but that only one component is active at a time. That is, the components are in *standby redundancy*. In this case, $\{X_n, n = 0, 1, \ldots\}$ *is* a Markov chain. Calculate its transition matrix if $\lambda_1 = \lambda_2$.

Question no. 54

Suppose that, in the preceding question, the components are placed in parallel and operate (independently from each other) *both at the same time*. We say that they are in *active redundancy*.

(a) The stochastic process $\{X_n, n = 0, 1, \ldots\}$ *is* a Markov chain. Calculate its matrix $\mathbf{P}$.

(b) Let

$$Y_n := \begin{cases} 0 & \text{if neither component is operating} \\ 1_1 & \text{if only component no. 1 is operating} \\ 1_2 & \text{if only component no. 2 is operating} \\ 2 & \text{if both components are operating} \end{cases}$$

after n time unit(s), for $n = 0, 1, \ldots$. Is the stochastic process $\{Y_n, n = 0, 1, \ldots\}$ a Markov chain? If it is, justify and calculate the transition matrix of the chain. If it's not, justify.

Question no. 55

A system comprises two components placed in parallel and operating (both at the same time) independently from each other. Component i has an exponential lifetime with parameter λ_i, for $i = 1, 2$. Let X_n, $n = 0, 1, \ldots$, be the number of active components after n time unit(s).

(a) Suppose that, after each time unit, we replace the failed component(s). Calculate the one-step transition probability matrix $\mathbf{P}$ of the Markov chain $\{X_n, n = 0, 1, \ldots\}$.

(b) It can be shown that the limiting probabilities π_j exist in part (a). Calculate these limiting probabilities.

(c) Suppose that $\lambda_1 = \lambda_2$ and that, after each time unit, we replace a single component, only when the system is down. Moreover, assume that $X_0 = 2$.

    (i) Calculate the matrix $\mathbf{P}$ of the Markov chain $\{X_n, n = 0, 1, \ldots\}$.

    (ii) What is the number of classes of the chain?

Question no. 56

We consider a *symmetric* random walk in two dimensions (see p. 75), whose state space is the set $\{(i, j): i = 0, 1, 2; j = 0, 1, 2\}$. Moreover, we suppose that the boundaries are *reflecting*. That is, when the process makes a transition that would take it outside the region defined by the state space, then it returns to the last position it occupied (on the boundary).

(a) Calculate the one-step transition probability matrix of the Markov chain.

(b) Show that the limiting probabilities exist and calculate them.

Question no. 57

A player has only one money unit and wishes to increase his fortune to five units. To do so, he plays independent repetitions of a game that, in case of a win, yields double the sum he betted. In case of a loss, he loses his bet. On each play, he bets an amount that, if he wins, enables him either to exactly reach his objective or to get as close as possible to it (for example, if he has three units, then he bets only one). We suppose that the game ends when the player either has achieved his target or has been ruined and that he has a probability equal to $1/2$ of winning an arbitrary play.

(a) What is the probability that he reaches his target?

(b) What is the average number of repetitions needed for the game to end?

Question no. 58

A particle moves in the plane according to a two-dimensional symmetric random walk (see p. 75). That is, the particle has a probability equal to $1/4$ of moving from its current position, (X_n, Y_n), to any of its four nearest neighbors. We suppose that the particle is at the origin at time $n = 0$, so that $X_0 = Y_0 = 0$. Thus, at time $n = 1$, the particle will be in one of the following states: $(0, 1)$, $(0, -1)$, $(1, 0)$, or $(-1, 0)$. Let

$$D_n^2 := X_n^2 + Y_n^2$$

be the square of the distance of the particle from the origin at time n. Calculate $E[D_n^2]$.

Question no. 59

Show that for a symmetric random walk of dimension $k = 2$ (see Question no. 58), the probability that the number of visits to already visited states will be infinite is equal to 1. Generalize this result to the case when $k \in \mathbb{N}$.

Question no. 60

A machine is composed of two identical components placed in series. The lifetime of a component is a random variable having an exponential distribution with parameter μ. We have at our disposal a stock of $n-2$ new components that we differentiate by numbering them from 3 to n (the components already installed bearing the numbers 1 and 2). When the machine fails, we immediately replace the component that caused the failure by the new component bearing the smallest number among those in stock. Let T be the total lifetime of the machine, and let N be the number of the only component that, at time T, will not be down. Find (a) the probability mass function of N, (b) the mathematical expectation of T, and (c) the distribution of T.

Question no. 61

We use k light bulbs to light an outside rink. The person responsible for the lighting of the rink does not keep spare light bulbs. Rather, he orders, at the beginning of each week, new light bulbs to replace the ones that burned out during the preceding week. These light bulbs are delivered the following week. Let X_n be the number of light bulbs in operation at the beginning of the nth week, and let Y_n be the number of light bulbs that will burn out during this nth week, for $n = 0, 1, \ldots$. We assume that, given that $X_n = i$, the random variable Y_n has a *discrete uniform* distribution over the set $\{0, 1, \ldots, i\}$:

$$P[Y_n = j \mid X_n = i] = \frac{1}{i+1} \quad \text{for } j = 0, 1, \ldots, i \text{ and } i = 0, 1, \ldots, k$$

(a) Calculate the one-step transition probability matrix of the Markov chain $\{X_n, n = 0, 1, \ldots\}$.

(b) Show that the limiting probabilities of the chain $\{X_n, n = 0, 1, \ldots\}$ exist and are given by

$$\pi_i = \frac{2(i+1)}{(k+1)(k+2)} \quad \text{for } i = 0, 1, \ldots, k$$

Question no. 62

Electric impulses are measured by a counter that only indicates the highest voltage it has registered up to the present time instant. We assume that the electric impulses are uniformly distributed over the set $\{1, 2, \ldots, N\}$.

(a) Let X_n be the voltage indicated after n electric impulses. The stochastic process $\{X_n, n = 1, 2, \ldots\}$ *is* a Markov chain. Find its one-step transition probability matrix.

(b) Let m_i be the average number of additional impulses needed for the counter to register the maximum voltage, N, when the voltage indicated is i, for $i = 1, 2, \ldots, N - 1$.

(i) Obtain a set of *difference equations* (as in the absorption problems, p. 100) for the m_i's, and solve these equations to determine m_i, for all i.

(ii) Calculate directly the value of the m_i's without making use of the difference equations.

Question no. 63

Let X_1, X_2, ... be an infinite sequence of independent and identically distributed random variables, such that $p_{X_1}(x) = 1/3$ if $x = -1, 0, 1$. The stochastic process $\{Y_n, n = 1, 2, \dots\}$ defined by

$$Y_n = \sum_{i=1}^{n} X_i \quad \text{for } n = 1, 2, \dots .$$

is a Markov chain (see p. 85).

(a) Calculate the one-step transition probability matrix $\mathbf{P}$ of the chain.

(b) Give an exact formula for $p_{0,0}^{(k)}$, for $k = 1, 2, \dots$.

(c) Use the central limit theorem to obtain an approximate formula for $p_{0,0}^{(k)}$ when k is large enough.

Question no. 64

A player has $900 at his disposal. His objective is reach the amount of $1000. To do so, he plays repetitions of a game for which the probability that he wins an arbitrary repetition is equal to 9/19, independently from one repetition to another.

(a) Calculate the probability that the player will reach his target if he bets $1 per repetition of the game.

(b) Calculate the probability that the player will reach his target if he adopts the following strategy: he bets

$$\begin{aligned}
\$(1000 - x) & \text{ if } 500 \leq x < 1000 \\
\$x & \text{ if } 0 < x < 500
\end{aligned}$$

where x is the amount of money at his disposal at the moment of betting.

(c) What is the expected gain with the strategy used in (b)?

Question no. 65

We consider a system made up of two components placed in parallel and operating independently. The lifetime T_1 of component no. 1 has an exponential distribution with parameter 1, while that of component no. 2 is a random variable $T_2 \sim \text{Exp}(1/2)$. When the system fails, 50% of the time the two components are replaced by new ones, and 50% of the time only the first (of the two components) that failed is replaced. Let $X_n = 1_1$ (respectively, 1_2) if only component no. 1 (resp., no. 2) is replaced at the moment of the nth failure, and $X_n = 2$ if both components are replaced. We can show that $\{X_n, n = 1, 2, \dots\}$ is a Markov chain.

(a) Find the one-step transition probability matrix of the chain.

(b) For each class of the chain, specify whether it is transient or recurrent. Justify.

(c) Calculate $p_{2,2}^{(k)}$, for $k = 1, 2, \ldots$.

(d) What is the period of state 1_2? Justify.

Question no. 66

In the gambler's ruin problem (see p. 101), suppose that the player has an infinite initial fortune and that $p = 1/2$. Let X_n be the gain (or the loss) of the player after n repetitions of the game. Suppose also that when $X_n < 0$, the player plays double or quits. That is, if $X_n = -1$, then $X_{n+1} = 0$ or -2; if $X_n = -2$, then $X_{n+1} = 0$ or -4, etc. The process $\{X_n, n = 0, 1, \ldots\}$ is a Markov chain whose state space is the set $\{\ldots, -4, -2, -1, 0, 1, 2, \ldots\}$.

(a) Find the one-step transition probability matrix of the chain.

(b) Suppose that the state space is $\{-4, -2, -1, 0, 1, 2\}$ and that $p_{-4,-4} = p_{2,2} = 1/2$ [the other probabilities being as in (a)]. Calculate, if they exist, the limiting probabilities.

(c) Suppose now that the player decides to stop playing if his losses or his profits reach four units. Calculate the probability that the player will stop playing because his losses have reached four units, given that he won the first repetition of the game.

Question no. 67

The state space of a Markov chain $\{X_n, n = 0, 1, 2 \ldots\}$ is the set of non-negative integers $\{0, 1, 2, \ldots\}$, and its one-step transition probability matrix is given by

$$\mathbf{P} = \begin{bmatrix} 0 & 1 & 0 & \cdots \\ 1/2 & 0 & 1/2 & 0 & \cdots \\ 1/3 & 1/3 & 0 & 1/3 & \cdots \\ 1/4 & 1/4 & 1/4 & 0 & 1/4 & 0 & \cdots \\ \cdots & \cdots & \cdots & \cdots & \cdots & \cdots \end{bmatrix}$$

(a) Calculate $p_{1,2}^{(3)}$.

(b) For each class of the chain, determine whether it is transient or recurrent. Justify.

(c) Find the period of each class.

(d) Let $T_{i,j}$ be the number of transitions needed for the process to move from state i to state j. Calculate $P[T_{1,0} < T_{1,3} \mid \{T_{1,0} \le 2\} \cup \{T_{1,3} \le 2\}]$.

Question no. 68

We consider a Markov chain whose state space is the set $\{0, 1, 2, \ldots\}$ and for which

$$p_{i,j} = p_j > 0 \quad \forall\, i, j \in \{0, 1, 2, \ldots\}$$

Calculate, assuming they exist, the limiting probabilities of the chain.

Question no. 69

Suppose that, in the gambler's ruin problem (see p. 101), $X_0 = 1$ and $k = 4$. Moreover, suppose that the value of p is not constant, but rather increases with X_n. More precisely, the one-step transition probability matrix of the chain $\{X_n, n = 0, 1, 2 \ldots\}$ is

$$\mathbf{P} = \begin{bmatrix} 1 & 0 & 0 & 0 & 0 \\ 1/2 & 0 & 1/2 & 0 & 0 \\ 0 & 1/3 & 0 & 2/3 & 0 \\ 0 & 0 & 1/4 & 0 & 3/4 \\ 0 & 0 & 0 & 0 & 1 \end{bmatrix}$$

Calculate the probability that the player will achieve his objective.

Question no. 70

Let $\{X_n, n = 0, 1, 2 \ldots\}$ be a branching process for which $X_0 = 1$. Suppose that there are two types of individuals, say A and B. Every individual (of type A or B) can give birth (independently from the others) to descendants of type A or B according to the formula

$$P[N_A = m, N_B = n] = 1/9 \quad \forall\, m, n \in \{0, 1, 2\}$$

where N_A (respectively, N_B) is the number of descendants of type A (resp., B).

(a) Calculate $E[X_1 \mid X_1 > 0]$.

(b) Show that the probability of eventual extinction of the population is $q_0 \simeq 0.15417$.

Section 3.3

Question no. 71

Let $Y := \min\{X_1, X_2\}$, where X_1 and X_2 are two independent exponential random variables, with parameters λ_1 and λ_2, respectively. We know (see p. 113) that $Y \sim \text{Exp}(\lambda_1 + \lambda_2)$. Find the probability density function of the random variable $Z := Y \mid \{X_1 < X_2\}$.

Remark. We can express Z as follows: $Z := Y \mid \{Y = X_1\}$. However, the random variables $Y \mid \{Y = X_1\}$ and X_1 are *not* identical, because

$$P[Y \leq y \mid Y = X_1] = P[X_1 \leq y \mid Y = X_1] = P[X_1 \leq y \mid X_1 < X_2]$$
$$\neq P[X_1 \leq y]$$

Note that the events $\{X_1 \leq y\}$ and $\{X_1 < X_2\}$ are not independent.

Question no. 72

Let $X_1, \ldots, X_n$ be independent random variables having an exponential distribution with parameter λ.

(a) Use the memoryless property of the exponential distribution to show that

$$P[X_1 > X_2 + X_3] = \frac{1}{4} = \frac{1}{2^{3-1}}$$

(b) Show, by mathematical induction, that

$$P\left[X_1 > \sum_{k=2}^{n} X_k\right] = \frac{1}{2^{n-1}} \quad \text{for } n = 2, 3, \ldots$$

(c) From the result in (b), calculate the probability $P[2Y > \sum_{k=1}^{n} X_k]$, where $Y := \max\{X_1, \ldots, X_n\}$.

Question no. 73

A birth and death process having parameters $\lambda_n \equiv 0$ and $\mu_n = \mu$, for all $n > 0$, is a pure death process with constant death rate. Find, without making use of the Kolmogorov equations, the transition function $p_{i,j}(t)$ for this process.

Question no. 74

Let $X_1 \sim \text{Exp}(\lambda_1)$ and $X_2 \sim \text{Exp}(\lambda_2)$ be two independent random variables. Show that, for all $x \geq 0$,

$$P[X_1 < X_2 \mid \min\{X_1, X_2\} = x] = P[X_1 < X_2] = \frac{\lambda_1}{\lambda_1 + \lambda_2}$$

Question no. 75

A system is made up of three components placed in standby redundancy: at first, only component no. 1 is active and when it fails, component no. 2 immediately relieves it. Next, at the moment when component no. 2 fails, component no. 3 becomes active at once. When the system breaks down, the three components are instantly replaced by new ones. Suppose that the lifetime T_k of component no. k has an exponential distribution with parameter λ_k, for $k = 1, 2, 3$, and that the random variables T_1, T_2, and T_3 are independent. Let the number of the component that is active at time t be the state of the system at this time instant. Write the Kolmogorov backward equations for this system.

Question no. 76

A university professor cannot receive more than two students at the same time in her office. On the day before an exam, students arrive according to a Poisson process with rate $\lambda = 3$ per hour to ask questions. The professor helps the students one at a time. There is a chair in her office where a person can wait his or her turn. However, if a student arrives when two other students are already in the professor's office, then this student must come back later. We suppose that the time that the professor takes to answer the questions of an arbitrary student is an exponential random variable with mean equal to 15 minutes, independently from one student to another. If we consider only the days preceding an exam, calculate (with the help of the limiting probabilities)

(a) the average number of students in the professor's office,

(b) the proportion of time, on the long run, when the professor is not busy answering questions.

(c) If the professor spent twice more time, on average, with each student, what would be the answer in (b)?

Question no. 77
Let $\{X(t), t \geq 0\}$ and $\{Y(t), t \geq 0\}$ be two continuous-time independent Markov chains. We consider the two-dimensional stochastic process $\{(X(t), Y(t)), t \geq 0\}$. Find the parameters $v_{(i,k)}$, $p_{(i,k),(j,k)}$, and $p_{(i,k),(i,l)}$ of this process.

Question no. 78
We consider a pure birth process for which, when there are n individuals in the population, the average time (in hours) needed for a birth to occur is equal to $1/n$, for $n \geq 0$.

(a) Knowing that at time t there are two individuals in the population and that at time $t + 1$ there are still two, what is the probability that the next birth will take place between $t + 2$ and $t + 3$?

(b) If, at the origin, the population is composed of a single individual, what is the probability that there will be exactly four births during the first two hours?

Question no. 79
A factory has m machines. Each machine fails at an exponential rate μ. When a machine fails, it remains down during a random time having an exponential distribution with parameter λ. Moreover, the machines are independent from one another. Let $X(t)$ be the number of machines that are in working order at time $t \geq 0$. It can be shown that the stochastic process $\{X(t), t \geq 0\}$ is a birth and death process.

(a) Find the birth and death rates of the process $\{X(t), t \geq 0\}$.

(b) Show that

$$\lim_{t \to \infty} P[X(t) = n] = P\left[B\left(m, \tfrac{\lambda}{\lambda+\mu}\right) = n\right]$$

for $n = 0, 1, \ldots, m$.

Question no. 80
Let $\{X(t), t \geq 0\}$ be a pure birth process such that $\lambda_j = j\lambda$, for $j = 0, 1, \ldots$, where $\lambda > 0$. We suppose that $X(0) = 1$.

(a) Let $T_n := \min\{t > 0 : X(t) = n \ (> 1)\}$. That is, T_n is the time needed for the number of individuals in the population to be equal to n. Show that the probability density function of T_n is given by

$$f_{T_n}(t) = \lambda(n-1)e^{-\lambda t}(1-e^{-\lambda t})^{n-2} \quad \text{for } t \geq 0$$

(b) Let $N(t)$ be the number of descendants of the ancestor of the population at time t, so that $N(t) = X(t) - 1$. Suppose that the random variable τ has an exponential distribution with parameter μ. Show that

$$P[N(\tau) = n] = \frac{\mu}{\lambda}B\left(\frac{\mu}{\lambda} + 1, n + 1\right) \quad \text{for } n = 0, 1, \ldots$$

where $B(\cdot, \cdot)$ is the *beta function* defined by

$$B(x,y) = \int_0^1 t^{x-1}(1-t)^{y-1}\, dt$$

for $x, y \in (0, \infty)$.

Remark. We can show that

$$B(x,y) = \frac{\Gamma(x)\Gamma(y)}{\Gamma(x+y)}$$

Question no. 81

A birth and death process, $\{X(t), t \geq 0\}$, has the following birth and death rates:

$$\lambda_n = \lambda \quad \text{for } n = 0, 1 \quad \text{and} \quad \mu_n = n\,\mu \quad \text{for } n = 1, 2$$

Moreover, the capacity of the system is equal to two individuals.

(a) Calculate, assuming that $\lambda = \mu$, the average number of individuals in the system at a time instant t (large enough), given that the system is not empty at this time.

(b) Calculate the probability that the process will spend more time in state 0 than in state 1 on two arbitrary visits to these states.

(c) Suppose that $\mu_1 = 0$ and that, when $X(t) = 2$, the next state visited will be 0, at rate 2μ. Write the balance equations of the system, and solve them to obtain the limiting probabilities.

Question no. 82

Let $\{X(t), t \geq 0\}$ be a birth and death process whose rates λ_n and μ_n are given by

$$\lambda_n = \mu_n = n\lambda \quad \text{for } n = 0, 1, 2, \ldots$$

We set $p_k(t) = P[X(t) = k]$, for all $k \in \{0, 1, 2, \ldots\}$ and for all $t \geq 0$. That is, $p_k(t)$ denotes the probability that the process will be in state k at time t. Suppose that $p_1(0) = 1$. It can be shown that

$$p_0(t) = \frac{\lambda t}{1 + \lambda t} \quad \text{and} \quad p_k(t) = \frac{(\lambda t)^{k-1}}{(1 + \lambda t)^{k+1}} \quad \text{for } k = 1, 2, \ldots$$

(a) Calculate $E[X(t) \mid X(t) > 0]$.

Indication. We have $E[X(t)] \equiv 1$.

(b) Calculate the limiting probabilities π_j and show that they satisfy the balance equations of the process.

(c) Use the Kolmogorov backward equation satisfied by $p_{1,0}(t)$ to obtain $p_{2,0}(t)$.

Indication. We have $p_{1,0}(t) = p_0(t)$ above.

Question no. 83

We consider a system composed of three components placed in parallel and operating independently. The lifetime X_i (in months) of component i has an exponential distribution with parameter λ, for $i = 1, 2, 3$. When the system breaks down, the three components are replaced in an exponential time (in months) with parameter μ. Let $X(t)$ be the number of components functioning at time t. Then $\{X(t), t \geq 0\}$ is a continuous-time Markov chain whose state space is the set $\{0, 1, 2, 3\}$.

(a) Calculate the average time that the process spends in each state.

(b) Is the process $\{X(t), t \geq 0\}$ a birth and death process? Justify.

(c) Write the Kolmogorov backward equation for $p_{0,0}(t)$.

(d) Calculate the limiting probabilities of the process if $\lambda = \mu$.

Question no. 84

Let $\{N(t), t \geq 0\}$ be a counting process (see p. 231) such that $N(0) = 0$. When the process is in state j, the next state visited will be $j + 1$, for all $j \geq 0$. Moreover, the time τ_j that the process spends in state j has the following probability density function:

$$f_{\tau_j}(s) = 2(j+1)\lambda s\, e^{-(j+1)\lambda s^2} \quad \text{for } s \geq 0$$

Finally, the τ_j's are independent random variables.

Now, consider the stochastic process $\{X(u), u \geq 0\}$ defined by

$$X(u) = \begin{cases} 0 \text{ if } u \leq \tau_0^2 \\ k \text{ if } \sum_{j=0}^{k-1} \tau_j^2 < u \leq \sum_{j=0}^{k} \tau_j^2 \end{cases}$$

for $k = 1, 2, \dots$.

(a) Show that the stochastic process $\{X(u), u \geq 0\}$ is a continuous-time Markov chain.

(b) Calculate $P[X(2) = 1]$.

(c) Calculate, if they exist, the limiting probabilities π_j of the stochastic process $\{X(u), u \geq 0\}$.

Question no. 85

We consider the particular case of the gambler's ruin problem (see p. 101) for which $k = 4$ and $p = 1/2$. Suppose that the length T (in minutes) of a play (the outcome of which is the player's winning or losing $1) has an exponential distribution with mean equal to 1/2. Moreover, when the player's fortune reaches $0 or $4, he waits for an exponential time S (in hours) with mean equal to 2 before starting to play again, and this time is independent of what happened before. Finally, suppose that the player has $1 when he starts to play again if he was ruined on the last play of the preceding game and that he has $3 if his fortune reached $4 on this last play.

Let $X(t)$, for $t \geq 0$, be the player's fortune at time t. The stochastic process $\{X(t), t \geq 0\}$ is a continuous-time Markov chain.

(a) (i) Is the process $\{X(t), t \geq 0\}$ a birth and death process? If it is, give its birth and death rates. If it's not, justify.

(ii) Answer the same question if the player always starts to play again with $1, whether his fortune reached $0 or $4 on the last play of the previous game.

(b) (i) Write, for each state j, the Kolmogorov backward equation satisfied by the function $p_{0,j}(t)$.

(ii) Use the preceding result to obtain the value of the sum $\sum_{j=0}^{4} p'_{0,j}(t)$.

(c) Calculate the limiting probabilities of the process $\{X(t), t \geq 0\}$, for all states j.

Question no. 86

Let $\{X_n, n = 0, 1, \ldots\}$ be a (discrete-time) Markov chain whose state space is $\{0, 1, 2\}$ and whose one-step transition probability matrix is

$$\mathbf{P} = \begin{bmatrix} \alpha & (1-\alpha)/2 & (1-\alpha)/2 \\ (1-\alpha)/2 & \alpha & (1-\alpha)/2 \\ 2(1-\alpha)/3 & (1-\alpha)/3 & \alpha \end{bmatrix}$$

where $\alpha \in [0, 1]$. Suppose that the process spends an exponential time with parameter λ in state i before making a transition, with probability $p_{i,j}$, to state j, for $i, j \in \{0, 1, 2\}$. Let $X(t)$ be the position, that is, the state in which the process is, at time t.

(a) For what value(s) of α is the stochastic process $\{X(t), t \geq 0\}$ a continuous-time Markov chain? Justify.

(b) For the value(s) of α in (a), calculate, assuming they exist, the limiting probabilities π_j, for $j = 0, 1, 2$.

Question no. 87

In the preceding question, suppose that the transition matrix $\mathbf{P}$ is instead the following:

$$\mathbf{P} = \begin{bmatrix} \alpha & \beta & 1-\alpha-\beta \\ \gamma & 0 & 1-\gamma \\ 1-\beta & \beta & 0 \end{bmatrix}$$

where α, β, and $\gamma \in [0, 1]$.

(a) For what values of α, β, and γ is the stochastic process $\{X(t), t \geq 0\}$ a birth and death process? Justify and give the value of the parameters v_j, λ_j, and μ_j, for $j = 0, 1, 2$.

(b) For the values of α and β found in (a), and for $\gamma \in (0, 1)$, calculate, assuming they exist, the limiting probabilities π_j, for all j.

Question no. 88

Let $\{X(t), t \geq 0\}$ be a (continuous-time) stochastic process whose state space is the set $\{0, 1\}$. Suppose that the process spends an exponential time with parameter Λ in a state before making a transition to the other state, where Λ is a discrete random variable taking the values 1 and 2 with probability 1/3 and 2/3, respectively.

(a) Is the stochastic process $\{X(t), t \geq 0\}$ a continuous-time Markov chain? Justify.

(b) Suppose that $X(0) = 0$. Let τ_0 be the time that the process spends in state 0 before making a first transition to state 1. Calculate the probability $P[\Lambda = 1 \mid \tau_0 < 1]$.

Question no. 89

Suppose that the continuous-time stochastic process $\{X(t), t \geq 0\}$, whose state space is $\{0, 1\}$, spends an exponential time with parameter 1 in a state the first time it visits this state. The second time it visits a state, it stays there an exponential time with parameter 2. When both states have been visited twice each, the process starts anew.

(a) Is the stochastic process $\{X(t), t \geq 0\}$ a birth and death process? Justify.

(b) Let N be the number of visits to state 0 from the initial time 0, and let τ_0 be the time that the process spends in state 0 on an arbitrary visit to this state. Calculate approximately $P[N$ is odd $\mid \tau_0 < 1]$ if we assume that the most recent visit to state 0 started at a very large time t.

Question no. 90

A system is composed of three components operating independently. Two active components are sufficient for the system to function. Calculate the failure rate of the system if the lifetime T_i of component i has an exponential distribution with parameter $\lambda = 2$, for $i = 1, 2, 3$.

Question no. 91

Let $\{N_1(t), t \geq 0\}$ and $\{N_2(t), t \geq 0\}$ be two independent Yule processes, with rates $\lambda_n = n\theta_1$ and $\lambda_n = n\theta_2$, for $n = 0, 1, \ldots$, respectively. We define

$$X(t) = N_1(t) + N_2(t) \quad \text{for } t \geq 0$$

(a) For what values of the constants θ_1 and θ_2 is the stochastic process $\{X(t), t \geq 0\}$ a continuous-time Markov chain? Justify and give the value of the parameters v_n of this process.

(b) For the values of θ_1 and θ_2 found in (a), calculate $p_{i,j}(t)$, for $j \geq i \geq 1$.

Question no. 92

We define

$$X(t) = |N_1(t) - N_2(t)| \quad \text{for } t \geq 0$$

where $\{N_1(t), t \geq 0\}$ and $\{N_2(t), t \geq 0\}$ are two independent Poisson processes, with rates $\lambda_1 = \lambda_2 = \lambda$.

(a) Show that $\{X(t), t \geq 0\}$ is a birth and death process, and give the rates λ_n and μ_n, for $n = 0, 1, \ldots$.

(b) Calculate, if they exist, the limiting probabilities π_j, for $j = 0, 1, \ldots$.

Question no. 93

Calculate $P[X_1 < X_2 < X_3]$ if X_1, X_2, and X_3 are independent random variables such that $X_i \sim \text{Exp}(\lambda_i)$, for $i = 1, 2, 3$.

Question no. 94

Let $\{X(t), t \geq 0\}$ be a birth and death process whose state space is the set $\{0, 1, 2\}$ and for which

$$\lambda_0 = \lambda, \quad \lambda_1 = \mu_1 = \lambda/2, \quad \text{and} \quad \mu_2 = \lambda$$

We consider two independent copies, $\{X_1(t), t \geq 0\}$ and $\{X_2(t), t \geq 0\}$, of this process, and we define

$$Y(t) = |X_1(t) - X_2(t)| \quad \text{for } t \geq 0$$

We can show that $\{Y(t), t \geq 0\}$ is also a birth and death process.

(a) Give the birth and death rates of the process $\{Y(t), t \geq 0\}$.

(b) Calculate the expected value of the random variable $Y(t)$ after two transitions if $X_1(0) = X_2(0) = 0$.

(c) Calculate the limiting probabilities of the process $\{Y(t), t \geq 0\}$.

Question no. 95

We consider a birth and death process, $\{X(t), t \geq 0\}$, whose state space is the set $\{0, 1, 2, \ldots\}$ and whose birth and death rates are given by

$$\lambda_n = n\lambda \quad \text{and} \quad \mu_n = n\mu \quad \text{for } n = 0, 1, \ldots$$

where $\lambda, \mu > 0$. Suppose that $X(0) = i \in \{1, 2, \ldots\}$. Calculate $E[X(t)]$.

Indication. We can use the Kolmogorov equations.

Question no. 96

The rates λ_n and μ_n of the birth and death process $\{X(t), t \geq 0\}$, whose state space is the set $\{0, 1, \ldots, c\}$, are

$$\lambda_n = (c - n)\lambda \quad \text{and} \quad \mu_n = n\mu$$

for $n = 0, 1, \ldots, c$. Suppose that $X(0) = k \in \{0, 1, \ldots, c\}$. Calculate the function $p_{k,c}(t)$.

Indication. In the case of a continuous-time Markov chain defined on the set $\{0, 1\}$, and for which

$$\lambda_0 = \lambda, \quad \lambda_1 = 0, \quad \mu_0 = 0, \quad \text{and} \quad \mu_1 = \mu$$

we have (see p. 131)

$$p_{0,0}(t) = \frac{\lambda}{\mu + \lambda} e^{-(\mu+\lambda)t} + \frac{\mu}{\mu + \lambda}$$

and

$$p_{1,0}(t) = \frac{\mu}{\mu + \lambda} - \frac{\mu}{\mu + \lambda} e^{-(\mu+\lambda)t}$$

Question no. 97

Let $\{X(t), t \geq 0\}$ be a birth and death process for which

$$\lambda_n = \frac{1}{n + 1} \quad \text{for } n = 0, 1, \dots \quad \text{and} \quad \mu_n = n \quad \text{for } n = 1, 2, \dots$$

and let $t_0 > 0$ be the time instant at which the first birth occurred.

(a) Suppose that we round off the time by taking the integer part. Calculate the probability that the first event from t_0 will be a birth.

Indication. If $X \sim \text{Exp}(\lambda)$, then $1 + \text{int}(X) \sim \text{Geom}(1 - e^{-\lambda})$, where *int* denotes the integer part.

(b) What is the probability that there will be at least two births among the first three events from t_0?

(c) Calculate, if they exist, the limiting probabilities of the process.

Question no. 98

The lifetime of a certain machine is a random variable having an exponential distribution with parameter λ. When the machine breaks down, there is a probability equal to p (respectively, $1 - p$) that the failure is of type I (resp., II). In the case of a type I failure, the machine is out of use for an exponential time, with mean equal to $1/\mu$ time unit(s). To repair a type II failure, two independent operations must be performed. Each operation takes an exponential time with mean equal to $1/\mu$.

(a) Define a state space such that the process $\{X(t), t \geq 0\}$, where $X(t)$ denotes the state of the system at time t, is a continuous-time Markov chain.

(b) Calculate, assuming the existence of the limiting probabilities, the probability that the machine will be functioning at a (large enough) given time instant.

Question no. 99

A person visits a certain Web site according to a Poisson process with rate λ per day. The site in question contains a main page and an internal link. The probability that the person visits only the main page is equal to 3/4 (independently from one visit to another). Moreover, when she clicks on the internal link (at most once per visit), the probability that she will return (afterward) to the main page is equal to 1/2. We define the states

0: the person is not visiting the site in question
1: the person is visiting the main page (coming from outside the site)
2: the person is visiting the internal link
3: the person is visiting the main page, after having visited the link

Let τ_k be the time (in hours) spent in state k, for $k = 0, 1, 2, 3$. We assume that the random variables τ_k are independent and that τ_k has an exponential distribution with parameter ν_k, for $k = 1, 2, 3$.

Remark. We suppose that when the process is in state 3, the internal link is highlighted and that this highlighting is removed when the person leaves the site.

Let $X(t)$ be the state the process is in at time $t \geq 0$. We can show that $\{X(t), t \geq 0\}$ is a continuous-time Markov chain.

(a) Give the probabilities $p_{i,j}$ of the process.

(b) Is the process $\{X(t), t \geq 0\}$ a birth and death process? If it's not, is it possible to rename the states so that $\{X(t), t \geq 0\}$ becomes a birth and death process? Justify.

(c) Calculate the average time spent on the site in question on an arbitrary visit, given that $\tau_1 = 1/4$ for this visit.

(d) Calculate the limiting probabilities π_j.

4

Diffusion Processes

4.1 The Wiener process

We already mentioned the Wiener process twice in Chapter 2. In this section, we will first present a classic way of obtaining this process from a random walk. Then we will give its main properties.

Consider the discrete-time Markov chain $\{X_n, n = 0, 1, \ldots\}$ whose state space is the set of all integers $\mathbf{Z} := \{0, \pm 1, \pm 2, \ldots\}$ and whose one-step transition probabilities are given by

$$p_{i,i+1} = p_{i,i-1} = 1/2 \quad \text{for all } i \in \mathbf{Z} \tag{4.1}$$

This Markov chain is a *symmetric random walk* (see p. 48). A possible interpretation of this process is the following: suppose that a particle moves randomly among all the integers. At each time unit, for example, each minute, a fair coin is tossed. If "tails" (respectively, "heads") appears, then the particle moves one integer (that is, one unit of distance) to the right (resp., left).

To obtain the stochastic process called the Brownian motion, we *accelerate* the random walk. The displacements are made every δ unit of time, and the distance traveled by the particle is equal to ϵ unit of distance to the left or to the right, where, by convention, $\delta > 0$ and $\epsilon > 0$ are real numbers that can be chosen as small as we want. As the Wiener process is a *continuous-time* and *continuous-state* process, we will take the limit as δ and ϵ decrease to 0, so that the particle will move continuously, but will travel an infinitesimal distance on each displacement. However, as will be seen subsequently, we cannot allow the constants δ and ϵ to decrease to 0 independently from each other; otherwise, the variance of the limiting process is equal either to zero or to infinity, so that this limiting process would be devoid of interest.

We denote by $X(t)$ the position of the particle at time t, and we suppose that $X(0) = 0$. That is, the particle is at the origin at the initial time. Let N be the number of transitions to the right that the particle has made after its first n displacements. We can then write that the position of the particle after $n\delta$ unit(s) of time is given by

$$X(n\delta) = (2N - n)\,\epsilon \tag{4.2}$$

Note that if all the displacements have been to the right, so that $N = n$, then we indeed have that $X(n\delta) = n\epsilon$. Similarly, if $N = 0$, then we obtain $X(n\delta) = -n\,\epsilon$, as it should be.

Remark. Since the particle only moves at time instants δ, 2δ, $\ldots$, we may write that its position at time t is given by

$$X(t) = X([t/\delta]\,\delta) \quad \text{for all } t \geq 0 \tag{4.3}$$

where $[\ ]$ denotes the *integer part*.

Because the tosses of the coin are independent, the random variable N has a *binomial distribution* with parameters n and $p = 1/2$. It follows that

$$E[X(n\delta)] == \left(2 \times \tfrac{n}{2} - n\right)\,\epsilon = 0 \tag{4.4}$$

and

$$V[X(n\delta)] = 4\,\epsilon^2\,V[N] = 4\,\epsilon^2 \times \frac{n}{4} = n\,\epsilon^2 \tag{4.5}$$

If we first let δ decrease to 0, then the random walk becomes a *continuous-time* process. However, given that

$$V[X(t)]|_{t=n\delta} = n\,\epsilon^2 = \frac{t}{\delta} \times \epsilon^2 \tag{4.6}$$

we find that ϵ must tend to 0 at the same time as δ; otherwise, the variance of $X(t)$ will be infinite. Actually, to obtain an interesting limiting process, δ and ϵ^2 must decrease to 0 at the same *speed*. Consequently, we assume that there exists a constant σ (> 0) such that

$$\epsilon = \sigma\sqrt{\delta} \quad \Longleftrightarrow \quad \epsilon^2 = \sigma^2\delta \tag{4.7}$$

Thus, when we let δ decrease to 0, we obtain a process that is also with *continuous-state space* and for which

$$E[X(t)] \equiv 0 \tag{4.8}$$

and

$$V[X(t)] \xrightarrow{\delta\downarrow 0} \sigma^2 t \quad \forall\, t \geq 0 \tag{4.9}$$

Remark. By choosing $\epsilon = \sigma\sqrt{\delta}$, we directly have that $V[X(t)] = \sigma^2 t$, for all $t \geq 0$. That is, the variance of $X(t)$ is actually *equal* to $\sigma^2 t$ for any positive value of δ, and not only in the limit as $\delta \downarrow 0$.

From the formula (1.113), we may write that

$$P[X(n\,\delta) \leq x] \simeq P[N(0, n\,\epsilon^2) \leq x] \tag{4.10}$$

from which we deduce that

$$P[W(t) \leq x] = P[N(0, \sigma^2 t) \leq x] \tag{4.11}$$

where

$$W(t) := \lim_{\delta \downarrow 0} X(t) \tag{4.12}$$

That is, the random variable $W(t)$ has a Gaussian distribution with zero mean and variance equal to $\sigma^2 t$. This result is the essential characteristic of the Wiener process. Moreover, since a random walk is a process with independent and stationary increments (see p. 50), we can assert that the process $\{W(t), t \geq 0\}$ has these two properties as well.

Based on what precedes, we now formally define the Wiener process.

Definition 4.1.1. *A stochastic process* $\{W(t), t \geq 0\}$ *is called a* **Wiener process**, *or a* **Brownian motion**, *if*
i) $W(0) = 0$,
ii) $\{W(t), t \geq 0\}$ *has independent and stationary increments,*
iii) $W(t) \sim N(0, \sigma^2 t)\ \forall\ t > 0$.

Remarks. i) The name *Brownian motion* is in honor of the Scottish botanist Robert Brown.[1] He observed through a microscope, in 1827, the purely random movement of grains of pollen suspended in water. This movement is due to the fact that the grains of pollen are bombarded by water molecules, which was only established in 1905, because the instruments Brown had at his disposal at the time did not enable him to observe the water molecules. The Brownian motion and the Poisson process (see Chapter 5) are the two most important processes for the applications. The Wiener process and processes derived from it are notably used extensively in financial mathematics.

ii) Let

$$B(t) := \frac{W(t)}{\sigma} \tag{4.13}$$

We have that $V[B(t)] = t$. The stochastic process $\{B(t), t \geq 0\}$ is named a *standard Brownian motion*. Moreover, if we sample a standard Brownian motion at regular intervals, we can obtain a symmetric random walk.

---

[1] Robert Brown, 1773–1858, was born in Scotland and died in England. He was a member of the Royal Society, in England.

iii) The Wiener process has been proposed as a model for the *position* of the particle at time t. Since the distance traveled in a time interval of length δ is proportional to $\sqrt{\delta}$, we then deduce that the order of magnitude of the *velocity* of the particle in this interval is given by

$$\frac{\sqrt{\delta}}{\delta} = \frac{1}{\sqrt{\delta}} \longrightarrow \infty \quad \text{when } \delta \downarrow 0 \qquad (4.14)$$

However, let $V(t)$ be the average velocity (from the initial time) of the particle at time $t > 0$. It was found experimentally that the model

$$V(t) := \frac{W(t)}{t} \quad \text{for } t > 0 \qquad (4.15)$$

was very good for values of t large enough (with respect to δ). We will see another model, in Subsection 4.2.5, which will be appropriate for the velocity of the particle even when t is small.

iv) We can replace condition iii) in Definition 4.1.1 by

$$W(t+s) - W(s) \sim \mathrm{N}(0, \sigma^2 t) \quad \forall \, s, t \geq 0 \qquad (4.16)$$

and then it is no longer necessary to assume explicitly that the process $\{W(t), t \geq 0\}$ has stationary increments, since it now follows from the new condition iii).

v) Let

$$W^*(t) := W(t) + c \qquad (4.17)$$

where c is a real constant, which is actually the value of $W^*(0)$. The process $\{W^*(t), t \geq 0\}$ is called a *Brownian motion starting from c*. We have that $W^*(t) \sim \mathrm{N}(c, \sigma^2 t)$, $\forall \, t \geq 0$. We could also consider the case when c is a random variable C independent of $W(t)$, for all $t \geq 0$. Then we would have

$$E[W^*(t)] = E[C] \quad \text{and} \quad V[W^*(t)] = \sigma^2 t + V[C] \qquad (4.18)$$

vi) Wiener proved the following very important result: $W(t)$ is a *continuous* function of t (with probability 1). Figure 4.1 shows a (simplified) example of the displacement of a particle that would follow a Brownian motion. In reality, the trajectory of the particle would be much more complicated, because there should be an *infinite* number of changes of direction of the particle in any interval of finite length. We can thus state that the function $W(t)$ is *nowhere differentiable* (see, however, Section 4.3).

In general, it is very difficult to explicitly calculate the kth-order density function (see p. 49) of a stochastic process. However, in the case of the Wiener process, $\{W(t), t \geq 0\}$, we only have to use the fact that this process has

$W(t)$

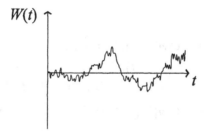

Fig. 4.1. (Simplified) example of the Brownian motion of a particle.

independent and *stationary increments*. We can thus write, for $t_1 < t_2 < \ldots < t_k$, that

$$f(w_1, \ldots, w_k; t_1, \ldots, t_k) = f(w_1; t_1) \prod_{j=2}^{k} f(w_j - w_{j-1}; t_j - t_{j-1}) \quad (4.19)$$

where

$$f(w; t) = \frac{1}{\sqrt{2\pi\sigma^2 t}} \exp\left\{-\frac{w^2}{2\sigma^2 t}\right\} \quad \text{for all } w \in \mathbb{R} \quad (4.20)$$

is the density function of a random variable having a Gaussian $N(0, \sigma^2 t)$ distribution. Indeed, we have

$$\bigcap_{j=1}^{k} \{W(t_j) = w_j\} \quad (4.21)$$

$$= \{W(t_1) = w_1\} \bigcap \left\{ \bigcap_{j=2}^{k} \{W(t_j) - W(t_{j-1}) = w_j - w_{j-1}\} \right\}$$

and the random variables $W(t_1)$, $W(t_2) - W(t_1)$, ... are independent and all have Gaussian distributions with zero means and variances given by $\sigma^2 t_1$, $\sigma^2(t_2 - t_1)$, ...

Remark. We deduce from what precedes that the Wiener process is a *Gaussian* process (see Section 2.4). It is also a *Markovian* process, because it is the limit of a Markov chain.

To calculate the *autocovariance* function (see p. 49) of the Wiener process, note first that

$$\begin{aligned} \text{Cov}[X, Y + Z] &= E[X(Y + Z)] - E[X]E[Y + Z] \\ &= \{E[XY] - E[X]E[Y]\} + \{E[XZ] - E[X]E[Z]\} \\ &= \text{Cov}[X, Y] + \text{Cov}[X, Z] \quad (4.22) \end{aligned}$$

Then we may write that

$$
\begin{aligned}
C_W(t, t+s) &\equiv \mathrm{Cov}[W(t), W(t+s)] \\
&= \mathrm{Cov}[W(t), W(t) + W(t+s) - W(t)] \\
&= \mathrm{Cov}[W(t), W(t)] + \mathrm{Cov}[W(t), W(t+s) - W(t)] \\
&\overset{\mathrm{ind.}}{=}\overset{\mathrm{incr.}}{} \mathrm{Cov}[W(t), W(t)] = V[W(t)] = \sigma^2 t \qquad (4.23)
\end{aligned}
$$

for all $s, t \geq 0$. This formula is equivalent to

$$
C_W(s, t) = \sigma^2 \min\{s, t\} \quad \text{for all } s, t \geq 0 \qquad (4.24)
$$

Remarks. i) The random variables $W(t)$ and $W(t+s)$ are *not* independent, because the intervals $[0, t]$ and $[0, t+s]$ are not disjoint. Thus, the larger $W(t)$ is, the larger we expect $W(t+s)$ to be. More precisely, we have

$$
W(t+s) \mid W(t) \sim \mathrm{N}(W(t), \sigma^2 s) \quad \forall \, s, t \geq 0 \qquad (4.25)
$$

ii) Since the function $C_W(t, t+s)$ is not a function of s alone, the Wiener process is *not* stationary, not even in the wide sense (see p. 52). As we already mentioned, the notion of processes with *stationary increments* and that of *stationary* processes must not be confounded.

iii) Since $E[W(t)] \equiv 0$, we also have

$$
R_W(s, t) = \sigma^2 \min\{s, t\} \quad \text{for all } s, t \geq 0 \qquad (4.26)
$$

where $R_W(\cdot, \cdot)$ is the *autocorrelation* function of the process $\{W(t), t \geq 0\}$ (see p. 49).

Example 4.1.1. If the random variables $W(t)$ and $W(t+s)$ were independent, we would have $W(t) + W(t+s) \sim \mathrm{N}(0, \sigma^2(2t+s))$, which is false. Indeed, we may write that

$$
W(t) + W(t+s) = 2 \, W(t) + [W(t+s) - W(t)] = X + Y
$$

where $X := 2 W(t)$ and $Y := W(t+s) - W(t)$ are *independent* random variables, because the Wiener process has independent increments. Moreover, we have

$$
E[X] = E[Y] \equiv 0
$$

and

$$
V[X] = 4 \, V[W(t)] = 4 \, \sigma^2 t \quad \text{and} \quad V[Y] = V[W(s)] = \sigma^2 s
$$

(using the fact that the Wiener process also has stationary increments). We thus have

$$
W(t) + W(t+s) \sim \mathrm{N}(0, \sigma^2(4t+s))
$$

which follows directly from the formula (1.108) as well.

Remark. As the increments of the Wiener process are stationary, the random variables $W(t + s) - W(t)$ and $W(s)$ $(= W(s) - W(0))$ have the same *distribution.* However, this does *not* mean that $W(t + s) - W(t)$ and $W(s)$ are *identical* variables. Indeed, suppose that $t = 1$, $s = 1$, and $W(1) = 0$. We cannot assert that $W(2) - W(1) = W(2) - 0 = 0$, since $P[W(2) = 0] = 0$, by continuity.

Given that the Brownian motion is Gaussian and that a Gaussian process is completely determined by its *mean* and its *autocovariance function* (see p. 59), we can give a second way of defining a Brownian motion: a continuous-time and continuous-state stochastic process, $\{X(t), t \geq 0\}$, is a Brownian motion if

i) $X(0) = 0$,

ii) $\{X(t), t \geq 0\}$ is a Gaussian process,

iii) $E[X(t)] \equiv 0$,

iv) $C_X(s, t) \equiv \text{Cov}[X(s), X(t)] = \sigma^2 \min\{s, t\} \ \forall \ s, t \geq 0$, where $\sigma > 0$ is a constant.

It is generally easier to check whether the process $\{X(t), t \geq 0\}$ possesses the four properties above, rather than trying to show that its increments are or are not independent and stationary. This second definition of the Brownian motion is particularly useful when $\{X(t), t \geq 0\}$ is some transformation of a Wiener process $\{W(t), t \geq 0\}$. As we saw in Section 2.4, any *affine* transformation of a Gaussian process is also a Gaussian process. That is, if

$$X(t) = c_1 W(t) + c_0 \tag{4.27}$$

where c_0 and $c_1 \neq 0$ are constants, then $\{X(t), t \geq 0\}$ is a Gaussian process. Moreover, if we only transform the variable t, for example, if

$$X(t) = W(t^2) \tag{4.28}$$

then the process $\{X(t), t \geq 0\}$ is also a Gaussian process.

Remark. We could drop the first condition above if we accept that a Brownian motion can start from any point $w_0 \in \mathbb{R}$. Similarly, we could replace the third condition by $E[X(t)] = \mu t$, where μ is a real constant. In this case, the stochastic process $\{X(t), t \geq 0\}$ would be a Brownian motion *with drift μ* (see Subsection 4.2.1).

Example 4.1.2. Let $\{W(t), t \geq 0\}$ be a Brownian motion. We set

$$X(0) = 0 \quad \text{and} \quad X(t) = t \, W(1/t) \quad \text{if } t > 0$$

At first sight, the stochastic process $\{X(t), t \geq 0\}$ does not seem to be a Wiener process. However, we have

$$E[X(t)] = tE[W(1/t)] = t \cdot 0 = 0 \quad \text{if } t > 0$$

because $E[W(t)] = 0$, for any value of $t > 0$, and then

$$
\begin{aligned}
C_X(s,t) &= E[X(s)X(t)] - 0 \times 0 \\
&= E[s\, W(1/s)\, t\, W(1/t)] = st\, C_W(1/s, 1/t) \\
&= st\, \sigma^2 \min\{1/s, 1/t\} = \sigma^2 \min\{s, t\}
\end{aligned}
$$

Moreover, we can assert that $\{X(t), t \geq 0\}$ is a Gaussian process, because here $X(t)$ is a *linear* transformation of $W(1/t)$. Since $X(0) = 0$ (by assumption), we can conclude that $\{X(t), t \geq 0\}$ is a Brownian motion having the same characteristics as $\{W(t), t \geq 0\}$.

Remark. We must not forget that the variable t is *deterministic*, and not random. Thus, we can consider it as a constant in the calculation of the moments of the process $\{X(t), t \geq 0\}$.

Example 4.1.3. We define the stochastic process $\{X(t), t \geq 0\}$ by

$$X(t) = B(t)\,|\{B(t) \geq 0\} \quad \text{for } t \geq 0$$

where $\{B(t), t \geq 0\}$ is a standard Brownian motion.
(a) Show that the probability density function of $X(t)$ is given by

$$f_{X(t)}(x) = 2\, f_{B(t)}(x) \quad \text{for } x \geq 0$$

(b) Calculate (i) $E[X(t)]$ and (ii) $V[X(t)]$, for $t \geq 0$.
(c) Is the stochastic process $\{X(t), t \geq 0\}$ (i) Gaussian? (ii) wide-sense stationary? Justify.
(d) Are the random variables $X(t)$ and $Y(t) := |B(t)|$ identically distributed?

Solution. (a) We may write that

$$f_{X(t)}(x) = \frac{f_{B(t)}(x)}{P[B(t) \geq 0]} = 2\, f_{B(t)}(x) \quad \text{for } x \geq 0$$

because $B(t) \sim N(0, t) \Rightarrow P[B(t) \geq 0] = 1/2$.
(b) (i) We calculate

$$
\begin{aligned}
E[X(t)] &\overset{(a)}{=} \int_0^\infty x\, \frac{2}{\sqrt{2\pi t}} e^{-x^2/2t}\, dx = -\left(\frac{2t}{\pi}\right)^{1/2} \int_0^\infty \left(-\frac{x}{t}\right) e^{-x^2/2t}\, dx \\
&= -(2t/\pi)^{1/2}\, e^{-x^2/2t}\Big|_0^\infty = (2t/\pi)^{1/2} \quad \text{for } t \geq 0
\end{aligned}
$$

(ii) Notice first that

$$E[B^2(t)] = E[B^2(t) \mid B(t) \geq 0]P[B(t) \geq 0] + E[B^2(t) \mid B(t) < 0] \underbrace{P[B(t) < 0]}_{1/2}$$

Since $E[B^2(t)] = V[B(t)] = t$ and $E[B^2(t) \mid B(t) \geq 0] = E[B^2(t) \mid B(t) < 0]$ (by symmetry and continuity), it follows that

$$V[X(t)] \overset{(i)}{=} E[B^2(t) \mid B(t) \geq 0] - \frac{2t}{\pi} = t - \frac{2t}{\pi} \quad \text{for } t \geq 0$$

(c) (i) Since $X(t) \geq 0 \;\forall\; t \geq 0$, the stochastic process $\{X(t), t \geq 0\}$ is *not* Gaussian.

(ii) It is not WSS either, because $E[X(t)]$ is not a constant.

(d) We have

$$P[Y(t) \leq y] \overset{y \geq 0}{=} P[-y \leq B(t) \leq y] = F_{B(t)}(y) - F_{B(t)}(-y)$$
$$\overset{\text{sym.}}{\Longrightarrow} \quad f_{Y(t)}(y) = 2 f_{B(t)}(y) \quad \text{for } y \geq 0$$

Thus, $X(t)$ and $Y(t)$ *are* identically distributed random variables.

4.2 Diffusion processes

Continuous-time and continuous-state *Markovian* processes are, under certain conditions, *diffusion processes*. The Wiener process is the archetype of this type of process. One way, which can be made even more rigorous, of defining a diffusion process is as follows.

Definition 4.2.1. *The continuous-time and continuous-state Markovian stochastic process $\{X(t), t \geq 0\}$, whose state space is an interval (a,b), is a* **diffusion process** *if*

$$\lim_{\epsilon \downarrow 0} \frac{1}{\epsilon} P[|X(t + \epsilon) - X(t)| > \delta \mid X(t) = x] = 0 \tag{4.29}$$

$\forall\; \delta > 0$ *and* $\forall\; x \in (a,b)$, *and if its infinitesimal parameters defined by (see p. 63)*

$$m(x;t) = \lim_{\epsilon \downarrow 0} \frac{1}{\epsilon} E[X(t + \epsilon) - X(t) \mid X(t) = x] \tag{4.30}$$

and

$$v(x;t) = \lim_{\epsilon \downarrow 0} \frac{1}{\epsilon} E[(X(t + \epsilon) - X(t))^2 \mid X(t) = x] \tag{4.31}$$

are **continuous** *functions of x and of t.*

Remarks. i) The condition (4.29) means that the probability that the process will travel a distance greater than a fixed constant δ during a sufficiently short period of time is very small. In practice, this condition implies that $X(t)$ is a continuous function of t.

ii) We assumed in the definition that the *infinitesimal mean* $m(x;t)$ and the *infinitesimal variance* $v(x;t)$ of the process exist.

iii) The state space $S_{X(t)}$ of the stochastic process $\{X(t), t \geq 0\}$ may actually be any interval: $[a, b]$, $(a, b]$, $[a, b)$, or (a, b). Moreover, if the interval does not contain the endpoint a, then a may be equal to $-\infty$. Similarly, we may have that $b = \infty$ if $S_{X(t)} = (a, b)$ or $[a, b)$. Finally, if $S_{X(t)} = [a, b]$, then the functions $m(x;t)$ and $v(x;t)$ must exist (and be continuous) for $a < x < b$ only, etc.

Example 4.2.1. In the case of the Wiener process, $\{W(t), t \geq 0\}$, we have that $W(t+\epsilon) \mid \{W(t) = w\} \sim N(w, \sigma^2\epsilon)$, for all $\epsilon > 0$. We calculate

$$\lim_{\epsilon \downarrow 0} \frac{1}{\epsilon} P[|W(t+\epsilon) - W(t)| > \delta \mid W(t) = w] = \lim_{\epsilon \downarrow 0} \frac{1}{\epsilon} P[|N(0, \sigma^2\epsilon)| > \delta]$$

$$= \lim_{\epsilon \downarrow 0} \frac{1}{\epsilon} P\left[|N(0, 1)| > \frac{\delta}{\sigma\sqrt{\epsilon}}\right] = \lim_{\epsilon \downarrow 0} \frac{2}{\epsilon} \left\{1 - \Phi\left(\frac{\delta}{\sigma\sqrt{\epsilon}}\right)\right\}$$

where

$$\Phi(x) := \int_{-\infty}^{x} \frac{1}{\sqrt{2\pi}} e^{-z^2/2} \, dz \tag{4.32}$$

is the distribution function of the $N(0, 1)$ distribution. We may write that

$$\Phi(x) = \frac{1}{2}\left(1 + \operatorname{erf}\left(\frac{x}{\sqrt{2}}\right)\right) \tag{4.33}$$

where $\operatorname{erf}(\cdot)$ is the *error function*.

Finally, making use of the formula (4.33) and of the asymptotic expansion

$$\operatorname{erf}(x) = 1 - \frac{e^{-x^2}}{\sqrt{\pi}} \left\{\frac{1}{x} - \frac{1}{2x^3} + \cdots\right\}$$

which is valid for $x > 1$, we find that

$$\lim_{\epsilon \downarrow 0} \frac{2}{\epsilon} \left\{1 - \Phi\left(\frac{\delta}{\sigma\sqrt{\epsilon}}\right)\right\} = 0$$

Next, we have

$$\lim_{\epsilon \downarrow 0} \frac{1}{\epsilon} E[W(t+\epsilon) - W(t) \mid W(t) = w] = \lim_{\epsilon \downarrow 0} \frac{1}{\epsilon} 0 = 0$$

and

$$\lim_{\epsilon \downarrow 0} \frac{1}{\epsilon} E[(W(t+\epsilon) - W(t))^2 \mid W(t) = w]$$

$$= \lim_{\epsilon \downarrow 0} \frac{1}{\epsilon} E[Z^2], \quad \text{where } Z \sim N(0, \sigma^2 \epsilon)$$

$$= \lim_{\epsilon \downarrow 0} \frac{1}{\epsilon} V[Z] = \lim_{\epsilon \downarrow 0} \frac{1}{\epsilon} \sigma^2 \epsilon = \sigma^2$$

Thus, we have $m(x; t) \equiv 0$ and $v(x; t) \equiv \sigma^2$. Because the infinitesimal parameters of the Wiener process are *constants*, the functions $m(x; t)$ and $v(x; t)$ are indeed continuous.

The most important case for the applications is the one when the diffusion process $\{X(t), t \geq 0\}$ is *time-homogeneous*, so that the infinitesimal moments of $\{X(t), t \geq 0\}$ are such that $m(x; t) \equiv m(x)$ and $v(x; t) \equiv v(x)$. We can then assert (see p. 64) that the process $\{Y(t), t \geq 0\}$ defined by

$$Y(t) = g[X(t)] \quad \text{for } t \geq 0 \tag{4.34}$$

where g is a strictly increasing or decreasing function on the interval $[a, b] \equiv S_{X(t)}$ and such that the second derivative $g''(x)$ exists and is continuous, for all $x \in (a, b)$, is also a diffusion process, whose infinitesimal parameters are given by

$$m_Y(y) = m(x)g'(x) + \frac{1}{2}v(x)g''(x) \quad \text{and} \quad v_Y(y) = v(x)[g'(x)]^2 \tag{4.35}$$

where the variable x is expressed in terms of y: $x = g^{-1}(y)$ (the *inverse* function of $g(x)$). Moreover, we have that $S_{Y(t)} = [g(a), g(b)]$ if g is strictly increasing, while $S_{Y(t)} = [g(b), g(a)]$ if g is strictly decreasing.

Remarks. i) The function g must *not* be a function of the variable t.

ii) We assume in what precedes that the process $\{X(t), t \geq 0\}$ can move from any state $x \in (a, b)$ to any other state $y \in (a, b)$ with a positive probability. We say that $\{X(t), t \geq 0\}$ is a *regular* diffusion process. Then the process $\{Y(t), t \geq 0\}$ is regular as well.

4.2.1 Brownian motion with drift

A first important transformation of the Wiener process is a generalization of this process. Let

$$Y(t) := \sigma B(t) + \mu t \tag{4.36}$$

where $\{B(t), t \geq 0\}$ is a *standard* Brownian motion, and μ and $\sigma \neq 0$ are real constants. Note that in this case the function g would be given by

$$g(x,t) = \sigma x + \mu t \qquad (4.37)$$

Thus, we cannot use the formulas (4.35) to calculate the infinitesimal parameters of the process $\{Y(t), t \geq 0\}$. However, we have

$$E[Y(t) \mid Y(t_0) = y_0] = y_0 + \mu(t - t_0) \qquad (4.38)$$

and

$$V[Y(t) \mid Y(t_0) = y_0] = \sigma^2(t - t_0) \qquad (4.39)$$

for all $t \geq t_0$. We then deduce from the formulas (2.58) and (2.59) that

$$m_Y(y) = \mu \quad \text{and} \quad v_Y(y) = \sigma^2 \quad \text{for all } y \qquad (4.40)$$

Remark. If we try to calculate the function $m_Y(y)$ from (4.35), treating t as a constant, we find that $m_Y(y) \equiv 0$, which is false, as we see in Eq. (4.40).

Definition 4.2.2. *Let $\{Y(t), t \geq 0\}$ be a diffusion process whose infinitesimal parameters are given by $m_Y(y) \equiv \mu$ and $v_Y(y) \equiv \sigma^2$. The process $\{Y(t), t \geq 0\}$ is called a* **Brownian motion** *(or* **Wiener process***) with drift μ.*

Remarks. i) The parameter μ is the *drift coefficient*, and σ^2 is the *diffusion coefficient* of the process. The term *parameter*, rather than *coefficient*, is used as well.

ii) If the random walk had not been symmetric in the preceding section, we would have obtained, under some conditions, a Wiener process with nonzero drift coefficient μ.

iii) Since the function $f[B(t)] := \sigma B(t) + \mu t$ is an *affine* transformation of the variable $B(t) \sim N(0, t)$, we may write

$$Y(t) \sim N(\mu t, \sigma^2 t) \qquad (4.41)$$

or, more generally,

$$Y(t) \mid \{Y(t_0) = y_0\} \sim N(y_0 + \mu(t - t_0), \sigma^2(t - t_0)) \quad \forall t > t_0 \qquad (4.42)$$

Moreover, the process $\{Y(t), t \geq 0\}$ is a *Gaussian* process having *independent and stationary increments*. It follows (with $Y(0) = 0$) that

$$
\begin{aligned}
E[Y(t+s)Y(t)] \quad &= \quad E[(Y(t+s) - Y(t) + Y(t))Y(t)] \\
&\overset{\text{ind. incr.}}{=} E[Y(t+s) - Y(t)]E[Y(t)] + E[Y^2(t)] \\
&\overset{\text{stat. incr.}}{=} E[Y(s)]E[Y(t)] + E[Y^2(t)] \\
&= \quad (\mu s)(\mu t) + (\sigma^2 t + \mu^2 t^2) \qquad (4.43)
\end{aligned}
$$

which implies that

$$C_Y(t+s,t) \equiv \text{Cov}[Y(t+s), Y(t)] = E[Y(t+s)Y(t)] - E[Y(t+s)]E[Y(t)]$$
$$= \mu^2 st + \sigma^2 t + \mu^2 t^2 - \mu(t+s)\mu t = \sigma^2 t \qquad \forall\, s,t \geq 0 \qquad (4.44)$$

Thus, the Brownian motion with drift has the same autocovariance function as the Wiener process.

iv) The conditional transition density function $p(y, y_0; t, t_0)$ (see p. 62) of the Brownian motion with drift coefficient μ and diffusion coefficient σ^2 satisfies the partial differential equation (see p. 64)

$$\frac{\partial p}{\partial t} + \mu \frac{\partial p}{\partial y} - \frac{\sigma^2}{2} \frac{\partial^2 p}{\partial y^2} = 0 \qquad (4.45)$$

as well as the equation

$$\frac{\partial p}{\partial t_0} + \mu \frac{\partial p}{\partial y_0} + \frac{\sigma^2}{2} \frac{\partial^2 p}{\partial y_0^2} = 0 \qquad (4.46)$$

We can check that

$$p(y, y_0; t, t_0) = \frac{1}{\sqrt{2\pi\sigma^2(t-t_0)}} \exp\left\{-\frac{1}{2} \frac{[y - (y_0 + \mu(t-t_0))]^2}{\sigma^2(t-t_0)}\right\} \qquad (4.47)$$

for $y, y_0 \in \mathbb{R}$ and $t > t_0 \geq 0$. We have

$$\lim_{t \downarrow t_0} p(y, y_0; t, t_0) = \delta(y - y_0) \qquad (4.48)$$

which is the appropriate initial condition.

4.2.2 Geometric Brownian motion

A diffusion process that is very important in financial mathematics is obtained by taking the exponential of a Brownian motion with drift.

Let $\{X(t), t \geq 0\}$ be a Wiener process with drift coefficient μ and diffusion coefficient σ^2. We set

$$Y(t) = e^{X(t)} \quad \text{for } t \geq 0 \qquad (4.49)$$

Since the function $g(x) = e^x$ does not depend on t, we deduce from (4.35) that

$$m_Y(y) = \mu e^x + \frac{1}{2}\sigma^2 e^x = \mu y + \frac{1}{2}\sigma^2 y \qquad (4.50)$$

and

$$v_Y(y) = \sigma^2 (e^x)^2 = \sigma^2 y^2 \qquad (4.51)$$

Definition 4.2.3. *The stochastic process* $\{Y(t), t \geq 0\}$ *whose infinitesimal parameters are given by* $m_Y(y) = (\mu + \frac{1}{2}\sigma^2)y$ *and* $v_Y(y) = \sigma^2 y^2$ *is called a* **geometric Brownian motion.**

Remarks. i) The state space of the geometric Brownian motion is the interval $(0, \infty)$, which follows directly from the definition $Y(t) = e^{X(t)}$. The origin is a *natural boundary* (see Section 4.4) for this process. It is used in financial mathematics as a model for the price of certain stocks.

ii) Since $Y(t) > 0$, for all $t \geq 0$, the geometric Brownian motion is *not* a Gaussian process. For a fixed t, the variable $Y(t)$ has a *lognormal distribution* with parameters μt and $\sigma^2 t$. That is,

$$f_{Y(t)}(y) = \frac{1}{\sqrt{2\pi\sigma^2 t}\, y} \exp\left\{-\frac{(\ln y - \mu t)^2}{2\sigma^2 t}\right\} \quad \text{for } y > 0 \qquad (4.52)$$

iii) We can generalize the definition of the geometric Brownian motion $\{Y(t), t \geq 0\}$ by setting

$$Y(t) = Y(0)e^{X(t)} \qquad (4.53)$$

where $Y(0)$ is a positive constant. As in the case of the Wiener process, the initial value $Y(0)$ could actually be a random variable.

iv) To obtain the conditional transition density function $p(y, y_0; t, t_0)$ of the process, we can solve the Kolmogorov forward equation

$$\frac{\partial p}{\partial t} + \left(\mu + \frac{1}{2}\sigma^2\right)\frac{\partial}{\partial y}(y\,p) - \frac{\sigma^2}{2}\frac{\partial^2}{\partial y^2}(y^2\,p) = 0 \qquad (4.54)$$

In the particular case when $t_0 = 0$, we find that the solution of this partial differential equation that satisfies the initial condition

$$\lim_{t\downarrow 0} p(y, y_0; t) = \delta(y - y_0) \qquad (4.55)$$

is

$$p(y, y_0; t) = \frac{1}{\sqrt{2\pi\sigma^2 t}\, y} \exp\left\{-\frac{(\ln \frac{y}{y_0} - \mu t)^2}{2\sigma^2 t}\right\} \qquad (4.56)$$

for $y, y_0 > 0$ and $t > 0$.

v) The geometric Brownian motion is appropriate to model the evolution of the value of certain stocks in financial mathematics when we assume that the ratios

$$\frac{X_1}{X_0}, \quad \frac{X_2}{X_1}, \quad \frac{X_3}{X_2}, \ldots \qquad (4.57)$$

where X_0 is the initial price of the stock and X_k is the price after k unit(s) of time, are independent and identically distributed random variables.

We have used the expression *diffusion process* at the beginning of this subsection. By definition, this means that the geometric Brownian motion must be a *Markovian* process, which we now show.

Proposition 4.2.1. *The geometric Brownian motion is a* Markovian *process, whose conditional transition density function is time-homogeneous.*

Proof. We have, in the general case when $Y(t) := \overset{\cdot}{Y}(0)e^{X(t)}$,

$$Y(t + s) = Y(0)e^{X(t+s)} = Y(0)e^{X(t+s)-X(t)+X(t)}$$
$$\implies \quad Y(t + s) = Y(t)e^{X(t+s)-X(t)} \tag{4.58}$$

We then deduce from the fact that the Wiener process has independent increments that $Y(t + s)$, given $Y(t)$, does not depend on the past. Thus, the process $\{Y(t), t \geq 0\}$ is *Markovian*.

We also deduce from the equation above and from the independent and stationary increments of the Wiener process that

$$P[Y(t_0 + s) \leq y \mid Y(t_0) = y_0] = P[Y(t_0)e^{X(t_0+s)-X(t_0)} \leq y \mid Y(t_0) = y_0]$$
$$= P[e^{X(t_0+s)-X(t_0)} \leq y/y_0]$$
$$= P[e^{X(s)} \leq y/y_0] \tag{4.59}$$

for all $y_0, y > 0$ and for all $t_0, s \geq 0$, from which we can assert that the function $p(y, y_0; t, t_0)$ of the geometric Brownian motion is such that

$$p(y, y_0; t, t_0) = p(y, y_0; t - t_0) \quad \square \tag{4.60}$$

Remarks. i) Equation (4.58) implies that

$$Y(t + s) \overset{\text{d}}{=} Y(t)e^{X(s)} \tag{4.61}$$

That is, the random variable $Y(t + s)$ has the *same distribution* as $Y(t)e^{X(s)}$.

ii) To prove that the conditional transition density function of the geometric Brownian motion is time-homogeneous, we can also check that the function [see (4.56)]

$$p(y, y_0; t, t_0) = \frac{1}{\sqrt{2\pi\sigma^2(t - t_0)}\, y} \exp\left\{-\frac{[\ln\frac{y}{y_0} - \mu(t - t_0)]^2}{2\sigma^2(t - t_0)}\right\} \tag{4.62}$$

for $y, y_0 > 0$ and $t > t_0 \geq 0$, satisfies the Kolmogorov forward equation (4.54), subject to

$$\lim_{t \downarrow t_0} p(y, y_0; t, t_0) = \delta(y - y_0) \tag{4.63}$$

From the formula $Y(t) = e^{X(t)}$, and making use of the formula for the *moment-generating function* (see p. 19) of a random variable X having a Gaussian $N(\mu, \sigma^2)$ distribution, namely

$$M_X(s) \equiv E[e^{sX}] = \exp\left\{ s\mu + \frac{1}{2}s^2\sigma^2 \right\} \tag{4.64}$$

we find that the mean and the variance of $Y(t)$ are given by

$$E[Y(t)] = \exp\left\{ \left(\mu + \frac{1}{2}\sigma^2 \right) t \right\} \tag{4.65}$$

and

$$V[Y(t)] = \exp\left\{ \left(\mu + \frac{1}{2}\sigma^2 \right) 2t \right\} \left(e^{\sigma^2 t} - 1 \right) \tag{4.66}$$

for all $t \geq 0$. Note that $E[Y(0)] = 1$, which is correct, since $Y(0) = 1$ (because $X(0) = 0$, by assumption). More generally, for all $t \geq \tau$, we have

$$E\left[\frac{Y(t)}{Y(\tau)} \right] = \exp\left\{ \left(\mu + \frac{1}{2}\sigma^2 \right) (t - \tau) \right\} \tag{4.67}$$

and

$$V\left[\frac{Y(t)}{Y(\tau)} \right] = \exp\left\{ \left(\mu + \frac{1}{2}\sigma^2 \right) 2(t - \tau) \right\} \left(e^{\sigma^2(t-\tau)} - 1 \right) \tag{4.68}$$

We also find, using the fact that the process is Markovian, that

$$E[Y(t) \mid Y(s), 0 \leq s \leq \tau] = E[Y(t) \mid Y(\tau)] = Y(\tau) \exp\left\{ \left(\mu + \frac{1}{2}\sigma^2 \right) (t - \tau) \right\} \tag{4.69}$$

and

$$E[Y^2(t) \mid Y(s), 0 \leq s \leq \tau] = E[Y^2(t) \mid Y(\tau)] = Y^2(\tau) \exp\{2(\mu + \sigma^2)(t - \tau)\} \tag{4.70}$$

for all $t \geq \tau$, from which we can calculate $V[Y(t) \mid Y(s), 0 \leq s \leq \tau]$.

Finally, we can write (see Ex. 4.1.1) that if $Y(0) = 1$, then

$$E[Y(t+s)Y(t)] = E\left[e^{X(t+s)} e^{X(t)} \right] = E\left[e^{X(t+s)+X(t)} \right]$$
$$= M_X(1), \quad \text{where } X \sim N(\mu(2t + s), \sigma^2(4t + s))$$

$$= \exp\left\{\mu(2t + s) + \frac{1}{2}\sigma^2(4t + s)\right\} \tag{4.71}$$

from which, $\forall\, s, t \geq 0$ we have

$$C_Y(t + s, t) = E[Y(t + s)Y(t)] - E[Y(t + s)]E[Y(t)]$$

$$= \exp\left\{\mu(2t + s) + \frac{1}{2}\sigma^2(4t + s)\right\}$$

$$- \exp\left\{\left(\mu + \frac{1}{2}\sigma^2\right)(t + s)\right\}\exp\left\{\left(\mu + \frac{1}{2}\sigma^2\right)t\right\}$$

$$= \exp\left\{\mu(2t + s) + \frac{1}{2}\sigma^2(4t + s)\right\} - \exp\left\{\left(\mu + \frac{1}{2}\sigma^2\right)(2t + s)\right\}$$

$$= \exp\left\{\left(\mu + \frac{1}{2}\sigma^2\right)(2t + s)\right\}\left(e^{\sigma^2 t} - 1\right) \tag{4.72}$$

Remarks. i) Contrary to the Wiener process, the geometric Brownian motion is *not* a process with independent and stationary increments. Indeed,

$$Y(t + s) - Y(s) \equiv Y(0)e^{X(t+s)} - Y(0)e^{X(s)} = Y(0)(e^{X(t+s)} - e^{X(s)}) \tag{4.73}$$

does not have the same distribution as

$$Y(t) - Y(0) \equiv Y(0)e^{X(t)} - Y(0) = Y(0)(e^{X(t)} - 1) \tag{4.74}$$

for all $s > 0$. Moreover, the random variables

$$Y(t + s) - Y(t) \equiv Y(0)e^{X(t+s)} - Y(0)e^{X(t)} = Y(0)(e^{X(t+s)} - e^{X(t)}) \tag{4.75}$$

and $Y(t) - Y(0)$ are not independent. To justify this assertion, we can calculate the covariance of these variables. Since $Y(0)$ is not random, we have

$$\mathrm{Cov}[Y(t + s) - Y(t), Y(t) - Y(0)]$$

$$= \{E[Y(t + s)Y(t)] - Y(0)E[Y(t + s)] - E[Y^2(t)] + Y(0)E[Y(t)]\}$$

$$- \{E[Y(t + s)] - E[Y(t)]\}\{E[Y(t)] - Y(0)\}$$

$$= \mathrm{Cov}[Y(t + s), Y(t)] - V[Y(t)] \tag{4.76}$$

Using the formulas (4.66) and (4.72), we may write that

$$\mathrm{Cov}[Y(t + s) - Y(t), Y(t) - Y(0)]$$

$$= \left(e^{\sigma^2 t} - 1\right)\left[\exp\left\{\left(\mu + \frac{1}{2}\sigma^2\right)(2t + s)\right\} - \exp\left\{\left(\mu + \frac{1}{2}\sigma^2\right)2t\right\}\right]$$

$$= \left(e^{\sigma^2 t} - 1\right)\exp\left\{\left(\mu + \frac{1}{2}\sigma^2\right)2t\right\}\left[\exp\left\{\left(\mu + \frac{1}{2}\sigma^2\right)s\right\} - 1\right] \tag{4.77}$$

which is different from zero if $s, t > 0$, so that the random variables considered are not independent.

ii) If the drift coefficient μ of the Wiener process is equal to $-\frac{1}{2}\sigma^2$, then the corresponding geometric Brownian motion is such that [see the formula (4.69)]

$$E[Y(t) \mid Y(s), 0 \le s \le \tau] = Y(\tau) \tag{4.78}$$

Thus, the process $\{Y(t), t \ge 0\}$ is a *martingale* (see p. 102). We can always obtain a martingale from an arbitrary Wiener process, by setting

$$Y^*(t) = Y(0) \exp\left\{ X(t) - \left(\mu + \frac{1}{2}\sigma^2\right) t \right\} = \exp\left\{ -\left(\mu + \frac{1}{2}\sigma^2\right) t \right\} Y(t) \tag{4.79}$$

where $Y(t)$ is defined in (4.53). Indeed, we then have

$$
\begin{aligned}
E[Y^*(t) \mid Y^*(s), 0 \le s \le \tau] & \\
&= \exp\left\{ -\left(\mu + \frac{1}{2}\sigma^2\right) t \right\} E[Y(t) \mid Y^*(s), 0 \le s \le \tau] \\
&= \exp\left\{ -\left(\mu + \frac{1}{2}\sigma^2\right) t \right\} E[Y(t) \mid Y(s), 0 \le s \le \tau] \\
&= \exp\left\{ -\left(\mu + \frac{1}{2}\sigma^2\right) t \right\} Y(\tau) \exp\left\{ \left(\mu + \frac{1}{2}\sigma^2\right)(t - \tau) \right\} \\
&= Y(\tau) \exp\left\{ -\left(\mu + \frac{1}{2}\sigma^2\right) \tau \right\} = Y^*(\tau) \tag{4.80}
\end{aligned}
$$

iii) There exists a discrete version of the geometric Brownian motion used, in particular, in financial mathematics. Let $Y_n := \ln X_n$, where X_n is the price of the shares of a certain company at time $n \in \{0, 1, \dots\}$. We assume that

$$Y_n = \mu + Y_{n-1} + \epsilon_n \tag{4.81}$$

where μ is a constant and the ϵ_n's are independent random variables. Then we have

$$Y_n = n\,\mu + \sum_{i=1}^{n} \epsilon_i + Y_0 \tag{4.82}$$

If we now assume that $\epsilon_n \sim N(0, \sigma^2)$, for all n, and that Y_0 is a constant, we obtain

$$E[Y_n] = n\,\mu + Y_0 \quad \text{and} \quad V[Y_n] = n\,\sigma^2 \tag{4.83}$$

The discrete-time and continuous-state processes $\{Y_n, n = 0, 1, \dots\}$ and $\{X_n, n = 0, 1, \dots\}$ are called a *discrete arithmetic Brownian motion* and a *discrete geometric Brownian motion*, respectively.

4.2.3 Integrated Brownian motion

Definition 4.2.4. *Let $\{Y(t), t \geq 0\}$ be a Brownian motion with drift coefficient μ and diffusion coefficient σ^2, and let*

$$Z(t) := Z(0) + \int_0^t Y(s)\,ds \tag{4.84}$$

The stochastic process $\{Z(t), t \geq 0\}$ is called an **integrated Brownian motion.**

Proposition 4.2.2. *The integrated Brownian motion is a* Gaussian *process.*

Proof. First, we use the definition of an integral as the limit of a sum:

$$Z(t) = Z(0) + \lim_{n \to \infty} \frac{t}{n} \sum_{k=1}^{n} Y\left(\frac{tk}{n}\right) \tag{4.85}$$

Since the Wiener process is a Gaussian process, the $Y(tk/n)$'s are Gaussian random variables. From this, it can be shown that the variable $Z(t)$ has a Gaussian distribution and also that the random vector $(Z(t_1), Z(t_2), \ldots, Z(t_n))$ has a multinormal distribution, for all $t_1, t_2, \ldots, t_n$ and for any n, so that the process $\{Z(t), t \geq 0\}$ is Gaussian. $\square$

Remark. We can write that

$$\begin{aligned}Z(t_k) = (Z(t_k) - Z(t_{k-1})) + (Z(t_{k-1}) - Z(t_{k-2})) + \cdots \\ + (Z(t_2) - Z(t_1)) + Z(t_1)\end{aligned} \tag{4.86}$$

for $k = 2, \ldots, n$, where we may assume that $t_1 < t_2 < \cdots < t_n$. We have

$$Z(t_k) - Z(t_{k-1}) = \int_{t_{k-1}}^{t_k} Y(s)\,ds \tag{4.87}$$

However, even though the increments of the Wiener process are independent, the random variables $(Z(t_2) - Z(t_1))$ and $Z(t_1)$, etc., are not independent, as will be seen further on. Consequently, we cannot proceed in this way to prove that the integrated Brownian motion is Gaussian.

Assuming that $Z(0)$ is a constant, we calculate

$$E[Z(t)] = Z(0) + \int_0^t E[Y(s)]\,ds = Z(0) + \int_0^t (Y(0) + \mu s)\,ds$$

$$= Z(0) + Y(0)\,t + \frac{\mu t^2}{2} \tag{4.88}$$

For the sake of simplicity, suppose now that $\{Y(t), t \geq 0\}$ is a standard Brownian motion (starting from 0) and that $Z(0) = 0$. Making use of the formula (see p. 178) $E[Y(s)Y(t)] = \min\{s,t\}$ (because $E[Y(t)] \equiv 0$ and $\sigma = 1$), we may write that

$$
\begin{aligned}
E[Z(t+s)Z(t)] &= E\left[\int_0^{t+s} Y(u)\,du \int_0^t Y(v)\,dv\right] \\
&= E\left[\int_0^t \int_0^{t+s} Y(u)Y(v)\,du\,dv\right] \\
&= \int_0^t \int_0^{t+s} E\left[Y(u)Y(v)\right]\,du\,dv \\
&= \int_0^t \left\{\int_0^v \min\{u,v\}\,du + \int_v^{t+s} \min\{u,v\}\,du\right\}\,dv \\
&= \int_0^t \left\{\int_0^v u\,du + \int_v^{t+s} v\,du\right\}\,dv \\
&= \int_0^t \left\{\frac{v^2}{2} + v(t+s-v)\right\}\,dv \\
&= t^2 \left(\frac{t}{3} + \frac{s}{2}\right) \tag{4.89}
\end{aligned}
$$

From the previous formula, we obtain, under the same assumptions as above, that

$$
\begin{aligned}
\mathrm{Cov}[Z(t+s) - Z(t), Z(t)] &= E\left\{[Z(t+s) - Z(t)]Z(t)\right\} - 0 \\
&= E[Z(t+s)Z(t)] - E[Z^2(t)] \\
&= t^2 \left(\frac{t}{3} + \frac{s}{2}\right) - \frac{t^3}{3} = \frac{t^2 s}{2} \neq 0 \tag{4.90}
\end{aligned}
$$

which implies that $Z(t+s) - Z(t)$ and $Z(t)$ are *not* independent random variables $\forall\, s > 0$, and

$$
\begin{aligned}
E\left[\{Z(t+s) - Z(s)\}^2\right] &= E[Z^2(t+s)] + E[Z^2(s)] - 2E[Z(t+s)Z(s)] \\
&= \frac{(t+s)^3}{3} + \frac{s^3}{3} - 2s^2\left(\frac{s}{3} + \frac{t}{2}\right) = \frac{t^3}{3} + t^2 s \\
&\neq \frac{t^3}{3} = E[Z^2(t)] \tag{4.91}
\end{aligned}
$$

from which we deduce that $Z(t+s) - Z(s)$ and $Z(t)$ are *not* identically distributed random variables.

We can generalize these results and state the following proposition.

Proposition 4.2.3. *The increments of the integrated Brownian motion are neither independent nor stationary.*

Next, by definition of $Z(t)$, we may write that

$$Z(t+s) = Z(0) + \int_0^{t+s} Y(\tau)\,d\tau = Z(0) + \int_0^t Y(\tau)\,d\tau + \int_t^{t+s} Y(\tau)\,d\tau$$

$$= Z(t) + \int_t^{t+s} Y(\tau)\,d\tau \tag{4.92}$$

Proposition 4.2.4. *The integrated Brownian motion is* not *a Markovian process. However, the two-dimensional stochastic process* $(Y(t), Z(t))$ *is Markovian.*

Proof. The value of the last integral above does not depend on the past if $Y(t)$ is known, because the Wiener process is Markovian. Now, if we only know the value of $Z(t)$, we do not know $Y(t)$. Thus, for the future to depend only on the present, we must consider the two processes at the same time. □

When $\{Y(t), t \geq 0\}$ is a *standard* Brownian motion, the joint probability density function of the random vector $(Y(t), Z(t))$, starting from (y_0, z_0), is given by

$$p(y, z, y_0, z_0; t)$$
$$:= \frac{P[Y(t) \in (y, y+dy], Z(t) \in (z, z+dz] \mid Y(0) = y_0, Z(0) = z_0]}{dy\,dz}$$
$$= \frac{\sqrt{3}}{\pi t^2} \exp\left\{ -\frac{2}{t}(y-y_0)^2 + \frac{6}{t^2}(y-y_0)(z-z_0-y_0 t) - \frac{6}{t^3}(z-z_0-y_0 t)^2 \right\}$$
$$\tag{4.93}$$

for $y, z, y_0, z_0 \in \mathbb{R}$ and $t > 0$.

Remarks. i) The density function $p(y, z, y_0, z_0; t)$ is a particular case of the joint conditional transition density function $p(y, z, y_0, z_0; t, t_0)$.

ii) To be more rigorous, we should take the limit as dy and dz decrease to 0 above. However, this notation is often used in research papers.

The function $p(y, z, y_0, z_0; t)$ is the solution of the partial differential equation (namely, the Kolmogorov forward equation)

$$\frac{\partial}{\partial t} p + y \frac{\partial}{\partial z} p - \frac{1}{2} \frac{\partial^2}{\partial y^2} p = 0 \tag{4.94}$$

which satisfies the initial condition

$$p(y, z, y_0, z_0; t=0) = \delta(y-y_0, z-z_0) = \begin{cases} 0 & \text{if } y \neq y_0 \text{ or } z \neq z_0 \\ \infty & \text{if } y = y_0 \text{ and } z = z_0 \end{cases} \tag{4.95}$$

Indeed, taking the limit as t decreases to zero in the formula (4.93), we obtain

$$\lim_{t \downarrow 0} p(y, z, y_0, z_0; t) = \delta(y-y_0, z-z_0) \tag{4.96}$$

Remark. Actually, it is preferable to define the *two-dimensional Dirac delta function* by

$$\delta(y - y_0, z - z_0) = 0 \quad \text{if } (y, z) \neq (y_0, z_0) \tag{4.97}$$

and

$$\int_{-\infty}^{\infty} \int_{-\infty}^{\infty} \delta(y - y_0, z - z_0)\, dy\, dz = 1 \tag{4.98}$$

We can then write that

$$\delta(y - y_0, z - z_0) = \delta(y - y_0)\delta(z - z_0) \tag{4.99}$$

Note that we also obtain the formula above, even if we set $\delta(0,0) = \delta(0) = \infty$, by assuming that

$$\delta(0)\delta(z - z_0) = \delta(y - y_0)\delta(0) = 0 \quad \text{if } z \neq z_0 \text{ and } y \neq y_0 \tag{4.100}$$

The formula for the function $p(y, z, y_0, z_0; t)$ is easily obtained by using the fact that it can be shown that the random vector $(Y(t), Z(t))$ has a bivariate normal distribution. Moreover, the bivariate normal distribution is completely characterized by the means, the variances, and the covariance of the two variables that constitute the random vector. Here, when $\{Y(t), t \geq 0\}$ is a standard Brownian motion starting from $Y(0) = y_0$, and $Z(0) = z_0$, we deduce from what precedes that

$$E[Y(t)] = y_0, \quad V[Y(t)] = t, \quad \text{and} \quad E[Z(t)] = z_0 + y_0 t \tag{4.101}$$

Furthermore, since $E[Y(s)Y(t)] = \min\{s, t\} + y_0^2$ when $Y(0) = y_0$, we find that the generalization of the formula (4.89) is

$$E[Z(t + s)Z(t)] = t^2 \left(\frac{t}{3} + \frac{s}{2} \right) + y_0^2 t^2 + z_0^2 + 2z_0 y_0 t \tag{4.102}$$

so that

$$V[Z(t)] = E[Z^2(t)] - \{E[Z(t)]\}^2$$
$$= \frac{t^3}{3} + y_0^2 t^2 + z_0^2 + 2z_0 y_0 t - (z_0 + y_0 t)^2 = \frac{t^3}{3} \tag{4.103}$$

Remark. The variance of $Z(t)$ is thus independent of the initial values $Z(0)$ and $Y(0)$, as we could have guessed.

Next, we calculate

$$E[Y(t)Z(t)] = E\left[Y(t) \left\{ z_0 + \int_0^t Y(\tau)\, d\tau \right\} \right]$$

$$= y_0 z_0 + \int_0^t E\left[Y(t)Y(\tau)\right] d\tau$$

$$= y_0 z_0 + \int_0^t (\tau + y_0^2) d\tau = y_0 z_0 + \frac{t^2}{2} + y_0^2 t \qquad (4.104)$$

which implies that

$$\mathrm{Cov}[Y(t), Z(t)] = E[Y(t)Z(t)] - E[Y(t)]E[Z(t)]$$

$$= \frac{t^2}{2} + y_0 z_0 + y_0^2 t - y_0(z_0 + y_0 t) = \frac{t^2}{2} \qquad (4.105)$$

and

$$\rho_{Y(t),Z(t)} := \frac{\mathrm{Cov}[Y(t), Z(t)]}{\sqrt{V[Y(t)]V[Z(t)]}} = \frac{t^2/2}{t^2/\sqrt{3}} = \frac{\sqrt{3}}{2} \qquad (4.106)$$

The expression for the function $p(y, z, y_0, z_0; t)$ is obtained by substituting these quantities into the formula (1.82).

We can check that if we replace t by $t - t_0$ in the formula (4.93), then the function $p(y, z, y_0, z_0; t, t_0)$ thus obtained is also a solution of Eq. (4.94), such that

$$\lim_{t \downarrow t_0} p(y, z, y_0, z_0; t, t_0) = \delta(y - y_0, z - z_0) \qquad (4.107)$$

Consequently, we can state the following proposition.

Proposition 4.2.5. *The integrated Brownian motion is* time-homogeneous. *That is,*

$$p(y, z, y_0, z_0; t, t_0) = p(y, z, y_0, z_0; t - t_0) \quad \forall\, t > t_0 \geq 0 \qquad (4.108)$$

Finally, in some applications, we consider as a model the Brownian motion integrated more than once. For example, the *doubly integrated Brownian motion*

$$D(t) := D(0) + \int_0^t Z(s)\, ds = D(0) + Z(0)\, t + \int_0^t \int_0^s Y(\tau)\, d\tau\, ds \qquad (4.109)$$

where $\{Y(t), t \geq 0\}$ is a Wiener process with drift coefficient μ and diffusion coefficient σ^2, is a *Gaussian* process for which

$$E[D(t)] = D(0) + Z(0)\, t + \int_0^t \int_0^s (Y(0) + \mu\tau)\, d\tau\, ds$$

$$= D(0) + Z(0)\, t + Y(0)\frac{t^2}{2} + \mu\frac{t^3}{6} \qquad (4.110)$$

We also find that

$$V[D(t)] = \sigma^2 \frac{t^5}{20} \qquad (4.111)$$

Moreover, the three-dimensional process $(Y(t), Z(t), D(t))$ has a multinormal (namely, trinormal or three-variate normal) distribution whose vector of means $\mathbf{m}$ and covariance matrix $\mathbf{K}$ (see p. 59) are given by

$$\mathbf{m} = \begin{bmatrix} Y(0) + \mu\, t \\ Z(0) + Y(0)\, t + \mu \frac{t^2}{2} \\ D(0) + Z(0)\, t + Y(0) \frac{t^2}{2} + \mu \frac{t^3}{6} \end{bmatrix} \qquad (4.112)$$

and

$$\mathbf{K} = \sigma^2 t \begin{bmatrix} 1 & t/2 & t^2/6 \\ t/2 & t^2/3 & t^3/8 \\ t^2/6 & t^3/8 & t^4/20 \end{bmatrix} \qquad (4.113)$$

Example 4.2.2. In Example 4.1.1, we calculated the distribution of the sum $W(t) + W(t+s)$ by using the fact that the increments of the Wiener process are independent. Since the integrated Brownian motion does not have this property, we cannot proceed in the same way to obtain the distribution of $Z :=$ $Z(t) + Z(t+s)$. However, we can assert that Z has a Gaussian distribution, with mean (if $\mu = 0$ and $Y(0) = Z(0) = 0$)

$$E[Z] = 0 + 0 = 0$$

and with variance (if $\sigma^2 = 1$)

$$V[Z] = V[Z(t)] + V[Z(t+s)] + 2\,\mathrm{Cov}[Z(t), Z(t+s)]$$
$$= \frac{t^3}{3} + \frac{(t+s)^3}{3} + 2\,t^2 \left(\frac{t}{3} + \frac{s}{2}\right)$$
$$= \frac{4t^3}{3} + 2t^2 s + ts^2 + \frac{s^3}{3}$$

4.2.4 Brownian bridge

The processes that we studied so far in this chapter were all defined for values of the variable t in the interval $[0, \infty)$. However, an interesting process that is based on a standard Brownian motion, $\{B(t), t \geq 0\}$, and that has been the subject of many research papers in the last 20 years or so, is defined for $t \in [0, 1]$ only. Moreover, it is a *conditional* diffusion process, because we suppose that $B(1) = 0$. Since it is as if the process thus obtained were tied at both ends, it is sometimes called the *tied Wiener process*, but most often the expression *Brownian bridge* is used.

Definition 4.2.5. *Let* $\{B(t), t \geq 0\}$ *be a standard Brownian motion. The* conditional *stochastic process* $\{Z(t), 0 \leq t \leq 1\}$, *where*

$$Z(t) := B(t) \mid \{B(1) = 0\} \tag{4.114}$$

is called a **Brownian bridge**.

Remark. We deduce from the properties of the Brownian motion that the Brownian bridge is a *Gaussian* diffusion process (thus, it is also *Markovian*).

Suppose that the standard Brownian motion $\{B(t), t \geq 0\}$ starts in fact from $B(0) = b_0$ and that $B(s) = b_s$. Then, using the formula (which follows from the (independent and) stationary increments of the Brownian motion)

$$f_{B(t) \mid \{B(s) = b_s\}}(b) = \frac{f_{B(t), B(s)}(b, b_s)}{f_{B(s)}(b_s)} = \frac{f_{B(t)}(b) f_{B(s)-B(t)}(b_s - b)}{f_{B(s)}(b_s)} \tag{4.115}$$

for $0 < t < s$, we find that

$$B(t) \mid \{B(0) = b_0, B(s) = b_s\} \sim \mathrm{N}\left(\frac{b_s t}{s} + \frac{b_0(s-t)}{s}, \frac{t(s-t)}{s}\right) \quad \forall\, t \in (0, s) \tag{4.116}$$

Thus, when $b_0 = 0$, $s = 1$, and $b_s = 0$, we obtain that

$$B(t) \mid \{B(0) = 0, B(1) = 0\} \sim \mathrm{N}\left(0, t(1-t)\right) \quad \forall\, t \in (0, 1) \tag{4.117}$$

so that

$$E[Z(t)] = 0 \quad \forall\, t \in (0, 1) \tag{4.118}$$

and

$$E[Z^2(t)] = V[Z(t)] = t(1-t) \quad \forall\, t \in (0, 1) \tag{4.119}$$

With the help of the formulas (4.116), (4.118), and (4.119), we calculate the autocovariance function of the Brownian bridge, for $0 < t \leq \tau < 1$, as follows:

$$\begin{aligned}
C_Z(t, \tau) &\equiv \mathrm{Cov}[Z(t), Z(\tau)] = E[Z(t)Z(\tau)] - 0 \times 0 \\
&= E\left[E[Z(t)Z(\tau) \mid Z(\tau)]\right] = E\left[Z(\tau)E[Z(t) \mid Z(\tau)]\right] \\
&= E\left[Z(\tau)\frac{t}{\tau}Z(\tau)\right] = \frac{t}{\tau}\tau(1-\tau) \\
&= t(1-\tau) \quad \text{if } 0 < t \leq \tau < 1
\end{aligned} \tag{4.120}$$

In general, we have

$$C_Z(t, \tau) = \min\{t, \tau\} - t\tau \quad \text{if } t, \tau \in (0, 1) \tag{4.121}$$

We also deduce from the formula (4.116) that

$$E[Z(t + \epsilon) - Z(t) \mid Z(t) = x] = -\frac{x\epsilon}{1 - t} \qquad (4.122)$$

and

$$E[(Z(t + \epsilon) - Z(t))^2 \mid Z(t) = x] = \epsilon + o(\epsilon) \qquad (4.123)$$

It follows that the *infinitesimal parameters* of the Brownian bridge are given by (see p. 63)

$$m(x; t) = \lim_{\epsilon \downarrow 0} -\frac{x}{1 - t} = -\frac{x}{1 - t} \qquad (4.124)$$

and

$$v(x; t) = \lim_{\epsilon \downarrow 0} 1 + \frac{o(\epsilon)}{\epsilon} = 1 \qquad (4.125)$$

for $x \in \mathbb{R}$ and $0 < t < 1$.

Remarks. i) The Brownian bridge thus has the same infinitesimal variance as the standard Brownian motion. However, its infinitesimal mean depends on t and tends to infinity (in absolute value if $x \neq 0$) as t increases to 1.

ii) Since $m(x; t)$ is a function of t, the Brownian bridge is *not* a time-homogeneous process.

As we did in the case of the Brownian motion, we can use the fact that the Brownian bridge is a *Gaussian* process to give a second definition of this process as being a Gaussian process with zero mean and whose autocovariance function is given by the formula (4.121).

Proposition 4.2.6. *Let $\{B(t), t \geq 0\}$ be a standard Brownian motion. The stochastic process $\{Z_1(t), 0 \leq t \leq 1\}$ defined by*

$$Z_1(t) = B(t) - tB(1) \qquad (4.126)$$

is a Brownian bridge.

Proof. We can assert that $\{Z_1(t), 0 \leq t \leq 1\}$ is a Gaussian process. Moreover, we calculate

$$E[Z_1(t)] = E[B(t)] - tE[B(1)] = 0 - t \times 0 = 0 \qquad (4.127)$$

and then,

$$\begin{aligned} \text{Cov}[Z_1(t), Z_1(\tau)] &= E[Z_1(t)Z_1(\tau)] = E[(B(t) - tB(1))(B(\tau) - \tau B(1))] \\ &= E[B(t)B(\tau)] - \tau \, E[B(t)B(1)] - t \, E[B(1)B(\tau)] \\ &\quad + t\tau E[B^2(1)] \end{aligned}$$

$$= \min\{t, \tau\} - \tau \times t - t \times \tau + t\tau \times 1$$
$$= \min\{t, \tau\} - t\tau \qquad (4.128)$$

Since the mean and the autocovariance function of the Gaussian process $\{Z_1(t), 0 \le t \le 1\}$ are identical to those of the Brownian bridge, we can conclude that it is indeed a Brownian bridge. □

Remark. In the same way, we can show that the process $\{Z_2(t), 0 \le t \le 1\}$ defined by

$$Z_2(t) = (1-t)B\left(\frac{t}{1-t}\right) \qquad \text{for } 0 < t < 1 \qquad (4.129)$$

and $Z_2(1) = 0$ is a Brownian bridge as well.

4.2.5 The Ornstein–Uhlenbeck process

As we mentioned in Section 4.1, the use of a Wiener process to model the displacement of a particle can be criticized. Indeed, for small values of the variable t, the Wiener process is not appropriate to represent the average velocity of the particle in the interval $[0, t]$. Moreover, the instantaneous velocity cannot be calculated, because the Brownian motion is nowhere differentiable.

To remedy this problem, in 1930 Uhlenbeck[2] and Ornstein[3] proposed a model in which they supposed that it is the *velocity* of the particle that is influenced, in part, by the shocks with the neighboring particles. The velocity also depends on the frictional resistance of the surrounding medium. The effect of this resistance is proportional to the velocity.

As in the case of the geometric Brownian motion, we can define the *Ornstein–Uhlenbeck (O.-U.) process* from a Wiener process. Let $\{B(t), t \ge 0\}$ be a standard Brownian motion. We set

$$U(t) = e^{-\alpha t} B\left(\frac{\sigma^2 e^{2\alpha t}}{2\alpha}\right) \qquad \text{for } t \ge 0 \qquad (4.130)$$

where α is a positive constant. It is thus a particular case of the transformation

$$X(t) = g(t)\, B\left(f(t)\right) \qquad (4.131)$$

where $f(t)$ is a nonnegative, continuous, and strictly increasing function, for $t \ge 0$, and $g(t)$ is a (real) continuous function. Indeed, the exponential function

---

[2] George Eugene Uhlenbeck, 1900–1988, was born in Indonesia and died in the United States. He was a physicist and mathematician whose family, coming from the Netherlands, returned there when he was six years old. His main research subject was statistical physics. He also worked on quantum mechanics and wrote two important papers on Brownian motion.

[3] Leonard Salomon Ornstein, 1880–1941, was born and died in the Netherlands. He was a physicist who worked on quantum mechanics. He applied statistical methods to problems in theoretical physics.

is continuous and $(\sigma^2/2\alpha)e^{2\alpha t}$ is nonnegative and strictly increasing, because $\alpha > 0$. For any transformation of this type, the following proposition can be shown.

Proposition 4.2.7. *Under the conditions mentioned above, the stochastic process* $\{X(t), t \geq 0\}$, *where* $X(t)$ *is defined in (4.131), is a* Gaussian *process, whose infinitesimal parameters are given by*

$$m(x;t) = \frac{g'(t)}{g(t)}x \quad and \quad v(x;t) = g^2(t)f'(t) \tag{4.132}$$

for $x \in \mathbb{R}$ *and* $t \geq 0$.

Remark. The process $\{X(t), t \geq 0\}$ is Gaussian, because the transformation of the Brownian motion is only with respect to the variable t (see p. 179). That is, we only change the time scale.

From the formulas in (4.132), we calculate

$$m(x;t) = \frac{-\alpha e^{-\alpha t}}{e^{-\alpha t}}x = -\alpha x \quad and \quad v(x;t) = (e^{-\alpha t})^2 \frac{\sigma^2}{2\alpha}e^{2\alpha t}2\alpha = \sigma^2 \tag{4.133}$$

Definition 4.2.6. *The stochastic process* $\{U(t), t \geq 0\}$ *whose infinitesimal parameters are given by* $m_U(u;t) = -\alpha u$, *for all* $t \geq 0$, *where* $\alpha > 0$, *and* $v_U(u;t) \equiv \sigma^2$ *is called an* **Ornstein–Uhlenbeck process.**

Remarks. i) We see that the Wiener process can be considered as the particular case of the O.–U. process obtained by taking the limit as α decreases to zero. Conversely, if $\{U(t), t \geq 0\}$ is an O.–U. process and if we set $B(0) = 0$ and

$$B(t) = \left(\frac{\sigma^2}{2\alpha t}\right)^{1/2} U\left[\frac{1}{2\alpha} \ln\left(\frac{2\alpha t}{\sigma^2}\right)\right] \quad \text{for } t > 0 \tag{4.134}$$

then $\{B(t), t \geq 0\}$ is a standard Brownian motion.

ii) We deduce from the definition, given in (4.130), of an O.–U. process in terms of $\{B(t), t \geq 0\}$ that $\{U(t), t \geq 0\}$ is a *Markovian* process. It is also a diffusion process.

iii) Note that the initial value of an O.–U. process, as defined above, is a random variable, since

$$U(0) = e^0 B\left(\frac{\sigma^2 e^0}{2\alpha}\right) = B\left(\frac{\sigma^2}{2\alpha}\right) \sim N\left(0, \frac{\sigma^2}{2\alpha}\right) \tag{4.135}$$

We can arrange things so that $U(0) = u_0$, a constant (see Ex. 4.2.3).

We calculate the mean and the variance of the O.–U. process from (4.130). We have

$$E[U(t)] = E\left[e^{-\alpha t}B\left(\frac{\sigma^2 e^{2\alpha t}}{2\alpha}\right)\right] \equiv 0 \qquad (4.136)$$

and

$$V[U(t)] = e^{-2\alpha t}V\left[B\left(\frac{\sigma^2 e^{2\alpha t}}{2\alpha}\right)\right] = e^{-2\alpha t}\frac{\sigma^2 e^{2\alpha t}}{2\alpha} = \frac{\sigma^2}{2\alpha} \quad \text{for } t \geq 0 \quad (4.137)$$

Next, using the formula $\text{Cov}[B(s), B(t)] = \min\{s, t\}$, we calculate

$$\text{Cov}[U(t), U(t+s)] = e^{-\alpha t}e^{-\alpha(t+s)}\text{Cov}\left[B\left(\frac{\sigma^2 e^{2\alpha t}}{2\alpha}\right), B\left(\frac{\sigma^2 e^{2\alpha(t+s)}}{2\alpha}\right)\right]$$

$$= e^{-2\alpha t}e^{-\alpha s}\frac{\sigma^2 e^{2\alpha t}}{2\alpha} = \frac{\sigma^2 e^{-\alpha s}}{2\alpha} \quad \text{for } s, t \geq 0 \qquad (4.138)$$

Given that the mean of the process is a *constant* and that its autocovariance function $C_U(t, t+s)$ does not depend on t, we can assert that the O.–U. process is a *wide-sense stationary* process (see p. 53). Moreover, since it is a Gaussian process, we can state the following proposition.

Proposition 4.2.8. *The Ornstein–Uhlenbeck process is a* strict-sense station-ary *process.*

Contrary to the Brownian motion, the increments of the O.–U. process are *not independent*. Indeed, we deduce from the formula (4.138) [and from (4.136)] that

$\text{Cov}[U(t+s) - U(t), U(t) - U(0)]$

$$= E\left[\{U(t+s) - U(t)\}\{U(t) - U(0)\}\right] - 0 \times 0$$
$$= E[U(t+s)U(t)] - E[U(t+s)U(0)] - E[U^2(t)] + E[U(t)U(0)]$$
$$= \frac{\sigma^2}{2\alpha}\left(e^{-\alpha s} - e^{-\alpha(t+s)} - 1 + e^{-\alpha t}\right) \neq 0 \qquad (4.139)$$

(if $s > 0$) so that we can assert that the random variables $U(t+s) - U(t)$ and $U(t) - U(0)$ are not independent for all $s, t \geq 0$.

However, as the O.–U. process is (strict-sense) stationary, its increments are *stationary*. To check this assertion, we use the fact that the O.–U. process is Gaussian, which implies that the random variable $U(t) - U(s)$, where $0 \leq s < t$, has a Gaussian distribution, with zero mean, and whose variance is given by

$$V[U(t) - U(s)] = V[U(t)] + V[U(s)] - 2\,\text{Cov}[U(t), U(s)]$$
$$= 2\left(\frac{\sigma^2}{2\alpha} - \frac{\sigma^2 e^{-\alpha(t-s)}}{2\alpha}\right) \qquad (4.140)$$

Now, this variance is identical to that of the variable $U(t+\tau) - U(s+\tau)$, which also has a Gaussian distribution with zero mean, from which we can conclude that the random variables $U(t) - U(s)$ and $U(t + \tau) - U(s + \tau)$ are identically distributed for all $\tau \geq 0$.

Let us now return to the problem of modeling the displacement of a particle, for which we proposed to use a Brownian motion. If we suppose that $U(t)$ is the velocity of the particle, then we may write that its position $X(t)$ at time t is given by

$$X(t) = X(0) + \int_0^t U(s) \, ds \qquad (4.141)$$

We can therefore assert that $X(t) - X(0)$ has a Gaussian distribution (because the O.–U. process is Gaussian), with mean [see (4.136)]

$$E[X(t) - X(0)] = E\left[\int_0^t U(s) \, ds\right] = \int_0^t E[U(s)] \, ds = \int_0^t 0 \, ds = 0 \quad (4.142)$$

and variance [see (4.138)]

$$
\begin{aligned}
V[X(t) - X(0)] &= E[(X(t) - X(0))^2] - 0^2 \\
&= E\left[\int_0^t U(s) \, ds \int_0^t U(\tau) \, d\tau\right] \\
&= \int_0^t \int_0^t E[U(s)U(\tau)] \, ds \, d\tau \\
&= \int_0^t \int_0^t \frac{\sigma^2}{2\alpha} e^{-\alpha|s-\tau|} \, ds \, d\tau \\
&= \frac{\sigma^2}{2\alpha} \int_0^t \left\{\int_0^\tau e^{\alpha(s-\tau)} \, ds + \int_\tau^t e^{\alpha(\tau-s)} \, ds\right\} d\tau \\
&= \frac{\sigma^2}{2\alpha^2} \int_0^t \left\{2 - e^{-\alpha\tau} - e^{\alpha(\tau-t)}\right\} d\tau \\
&= \frac{\sigma^2}{\alpha^3}\left(\alpha t - 1 + e^{-\alpha t}\right) \qquad (4.143)
\end{aligned}
$$

From the series expansion of $e^{-\alpha t}$:

$$e^{-\alpha t} = 1 - \alpha t + \frac{1}{2}(\alpha t)^2 - \frac{1}{6}(\alpha t)^3 + \ldots \qquad (4.144)$$

we may write that

$$
V[X(t) - X(0)] \sim
\begin{cases}
\dfrac{\sigma^2}{2\alpha} t^2 & \text{if } t \text{ is small} \\[2ex]
\dfrac{\sigma^2}{\alpha^2} t & \text{if } t \text{ is large}
\end{cases}
\qquad (4.145)
$$

Thus, the variance of the integral of the O.–U. process tends to that of a Brownian motion when t tends to infinity. However, for small values of t, the variance of the integral is proportional to the square of the time t elapsed since the initial time instant. Now, this result is more realistic than assuming that the variance is always proportional to t. Indeed, let $X \sim N(0, \sigma_X^2)$. By symmetry, we may write that

$$E[|X|] = 2 \int_0^\infty \frac{x}{\sqrt{2\pi}\sigma_X} e^{-x^2/(2\sigma_X^2)} \, dx$$

$$= -2 \frac{1}{\sqrt{2\pi}\sigma_X} \sigma_X^2 e^{-x^2/(2\sigma_X^2)} \Big|_0^\infty = \frac{\sqrt{2}}{\sqrt{\pi}} \sigma_X \tag{4.146}$$

Since the standard deviation of $X(t) - X(0)$ is proportional to t for small values of t, we deduce from (4.146) that the order of magnitude of the velocity of the particle in a small interval of length δ, from the initial time, is a constant:

$$\frac{\sqrt{2}}{\sqrt{\pi}} \frac{(\sigma/\sqrt{2\alpha})\delta}{\delta} = \frac{\sigma}{\sqrt{\pi\alpha}} \quad \text{for all } \delta > 0 \tag{4.147}$$

whereas the order of magnitude of this velocity tends to infinity in the case of the Brownian motion (see p. 176).

The Kolmogorov forward equation corresponding to the O.–U. process is the following:

$$\frac{\partial p}{\partial t} = \alpha \frac{\partial}{\partial u}(u\, p) + \frac{\sigma^2}{2} \frac{\partial^2 p}{\partial u^2} \tag{4.148}$$

The solution that satisfies the initial condition

$$\lim_{t \downarrow t_0} p(u, u_0; t, t_0) = \delta(u - u_0) \tag{4.149}$$

is

$$p(u, u_0; t, t_0) = \frac{1}{\sqrt{2\pi\sigma_{OU}^2}} \exp\left\{-\frac{1}{2\sigma_{OU}^2}(u - \mu_{OU})^2\right\} \tag{4.150}$$

for $u_0, u \in \mathbb{R}$ and $t > t_0 \geq 0$, where

$$\mu_{OU} := u_0\, e^{-\alpha\,(t-t_0)} \quad \text{and} \quad \sigma_{OU}^2 := \frac{\sigma^2}{2\alpha}(1 - e^{-2\alpha\,(t-t_0)}) \tag{4.151}$$

That is, $U(t) \mid \{U(t_0) = u_0\} \sim N(\mu_{OU}, \sigma_{OU}^2)$.

We deduce from what precedes that the stochastic process defined from a standard Brownian motion in (4.130) is actually the *stationary version* of the Ornstein–Uhlenbeck process, obtained by taking the limit as t tends to infinity of the solution above. Indeed, we have

$$\lim_{t \to \infty} \mu_{OU} = 0 \quad \text{and} \quad \lim_{t \to \infty} \sigma_{OU}^2 = \frac{\sigma^2}{2\alpha} \tag{4.152}$$

as in (4.136) and (4.137).

Remark. The solution (4.150) enables us to state that the O.–U. process is *time-homogeneous*. However, this version of the process is not even wide-sense stationary, since its mean depends on t.

Example 4.2.3. For the initial value of the O.–U. process $\{U(t), t \geq 0\}$ to be deterministic, we can set, when $\alpha = 1$ and $\sigma^2 = 2$,

$$U(t) = u_0 + e^{-t}B(e^{2t}) - B(1)$$

We then have

$$E[U(t)] = u_0 + e^{-t}E[B(e^{2t})] - E[B(1)] = u_0$$

and, for all $t \geq 0$,

$$
\begin{aligned}
V[U(t)] &= V[e^{-t}B(e^{2t}) - B(1)] \\
&= e^{-2t}V[B(e^{2t})] + V[B(1)] - 2e^{-t}\mathrm{Cov}[B(e^{2t}), B(1)] \\
&= e^{-2t}\,e^{2t} + 1 - 2e^{-t}\min\{e^{2t}, 1\} \\
&= 2 - 2e^{-t} \times 1 = 2(1 - e^{-t})
\end{aligned}
$$

4.2.6 The Bessel process

Let $\{B_k(t), t \geq 0\}$, for $k = 1, \ldots, n$, be independent standard Brownian motions. We set

$$X(t) = B_1^2(t) + \ldots + B_n^2(t) \quad \forall\, t \geq 0 \tag{4.153}$$

The random variable $X(t)$ can be interpreted as the square of the distance from the origin of a standard Brownian motion in n dimensions. The following proposition can be shown.

Proposition 4.2.9. *The stochastic process $\{X(t), t \geq 0\}$ is a diffusion process whose infinitesimal parameters are given by*

$$m(x; t) = n \quad and \quad v(x; t) = 4x \quad for\ x \geq 0\ and\ t \geq 0 \tag{4.154}$$

Remark. Note that the infinitesimal parameters of the process do not depend on the variable t.

We now define the process $\{Y(t), t \geq 0\}$ by

$$Y(t) = g[X(t)] = X^{1/2}(t) \quad \text{for } t \geq 0 \tag{4.155}$$

Since the transformation $g(x) = x^{1/2}$ is strictly increasing, and its second derivative $g''(x) = -\frac{1}{4}x^{-3/2}$ exists and is continuous for $x \in (0, \infty)$, we can use the formulas in (4.35) to calculate the infinitesimal parameters of the *diffusion process* $\{Y(t), t \geq 0\}$:

$$m_Y(y) = n \left(\frac{1}{2x^{1/2}}\right) + \frac{4x}{2} \left(-\frac{1}{4x^{3/2}}\right)$$

$$= n \left(\frac{1}{2y}\right) + \frac{4y^2}{2} \left(-\frac{1}{4y^3}\right) = \frac{n-1}{2y} \qquad (4.156)$$

and

$$v_Y(y) = 4x \left(\frac{1}{2x^{1/2}}\right)^2 = 1 \qquad (4.157)$$

It can be shown that this process is indeed *Markovian*. Moreover, we can generalize the definition of the process $\{Y(t), t \geq 0\}$ by replacing n by a real parameter $\alpha > 0$.

Definition 4.2.7. *The diffusion process* $\{Y(t), t \geq 0\}$ *whose state space is the interval* $[0, \infty)$ *and whose infinitesimal mean and variance are given, respectively, by*

$$m(y; t) = \frac{\alpha - 1}{2y} \quad and \quad v(y; t) = 1 \quad for \ t \geq 0 \ and \ y \geq 0 \qquad (4.158)$$

is called a **Bessel**[4] *process of dimension* $\alpha > 0$.

Remarks. i) The term *dimension* used for the parameter α comes from the interpretation of the process when $\alpha = n \in \mathbb{N}$.

ii) When $\alpha = 1$, the infinitesimal parameters of the process are the same as those of the standard Brownian motion. However, we can represent it as follows:

$$Y(t) = |B(t)| \quad for \ t \geq 0 \qquad (4.159)$$

where $\{B(t), t \geq 0\}$ is a standard Brownian motion. It is as if the origin were a *reflecting boundary* for $\{B(t), t \geq 0\}$.

The conditional transition density function of the Bessel process satisfies the Kolmogorov forward equation

$$\frac{\partial p}{\partial t} = \frac{1-\alpha}{2} \frac{\partial}{\partial y} \left(\frac{p}{y}\right) + \frac{1}{2} \frac{\partial^2 p}{\partial y^2} \qquad (4.160)$$

We can check that

$$p(y, y_0; t, t_0 = 0) = \frac{1}{t} \left(\frac{y}{y_0}\right)^\nu y \exp\left\{-\frac{y_0^2 + y^2}{2t}\right\} I_\nu \left(\frac{y_0 y}{t}\right) \qquad (4.161)$$

---

[4] See p. 114.

for y_0, y, and $t > 0$, where

$$\nu := \frac{\alpha}{2} - 1 \tag{4.162}$$

and $I_\nu(\cdot)$ is a *modified Bessel function of the first kind* (of order ν), defined by (see p. 375 of Ref. [1])

$$I_\nu(z) = (z/2)^\nu \sum_{k=0}^\infty \frac{(z^2/4)^k}{k!\Gamma(\nu + k + 1)} \tag{4.163}$$

Remark. The quantity ν defined in (4.162) is called the *index* of the Bessel process.

Definition 4.2.8. *The diffusion process $\{X(t), t \geq 0\}$ defined by*

$$X(t) = Y^2(t) \quad for\ t \geq 0 \tag{4.164}$$

is called a **squared Bessel process** *of* **dimension** $\alpha > 0$. *Its infinitesimal parameters are given by*

$$m(x;t) \equiv \alpha \quad and \quad v(x;t) = 4x \quad for\ t \geq 0\ and\ x \geq 0 \tag{4.165}$$

Remarks. i) We deduce from the representation of the process $\{X(t), t \geq 0\}$ given in (4.153) that if $\{X_1(t), t \geq 0\}$ and $\{X_2(t), t \geq 0\}$ are two independent squared Bessel processes, of dimensions $\alpha_1 = n_1$ and $\alpha_2 = n_2$, respectively, then the process $\{S(t), t \geq 0\}$, where

$$S(t) := X_1(t) + X_2(t) \quad for\ t \geq 0 \tag{4.166}$$

is also a squared Bessel process, of dimension $\alpha := n_1 + n_2$. Actually, this result is valid for all $\alpha_1 > 0$ and $\alpha_2 > 0$, and not only when α_1 and α_2 are integers.

ii) The function $p(x, x_0; t, t_0 = 0)$ of the squared Bessel process of dimension α is given by

$$p(x, x_0; t, t_0 = 0) = \frac{1}{t}\left(\frac{x}{x_0}\right)^{\nu/2} \exp\left\{-\frac{x_0 + x}{2t}\right\} I_\nu\left(\frac{\sqrt{x_0 x}}{t}\right) \tag{4.167}$$

for x_0, x, and $t > 0$, where $\nu = (\alpha/2) - 1$.

A diffusion process used in financial mathematics to model the variations of interest rates, and which can be expressed in terms of a squared Bessel process, is named the **Cox–Ingersoll–Ross**[5]- **(CIR) process** (see Ref. [4]). We set

---

[5] John C. Cox and Stephen A. Ross are professors at the MIT Sloan School of Management. Jonathan E. Ingersoll, Jr. is a professor at the Yale School of Management.

$$R(t) = e^{-bt} X \left(\frac{\sigma^2}{4b} (e^{bt} - 1) \right) \quad \text{for } t \geq 0 \tag{4.168}$$

where $b \in \mathbb{R}$ and $\sigma > 0$ are constants, and $\{X(t), t \geq 0\}$ is a squared Bessel process of dimension $\alpha = 4a/\sigma^2$ (with $a > 0$). We find that the infinitesimal parameters of the process $\{R(t), t \geq 0\}$ are given by

$$m(r; t) = a - br \quad \text{and} \quad v(r; t) = \sigma^2 r \quad \text{for } t \geq 0 \text{ and } r \geq 0 \tag{4.169}$$

From the formula (4.167), we find that

$$p(r, r_0; t, t_0 = 0) = \frac{e^{bt}}{2k(t)} \left(\frac{re^{bt}}{r_0} \right)^{\nu/2} \exp \left\{ -\frac{r_0 + re^{bt}}{2k(t)} \right\} I_\nu \left(\frac{\sqrt{r_0 re^{bt}}}{k(t)} \right) \tag{4.170}$$

for r_0, r, and $t > 0$, where

$$k(t) := \frac{\sigma^2}{4b} (e^{bt} - 1) \quad \text{and} \quad \nu := \frac{2a}{\sigma^2} - 1 \tag{4.171}$$

4.3 White noise

In Section 2.4, we defined the stochastic process called *white noise* as being a process (that we now denote by) $\{X(t), t \geq 0\}$ with zero mean and whose autocovariance function is of the form

$$C_X(t_1, t_2) = q(t_1)\delta(t_2 - t_1) \tag{4.172}$$

where $q(t_1)$ is a positive function and $\delta(\cdot)$ is the Dirac delta function (see p. 63).

When $q(t_1) \equiv \sigma^2$, the function $C_X(t_1, t_2)$ is the second mixed derivative of the autocovariance function $C_W(t_1, t_2) = \sigma^2 \min\{t_1, t_2\}$ of a Brownian motion (without drift and) with diffusion coefficient σ^2. Indeed, we have

$$\frac{\partial}{\partial t_2} C_W(t_1, t_2) = \begin{cases} 0 & \text{if } t_1 < t_2 \\ \sigma^2 & \text{if } t_1 \geq t_2 \end{cases} \tag{4.173}$$

so that

$$\frac{\partial^2}{\partial t_1 \partial t_2} C_W(t_1, t_2) = \frac{\partial}{\partial t_1} \sigma^2 u(t_1 - t_2) = \sigma^2 \delta(t_1 - t_2) = \sigma^2 \delta(t_2 - t_1) \tag{4.174}$$

where $u(\cdot)$ is the Heaviside function (see p. 11).

Definition 4.3.1. *The (generalized) stochastic process $\{X(t), t \geq 0\}$ with zero mean and autocovariance function*

$$C_X(t_1, t_2) = \sigma^2 \delta(t_2 - t_1) \tag{4.175}$$

is called a **Gaussian white noise** *(or white Gaussian noise).*

Remarks. i) Since $E[X(t)] \equiv 0$, we also have

$$R_X(t_1, t_2) = \sigma^2 \delta(t_2 - t_1) \tag{4.176}$$

The second mixed derivative $\frac{\partial^2}{\partial t_1 \partial t_2} R_Y(t_1, t_2)$ is called the *generalized mean-square derivative* of the stochastic process $\{Y(t), t \geq 0\}$. We say that the process $\{X(t), t \geq 0\}$ having the autocorrelation function $\frac{\partial^2}{\partial t_1 \partial t_2} R_Y(t_1, t_2)$ is a *generalized stochastic process*.

ii) A white noise (Gaussian or not) is such that the random variables $X(t_1)$ and $X(t_2)$ are *uncorrelated* if $t_1 \neq t_2$. If the variables $X(t_1)$ and $X(t_2)$ are *independent*, the expression *strict white noise* is used to designate the corresponding process.

Now, we mentioned that the Brownian motion is nowhere differentiable. Consider, however, the process $\{X(t), t \geq 0\}$ defined (symbolically) by

$$X(t) = \lim_{\epsilon \downarrow 0} \frac{W(t + \epsilon) - W(t)}{\epsilon} \tag{4.177}$$

where $\epsilon > 0$ is a constant and $\{W(t), t \geq 0\}$ is a Brownian motion with coefficients $\mu = 0$ and $\sigma^2 > 0$. Assume that we can interchange the limit and the mathematical expectation. Then

$$E[X(t)] = \lim_{\epsilon \downarrow 0} E\left[\frac{W(t + \epsilon) - W(t)}{\epsilon}\right] = \lim_{\epsilon \downarrow 0} 0 = 0 \tag{4.178}$$

so that

$$C_X(t_1, t_2) = \lim_{\epsilon \downarrow 0} E\left[\left(\frac{W(t_1 + \epsilon) - W(t_1)}{\epsilon}\right)\left(\frac{W(t_2 + \epsilon) - W(t_2)}{\epsilon}\right)\right]$$

$$= \lim_{\epsilon \downarrow 0} \frac{1}{\epsilon^2}\Big\{E\left[W(t_1 + \epsilon)W(t_2 + \epsilon)\right] - E\left[W(t_1 + \epsilon)W(t_2)\right]$$

$$- E\left[W(t_1)W(t_2 + \epsilon)\right] + E\left[W(t_1)W(t_2)\right]\Big\} \tag{4.179}$$

Suppose first that $t_1 < t_2$. Then, for an ϵ small enough, we will have that $t_1 + \epsilon < t_2$. It follows that

$$C_X(t_1, t_2) = \lim_{\epsilon \downarrow 0} \frac{\sigma^2}{\epsilon^2}\{(t_1 + \epsilon) - (t_1 + \epsilon) - t_1 + t_1\} = \lim_{\epsilon \downarrow 0} \frac{\sigma^2}{\epsilon^2} 0 = 0 \tag{4.180}$$

When $t_1 = t_2$, we obtain

$$C_X(t_1, t_2) = \lim_{\epsilon \downarrow 0} \frac{\sigma^2}{\epsilon^2}\{(t_1 + \epsilon) - t_1 - t_1 + t_1\}$$

$$= \lim_{\epsilon \downarrow 0} \frac{\sigma^2}{\epsilon^2}\epsilon = \lim_{\epsilon \downarrow 0} \frac{\sigma^2}{\epsilon} = \infty \tag{4.181}$$

We can therefore write that

$$C_X(t_1, t_2) = \sigma^2 \delta(t_2 - t_1) \tag{4.182}$$

Thus, we interpret a Gaussian white noise as being, in a certain way, the *derivative* of a Brownian motion. From now on, we will denote the Gaussian white noise by $\{dW(t), t \geq 0\}$. We also have that $dW(t) = W'(t) \, dt$. In engineering, the notation $dW(t)/dt = \epsilon(t)$ is often used.

Even though the Brownian motion is not differentiable, we can use the notion of a *generalized derivative*, which is defined as follows.

Definition 4.3.2. *Let f be a function whose derivative exists and is continuous in the interval $[0, t]$. The **generalized derivative** of the function $W(t)$, where $\{W(t), t \geq 0\}$ is a Brownian motion, is given by*

$$\int_0^t f(s) W'(s) \, ds = f(t) W(t) - \int_0^t W(s) f'(s) \, ds \tag{4.183}$$

Remarks. i) This definition of the generalized derivative of $W(t)$ is actually a more rigorous way of defining a *Gaussian white noise* $\{dW(t), t \geq 0\}$.

ii) For a deterministic function g, its *generalized derivative* is defined by

$$\int_0^\infty f(s) g'(s) \, ds = -\int_0^\infty g(s) f'(s) \, ds \tag{4.184}$$

where we assume that $f(t) = 0$ for all $t \notin [a, b]$, with $a > 0$ and $b < \infty$.

The formula (4.183) leads us to the notion of a *stochastic integral*.

Definition 4.3.3. *Let f be a function whose derivative exists and is continuous in the interval $[a, b]$, where $a \geq 0$, and let $\{W(t), t \geq 0\}$ be a Wiener process. We define the **stochastic integral** $\int_a^b f(t) \, dW(t)$ by*

$$\int_a^b f(t) \, dW(t) = f(b) W(b) - f(a) W(a) - \int_a^b W(t) \, df(t) \tag{4.185}$$

Remarks. i) We can also define a stochastic integral as follows:

$$\int_a^b f(t) \, dW(t) = \lim_{\substack{n \to \infty \\ \max_{1 \leq i \leq n} \{t_i - t_{i-1}\} \downarrow 0}} \sum_{i=1}^n f(t_{i-1})[W(t_i) - W(t_{i-1})] \tag{4.186}$$

where $a = t_0 < t_1 < \ldots < t_n = b$ is a *partition* of $[a, b]$.

ii) The definition of $\int_a^b f(t) \, dW(t)$ follows from the formula (4.177) as well, by setting

$$\int_a^b f(t)\, dW(t) = \lim_{\epsilon\downarrow 0} \int_a^b f(t)\left(\frac{W(t+\epsilon)-W(t)}{\epsilon}\right)\, dt \qquad (4.187)$$

Indeed, using the formula

$$\frac{W(t+\epsilon)-W(t)}{\epsilon} = \frac{d}{dt}\left(\frac{1}{\epsilon}\int_t^{t+\epsilon} W(s)\, ds\right) \qquad (4.188)$$

and integrating by parts, we obtain

$$\int_a^b f(t)\, dW(t) = \lim_{\epsilon\downarrow 0}\left\{\left[f(t)\frac{1}{\epsilon}\int_t^{t+\epsilon} W(s)\, ds\right]_a^b \right.$$
$$\left. - \int_a^b f'(t)\left(\frac{1}{\epsilon}\int_t^{t+\epsilon} W(s)\, ds\right)\, dt\right\} \qquad (4.189)$$

Finally, since $W(t)$ is a continuous function, we have (making use of l'Hospital's[6] rule)

$$\lim_{\epsilon\downarrow 0}\frac{1}{\epsilon}\int_t^{t+\epsilon} W(s)\, ds = \lim_{\epsilon\downarrow 0}\frac{d}{d\epsilon}\int_t^{t+\epsilon} W(s)\, ds = \lim_{\epsilon\downarrow 0} W(t+\epsilon) = W(t) \quad (4.190)$$

from which we retrieve the formula (4.185).

Properties. i) From the formula (4.185), we can assert that a stochastic integral has a *Gaussian distribution*, because it is a linear combination of Gaussian random variables.

ii) We have

$$E\left[\int_a^b f(t)\, dW(t)\right] = f(b)E[W(b)] - f(a)E[W(a)] - E\left[\int_a^b W(t)\, df(t)\right] = 0$$
$$(4.191)$$

where we assumed that we can interchange the mathematical expectation and the integral.

iii) To calculate the variance of a stochastic integral, we can use the formula (4.186) and the fact that the increments of the Brownian motion are independent and stationary. We have

$$V\left[\sum_{i=1}^n f(t_{i-1})[W(t_i)-W(t_{i-1})]\right] = \sum_{i=1}^n f^2(t_{i-1})V[W(t_i)-W(t_{i-1})]$$
$$= \sum_{i=1}^n f^2(t_{i-1})V[W(t_i - t_{i-1})]$$

---

[6] Guillaume François Antoine de l'Hospital, 1661–1704, was born and died in France. He published the first textbook on differential calculus, in which the rule that bears his name can be found.

$$= \sum_{i=1}^{n} f^2(t_{i-1})\sigma^2(t_i - t_{i-1}) \quad (4.192)$$

It follows (interchanging the limit and the calculation of the variance) that

$$V\left[\int_a^b f(t)\, dW(t)\right] = \lim_{\substack{n \to \infty \\ \max_{1 \le i \le n}\{t_i - t_{i-1}\} \downarrow 0}} \sum_{i=1}^{n} f^2(t_{i-1})\sigma^2(t_i - t_{i-1})$$

$$= \sigma^2 \int_a^b f^2(t)\, dt \quad (4.193)$$

iv) We can generalize the preceding formula as follows:

$$E\left[\int_a^b f(t)\, dW(t) \int_c^d g(t)\, dW(t)\right] = \begin{cases} 0 & \text{if } a \le b \le c \le d \\ \sigma^2 \int_a^b f(t)g(t)\, dt & \text{if } a = c \le b \le d \end{cases} \quad (4.194)$$

where g is a function whose derivative exists and is continuous in the interval $[c, d]$.

Let $\{X(t), t \ge 0\}$ be a continuous-time and continuous-state stochastic process whose infinitesimal parameters are $m(x; t)$ and $v(x; t)$. This process can be represented in the following way (see Ref. [16], for instance):

$$X(t) = X(0) + \int_0^t m[X(s); s]\, ds + \int_0^t v^{1/2}[X(s); s]\, dB(s) \quad (4.195)$$

where $\{B(t), t \ge 0\}$ is a standard Brownian motion. It follows that $X(t)$ is a solution of the *stochastic differential equation*

$$dX(t) = m[X(t); t]\, dt + v^{1/2}[X(t); t]\, dB(t) \quad (4.196)$$

Under the condition $X(t) = x$, the equation above becomes

$$dX(t) = m(x; t)\, dt + v^{1/2}(x; t)\, dB(t) \quad (4.197)$$
$$\Longleftrightarrow \quad \dot{X}(t) = m(x; t) + v^{1/2}(x; t)\, \dot{B}(t) \quad (4.198)$$

(with the notation $\dot{X}(t) = \frac{d}{dt}X(t)$).

We can consider stochastic differential equations in n dimensions. A useful result is given in the following proposition.

Proposition 4.3.1. *Let $\{\mathbf{X}(t), t \ge 0\}$ be an n-dimensional stochastic process defined by*

$$dX(t) = (AX(t) + a) \, dt + N^{1/2} \, dB(t) \tag{4.199}$$

where $\{B(t), t \geq 0\}$ is an n-dimensional standard Brownian motion, A is a square matrix of order n, a is an n-dimensional vector, and $N^{1/2}$ is a positive definite square matrix of order n. Then, given that $X(t_0) = x$, we may write that

$$X(t) \sim N(m(t), K(t)) \quad \text{for } t \geq t_0 \tag{4.200}$$

where

$$m(t) := \Phi(t) \left(x + \int_{t_0}^t \Phi^{-1}(u) \, a \, du \right) \tag{4.201}$$

and

$$K(t) := \Phi(t) \left(\int_{t_0}^t \Phi^{-1}(u) N [\Phi^{-1}(u)]' \, du \right) \Phi'(t) \tag{4.202}$$

where the symbol prime denotes the transpose of the matrix, and the function $\Phi(t)$ is given by

$$\Phi(t) := e^{A(t-t_0)} = \sum_{n=0}^{\infty} A^n \frac{(t-t_0)^n}{n!} \tag{4.203}$$

Remarks. i) A matrix $M = (m_{i,j})_{i,j=1,\ldots,n}$ is positive definite if

$$c'Mc = \sum_{i=1}^{n} \sum_{j=1}^{n} c_i c_j m_{i,j} > 0 \tag{4.204}$$

for any vector $c' := (c_1, \ldots, c_n)$ that is not the null vector $(0, \ldots, 0)$.

ii) When $n = 1$, the formulas for the mean and the variance become

$$m(t) = \begin{cases} x + a(t - t_0) & \text{if } A = 0 \\ \left(x + \dfrac{a}{A} \right) e^{A(t-t_0)} - \dfrac{a}{A} & \text{if } A \neq 0 \end{cases} \tag{4.205}$$

and

$$K(t) = \begin{cases} N(t - t_0) & \text{if } A = 0 \\ \dfrac{N}{2A} \left(e^{2A(t-t_0)} - 1 \right) & \text{if } A \neq 0 \end{cases} \tag{4.206}$$

iii) We can generalize the proposition to the case where $A = A(t)$, $a = a(t)$, and $N^{1/2} = N^{1/2}(t)$. The function $\Phi(t)$ is then obtained by solving the matrix differential equation

$$\frac{d}{dt} \Phi(t) = A(t) \Phi(t) \tag{4.207}$$

with the initial condition $\Phi(t_0) = I_n$ (the identity matrix of order n).

Example 4.3.1. The Brownian motion $\{X(t), t \geq 0\}$ with drift coefficient $\mu \neq 0$ and diffusion coefficient σ^2 (> 0) is defined by the stochastic differential equation

$$dX(t) = \mu dt + \sigma dB(t)$$

According to the second remark above, the random variable $X(t)$, given that $X(0) = x$, has a Gaussian distribution with parameters

$$\mu_{X(t)} = m(t) = x + \mu t \quad \text{and} \quad \sigma^2_{X(t)} = K(t) = \sigma^2 t$$

which does correspond to the results mentioned in Subsection 4.2.2.

In the case of the Ornstein–Uhlenbeck process, $\{U(t), t \geq 0\}$, we have the following stochastic differential equation:

$$dU(t) = -\alpha \, U(t) \, dt + \sigma \, dB(t)$$

so that $A = -\alpha$, $a = 0$, and $N^{1/2} = \sigma$. It follows that the random variable $Y(t) := U(t) \mid \{U(t_0) = u_0\}$ has a Gaussian distribution whose parameters are (see p. 203)

$$\mu_{Y(t)} = u_0 \, e^{-\alpha(t-t_0)} \quad \text{and} \quad \sigma^2_{Y(t)} = \frac{\sigma^2}{-2\alpha} \left(e^{-2\alpha(t-t_0)} - 1 \right)$$

Example 4.3.2. Consider now the two-dimensional diffusion process $\mathbf{X}(t) = (Z(t), Y(t))$ defined by the system of equations

$$dZ(t) = Y(t) \, dt$$
$$dY(t) = \mu \, dt + \sigma dB(t)$$

That is, $\{Y(t), t \geq 0\}$ is a Brownian motion with drift coefficient μ $(\in \mathbb{R})$ and diffusion coefficient σ^2 (> 0), and $\{Z(t), t \geq 0\}$ is its integral. Suppose that $(Z(0), Y(0)) = (z, y)$. We may write that

$$\mathbf{A} = \begin{bmatrix} 0 & 1 \\ 0 & 0 \end{bmatrix}, \quad \mathbf{a} = \begin{bmatrix} 0 \\ \mu \end{bmatrix}, \quad \text{and} \quad \mathbf{N}^{1/2} = \begin{bmatrix} 0 & 0 \\ 0 & \sigma \end{bmatrix}$$

Let us first calculate the function $\Phi(t)$. We have

$$\mathbf{A}^2 = \begin{bmatrix} 0 & 1 \\ 0 & 0 \end{bmatrix} \begin{bmatrix} 0 & 1 \\ 0 & 0 \end{bmatrix} = \begin{bmatrix} 0 & 0 \\ 0 & 0 \end{bmatrix}$$

It follows that

$$\Phi(t) = \mathbf{I}_2 + \mathbf{A}t = \begin{bmatrix} 1 & t \\ 0 & 1 \end{bmatrix}$$

so that

$$\Phi'(t) = \begin{bmatrix} 1 & 0 \\ t & 1 \end{bmatrix} \quad \text{and} \quad \Phi^{-1}(t) = \begin{bmatrix} 1 & -t \\ 0 & 1 \end{bmatrix}$$

We then have

$$\int_0^t \Phi^{-1}(u)\, \mathbf{a}\, du = \int_0^t \begin{bmatrix} -\mu u \\ \mu \end{bmatrix} du = \begin{bmatrix} -\frac{1}{2}\mu t^2 \\ \mu t \end{bmatrix}$$

which implies that

$$\mathbf{m}(t) = \begin{bmatrix} 1 & t \\ 0 & 1 \end{bmatrix} \left\{ \begin{bmatrix} z \\ y \end{bmatrix} + \begin{bmatrix} -\frac{1}{2}\mu t^2 \\ \mu t \end{bmatrix} \right\} = \begin{bmatrix} z + yt + \frac{1}{2}\mu t^2 \\ y + \mu t \end{bmatrix}$$

Next, we have

$$\mathbf{N} = \begin{bmatrix} 0 & 0 \\ 0 & \sigma^2 \end{bmatrix}$$

and

$$\int_0^t \Phi^{-1}(u)\mathbf{N}[\Phi^{-1}(u)]'\, du = \int_0^t \left\{ \begin{bmatrix} 1 & -u \\ 0 & 1 \end{bmatrix} \begin{bmatrix} 0 & 0 \\ 0 & \sigma^2 \end{bmatrix} \begin{bmatrix} 1 & 0 \\ -u & 1 \end{bmatrix} \right\} du$$

$$= \int_0^t \begin{bmatrix} u^2\sigma^2 & -u\sigma^2 \\ -u\sigma^2 & \sigma^2 \end{bmatrix} du$$

$$= \begin{bmatrix} \sigma^2 t^3/3 & -\sigma^2 t^2/2 \\ -\sigma^2 t^2/2 & \sigma^2 t \end{bmatrix}$$

from which we calculate

$$\mathbf{K}(t) = \begin{bmatrix} 1 & t \\ 0 & 1 \end{bmatrix} \begin{bmatrix} \sigma^2 t^3/3 & -\sigma^2 t^2/2 \\ -\sigma^2 t^2/2 & \sigma^2 t \end{bmatrix} \begin{bmatrix} 1 & 0 \\ t & 1 \end{bmatrix} = \begin{bmatrix} \sigma^2 t^3/3 & \sigma^2 t^2/2 \\ \sigma^2 t^2/2 & \sigma^2 t \end{bmatrix}$$

Note that these results agree with those mentioned in the case of the three-dimensional process $(Y(t), Z(t), D(t))$ (see p. 196).

4.4 First-passage problems

Let T_d be the random variable that designates the time needed for a standard Brownian motion to go from $B(0) = 0$ to $B(t) = d \neq 0$. Symbolically, we write

$$T_d := \min\{t > 0 \colon B(t) = d\} \tag{4.208}$$

Remark. To be more rigorous, we should write that T_d is the *infimum* (rather than the *minimum*) of the positive values of t for which $B(t) = d$. Indeed, in the case of the standard Brownian motion, we can show [see Eq. (4.218)] that the probability that the process will eventually hit the boundary at d is equal to 1, for any $d \in \mathbb{R}$. However, in other cases, it is not certain that the process will hit the boundary at d. If the stochastic process does not hit the boundary, the set whose minimum is to be found is then empty and this minimum does not exist. On the other hand, the infimum of the (empty) set in question is, by definition, equal to infinity.

Suppose first that d is a positive constant. We can obtain the distribution of the random variable T_d by using the fact that the distribution of $B(t)$ is known, and by conditioning on the possible values of T_d. We have

$$P[B(t) \geq d] = P[B(t) \geq d \mid T_d \leq t]P[T_d \leq t] + P[B(t) \geq d \mid T_d > t]P[T_d > t] \tag{4.209}$$

We deduce from the continuity of the process $\{B(t), t \geq 0\}$ that

$$P[B(t) \geq d \mid T_d > t] = 0 \tag{4.210}$$

Moreover, if $T_d < t$, then there exists a $t_0 \in (0, t)$ such that $B(t_0) = d$. Since

$$B(t) \mid \{B(t_0) = d\} \sim N(d, t - t_0) \quad \text{for } t \geq t_0 \tag{4.211}$$

we may write (by symmetry) that

$$P[B(t) \geq d \mid T_d < t] = \frac{1}{2} \tag{4.212}$$

Finally, $P[B(t) \geq d \mid T_d = t] = 1$. However, as T_d is a continuous random variable (which implies that $P[T_d = t] = 0$), we may conclude that

$$P[T_d \leq t] = 2\left[1 - \Phi(d/\sqrt{t})\right], \quad \text{where } \Phi(x) := P[N(0,1) \leq x] \tag{4.213}$$

Remarks. i) The preceding formula is called the *reflection principle* for the Brownian motion.

ii) We also deduce from the continuity of the Wiener process that

$$P\left[\max_{0 \leq s \leq t} B(s) \geq d \ (> 0)\right] = P[T_d \leq t] = 2\left[1 - \Phi(d/\sqrt{t})\right] \tag{4.214}$$

That is, the probability that the process takes on a value greater than or equal to $d > 0$ in the interval $[0, t]$ is the same as the probability that it reaches the boundary at d not later than at time t.

iii) By symmetry, T_d and T_{-d} are identically distributed random variables. It follows that

$$P[T_d \leq t] = 2\left[1 - \Phi(|d|/\sqrt{t})\right] \quad \forall \, d \neq 0 \tag{4.215}$$

iv) The density function of T_d is obtained by differentiating the function above:

$$f_{T_d}(t) = \frac{d}{dt}\left[2\left(1 - \int_{-\infty}^{|d|/\sqrt{t}} \frac{1}{\sqrt{2\pi}} e^{-z^2/2} \, dz\right)\right]$$

$$= -2\frac{1}{\sqrt{2\pi}} \exp\left\{-\frac{d^2}{2t}\right\}\left(-\frac{|d|}{2t^{3/2}}\right)$$

$$= \frac{|d|}{\sqrt{2\pi t^3}} \exp\left\{-\frac{d^2}{2t}\right\} \quad \text{for } t > 0 \tag{4.216}$$

This density is a particular case of the *inverse Gaussian* or *Wald*[7] *distribution*.

Remark. The density function of a random variable X having an *inverse Gaussian distribution* with parameters $\mu > 0$ and $\lambda > 0$ is given by

$$f_X(x; \mu, \lambda) = \frac{\sqrt{\lambda}}{\sqrt{2\pi x^3}} \exp\left\{-\frac{\lambda(x-\mu)^2}{2\mu^2 x}\right\} \quad \forall\, x > 0 \tag{4.217}$$

The distribution of T_d above is thus, in fact, obtained by setting $\lambda = d^2$ and by taking the limit as the parameter μ tends to infinity.

Note that from (4.215), we deduce that

$$P[T_d < \infty] = \lim_{t\to\infty} P[T_d \le t] = 2\,[1 - \Phi(0)] = 2\left(1 - \frac{1}{2}\right) = 1 \tag{4.218}$$

Therefore, the standard Brownian motion is certain to eventually reach any real value d. However, the formula (4.216) implies that

$$E[T_d] = \int_0^\infty \frac{|d|}{\sqrt{2\pi}t^{1/2}} \exp\left\{-\frac{d^2}{2t}\right\} dt = \infty \tag{4.219}$$

because

$$\frac{1}{t^{1/2}} \exp\left\{-\frac{d^2}{2t}\right\} \sim \frac{1}{t^{1/2}} \tag{4.220}$$

That is, the function that we integrate behaves like $t^{-1/2}$ as t tends to infinity. Given that

$$\int_{t_0}^\infty t^{-1/2}\, dt = \infty \quad \forall\, t_0 > 0 \tag{4.221}$$

we must conclude that the mathematical expectation of the random variable T_d is indeed infinite.

We also deduce from the formula (4.215) that

$$P[d^2 T_1 \le t] = 2\left[1 - \Phi\left(1\Big/\sqrt{t/d^2}\right)\right] = 2\left[1 - \Phi\left(|d|/\sqrt{t}\right)\right] \tag{4.222}$$

---

[7] Abraham Wald, 1902–1950, was born in Kolozsvár, Hungary (now Cluj, in Romania), and died (in a plane crash) in India. He first worked on geometry and then on econometrics. His main contributions were, however, to statistics, notably to *sequential analysis*.

Thus, the random variables T_d and $T_1^* := d^2 T_1$ are identically distributed.

Finally, let $X := 1/B^2(1)$. Since $B(1)$ has a Gaussian $N(0,1)$ distribution, we calculate, for $x > 0$,

$$P[X \leq x] = P[B^2(1) \geq 1/x] = 1 - P\left[-1/\sqrt{x} < B(1) < 1/\sqrt{x}\,\right]$$
$$= 1 - \left(2\,\Phi\left(1/\sqrt{x}\right) - 1\right) = 2\left(1 - \Phi\left(1/\sqrt{x}\right)\right) \qquad (4.223)$$

from which we can assert that T_1 and X are also identically distributed random variables.

Remark. The square of a standard Gaussian random variable Z has a *chi-square distribution* with 1 *degree of freedom*, which is also a particular gamma distribution: $Z^2 \sim G(1/2, 1/2)$. The random variable T_1 is thus the reciprocal of a variable having a gamma distribution.

A much more general technique to obtain the distribution of a *first-passage time T* for an arbitrary diffusion process consists in trying to solve the Kolmogorov backward equation satisfied by the density function of this random variable, under the appropriate conditions.

Let $\{X(t), t \geq 0\}$ be a time-homogeneous diffusion process whose infinitesimal parameters are $m(x)$ and $v(x)$. Its conditional transition density function, $p(x, x_0; t, t_0)$, satisfies

$$\frac{\partial p}{\partial t_0} + m(x_0)\frac{\partial p}{\partial x_0} + \frac{1}{2}v(x_0)\frac{\partial^2 p}{\partial x_0^2} = 0 \qquad (4.224)$$

If $t_0 = 0$, we can use the fact that the process is time-homogeneous, so that

$$\frac{\partial}{\partial t_0}p(x, x_0; t, t_0) = \frac{\partial}{\partial t_0}p(x, x_0; t - t_0) = -\frac{\partial}{\partial t}p(x, x_0; t - t_0) \qquad (4.225)$$

to write that

$$\frac{1}{2}v(x_0)\frac{\partial^2 p}{\partial x_0^2} + m(x_0)\frac{\partial p}{\partial x_0} = \frac{\partial p}{\partial t} \qquad (4.226)$$

Let $\rho(t; x_0)$ be the probability density function of

$$T_{c,d}\,(= T_{c,d}(x_0)) := \min\{t \geq 0 : X(t) \notin (c,d) \mid X(0) = x_0 \in [c,d]\} \qquad (4.227)$$

That is,

$$\rho(t; x_0)\,dt := P[T_{c,d} \in (t, t+dt] \mid X(0) = x_0 \in [c,d]\} \qquad (4.228)$$

We find that the function $\rho(t; x_0)$ satisfies the partial differential equation (4.226). Moreover, since the coefficients $m(x_0)$ and $v(x_0)$ do not depend on t, we can take the *Laplace transform* of the equation (with respect to t) to reduce it to an ordinary differential equation. Let

$$L(x_0; \alpha) := \int_0^\infty e^{-\alpha t} \rho(t; x_0) \, dt \qquad (4.229)$$

where α is a real positive constant.

Remark. The Laplace transform of the density function of the continuous and nonnegative random variable X is the moment-generating function $M_X(-\alpha)$ of this random variable (see p. 19).

We can check that the function $L(x_0; \alpha)$ satisfies the following differential equation:

$$\frac{1}{2} v(x_0) \frac{\partial^2 L}{\partial x_0^2} + m(x_0) \frac{\partial L}{\partial x_0} = \alpha L \qquad (4.230)$$

This equation is valid for $c < x_0 < d$. Indeed, we have

$$\int_0^\infty e^{-\alpha t} \frac{\partial}{\partial t} \rho(t; x_0) \, dt = e^{-\alpha t} \rho(t; x_0)\big|_0^\infty + \alpha \int_0^\infty e^{-\alpha t} \rho(t; x_0) \, dt$$

$$= 0 + \alpha L = \alpha L \qquad (4.231)$$

because $\rho(0; x_0) = 0$ if $x_0 \neq c$ or d. Since $T_{c,d} = 0$ if $x_0 = c$ or d, the boundary conditions are

$$L(x_0; \alpha) \equiv E\left[e^{-\alpha T_{c,d}} \mid X(0) = x_0 \right] = 1 \quad \text{if } x_0 = c \text{ or } d \qquad (4.232)$$

Once we have found the function $L(x_0; \alpha)$, we must invert the Laplace transform to obtain the density function of $T_{c,d}$.

Example 4.4.1. Let $\{X(t), t \geq 0\}$ be a Wiener process whose drift and diffusion coefficients are μ and σ^2, respectively. We must solve the following differential equation:

$$\frac{\sigma^2}{2} \frac{\partial^2 L}{\partial x_0^2} + \mu \frac{\partial L}{\partial x_0} = \alpha L$$

The general solution of this equation is given by

$$L(x_0; \alpha) = c_1 e^{x_0 r_1(\alpha)} + c_2 e^{x_0 r_2(\alpha)} \qquad (4.233)$$

where c_1 and c_2 are constants, and

$$r_1(\alpha) := \frac{1}{\sigma^2}\left(-\mu - \sqrt{\mu^2 + 2\alpha\sigma^2}\right) \quad \text{and} \quad r_2(\alpha) := \frac{1}{\sigma^2}\left(-\mu + \sqrt{\mu^2 + 2\alpha\sigma^2}\right)$$

The solution that satisfies the conditions $L(c; \alpha) = L(d; \alpha) = 1$ is

$$L(x_0; \alpha) = \frac{\left(e^{d\, r_2(\alpha)} - e^{c\, r_2(\alpha)}\right) e^{x_0 r_1(\alpha)} + \left(e^{c\, r_1(\alpha)} - e^{d\, r_1(\alpha)}\right) e^{x_0 r_2(\alpha)}}{e^{c\, r_1(\alpha) + d\, r_2(\alpha)} - e^{c\, r_2(\alpha) + d\, r_1(\alpha)}}$$

$$(4.234)$$

for $c \leq x_0 \leq d$.

Inverting the Laplace transform above is not easy. However, in this type of problem, we must often content ourselves with explicitly calculating the moment-generating function of the first-passage time considered.

Suppose, to simplify, that there is a single boundary, at d (which is tantamount to taking $c = -\infty$) and that $\mu = 0$. Since, from Eq. (4.232), $L(x_0; \alpha)$ must be in the interval $[0, 1]$, for any $x_0 \leq d$, we must discard the solution with $r_1(\alpha)$ in (4.233). Indeed, with $\mu = 0$, this solution is

$$c_1 e^{-\sqrt{2\alpha}x_0/\sigma} \longrightarrow c_1 \times \infty \quad \text{as } x_0 \longrightarrow -\infty$$

We must therefore choose the constant $c_1 = 0$. We then have

$$L(x_0; \alpha) = c_2 e^{\sqrt{2\alpha}x_0/\sigma}$$

and the condition $L(d; \alpha) = 1$ implies that

$$L(x_0; \alpha) = e^{\sqrt{2\alpha}(x_0-d)/\sigma} \quad \forall\, x_0 \leq d \qquad (4.235)$$

Remark. We obtain the same result by taking the limit as c decreases to $-\infty$ in the formula (4.234) (with $\mu = 0$).

We can check that the inverse Laplace transform of the function L in Eq. (4.235) is given by (see the inverse Gaussian distribution, p. 216)

$$f_{T_d}(t) = \frac{(d - x_0)}{\sqrt{2\pi\sigma^2 t^3}} \exp\left\{-\frac{(d - x_0)^2}{2\sigma^2 t}\right\} \quad \text{for } t > 0$$

The density function of the random variable T_d in the general case where $\mu \in \mathbb{R}$ is

$$f_{T_d}(t) = \frac{(d - x_0)}{\sqrt{2\pi\sigma^2 t^3}} \exp\left\{-\frac{(d - x_0 - \mu t)^2}{2\sigma^2 t}\right\} \quad \text{for } t > 0 \qquad (4.236)$$

and the function L becomes

$$L(x_0; \alpha) = \exp\left\{\frac{(d - x_0)}{\sigma^2}\left[\mu - (\mu^2 + 2\alpha\sigma^2)^{1/2}\right]\right\} \quad \text{for } x_0 \leq d \qquad (4.237)$$

After having obtained the moment-generating function of the random variable T_d, we can calculate the probability that this variable is finite, as follows:

$$P[T_d < \infty] = \lim_{\alpha \downarrow 0} L(x_0; \alpha) \qquad (4.238)$$

Example 4.4.2. In the preceding example, with a single boundary and $\mu = 0$, we obtain that

$$P[T_d < \infty] = \lim_{\alpha \downarrow 0} e^{\sqrt{2\alpha}(x_0-d)/\sigma} = 1$$

which generalizes (in the case when $d > 0$) the result already mentioned regarding the standard Brownian motion.

However, when $\mu \neq 0$, we deduce from the formula (4.237) that

$$P[T_d < \infty] = \lim_{\alpha \downarrow 0} \exp\left\{\frac{(d - x_0)}{\sigma^2}\left[\mu - (\mu^2 + 2\alpha\sigma^2)^{1/2}\right]\right\}$$

$$= \begin{cases} 1 & \text{if } \mu > 0 \\ e^{2\mu(d-x_0)/\sigma^2} & \text{if } \mu < 0 \end{cases}$$

To obtain the moments of the random variable T_d, we can use the formula (see p. 19)

$$E[T_d^n] = (-1)^n \frac{\partial^n}{\partial \alpha^n} L(x_0; \alpha)|_{\alpha=0} \tag{4.239}$$

for $n = 1, 2, \ldots$. Thus, in Example 4.4.1, if $c = -\infty$ and $\mu > 0$, we calculate [see the formula (4.237)]

$$E[T_d] = -\frac{\partial}{\partial \alpha} \exp\left\{\frac{(d - x_0)}{\sigma^2}\left[\mu - (\mu^2 + 2\alpha\sigma^2)^{1/2}\right]\right\}\Bigg|_{\alpha=0}$$

$$= -L(x_0; \alpha)\left(\frac{d - x_0}{\sigma^2}\right)(-1)(\mu^2 + 2\alpha\sigma^2)^{-1/2}\sigma^2\Bigg|_{\alpha=0}$$

$$= \frac{d - x_0}{\mu} \tag{4.240}$$

Remark. We can also try to solve the ordinary differential equation satisfied by the function

$$m_{n,d}(x_0) := E[T_d^n] \quad \text{for } n = 1, 2, \ldots \tag{4.241}$$

namely,

$$\frac{1}{2}v(x_0)\frac{d^2}{dx_0^2}m_{n,d}(x_0) + m(x_0)\frac{d}{dx_0}m_{n,d}(x_0) = -n\, m_{n-1,d}(x_0) \tag{4.242}$$

under the boundary condition $m_{n,d}(d) = 0$. In particular, we have

$$\frac{1}{2}v(x_0)\frac{d^2}{dx_0^2}m_{1,d}(x_0) + m(x_0)\frac{d}{dx_0}m_{1,d}(x_0) = -1 \tag{4.243}$$

When there are two boundaries, if we wish to calculate the probability that the process will hit the boundary at d before that at c, we can simply solve the ordinary differential equation

$$\frac{1}{2}v(x_0)\frac{d^2}{dx_0^2}p_d(x_0) + m(x_0)\frac{d}{dx_0}p_d(x_0) = 0 \tag{4.244}$$

where

$$p_d(x_0) := P[X(T_{c,d}(x_0)) = d] \qquad (4.245)$$

This differential equation is obtained from Eq. (4.230) by setting $\alpha = 0$. Indeed, we may write that

$$p(t; x_0) = \rho_c(t; x_0) + \rho_d(t; x_0) \qquad (4.246)$$

where $\rho_c(t; x_0)$ (respectively, $\rho_d(t; x_0)$) is the density function of the random variable T_c' (resp., T_d') denoting the time the process takes to go from x_0 to c (resp., d) without hitting d (resp., c). Let

$$L_d(x_0; \alpha) := \int_0^\infty e^{-\alpha t} \rho_d(t; x_0) \, dt \qquad (4.247)$$

We have

$$P[X(T_{c,d}(x_0)) = d] = P[T_d'(x_0) < \infty] = \int_0^\infty \rho_d(t; x_0) \, dt = L_d(x_0; 0) \quad (4.248)$$

Now, the function $L_d(x_0; \alpha)$ also satisfies Eq. (4.230).

The differential equation (4.244) is valid for $c < x_0 < d$. The boundary conditions are

$$p_d(d) = 1 \quad \text{and} \quad p_d(c) = 0 \qquad (4.249)$$

Example 4.4.3. If $\{W(t), t \geq 0\}$ is a Wiener process with diffusion coefficient σ^2, then we must solve

$$\frac{\sigma^2}{2} \frac{d^2}{dx_0^2} p_d(x_0) = 0$$

We find at once that

$$p_d(x_0) = c_1 x_0 + c_0$$

where c_0 and c_1 are constants. The boundary conditions imply that

$$p_d(x_0) = \frac{x_0 - c}{d - c} \quad \text{for } c \leq x_0 \leq d$$

Note that this formula does not depend on σ^2. Moreover, if $c = -d$, then we have that $p_d(0) = 1/2$, which could have been predicted, by symmetry.

Until now, we only considered the time a diffusion process takes to reach a boundary or either of two boundaries. We can define these boundaries as being *absorbing*. If the boundary at c is *reflecting*, it can be shown that the boundary condition becomes

$$\frac{\partial}{\partial x_0} L(x_0; \alpha)\Big|_{x_0=c} = 0 \tag{4.250}$$

Finally, we can also try to calculate first-passage time distributions in two or more dimensions. However, the problem of solving the appropriate partial differential equation is generally very difficult.

Example 4.4.4. If $\{B(t), t \geq 0\}$ is a standard Brownian motion, starting from b_0, and if $\{X(t), t \geq 0\}$ is its integral, we find that the Kolmogorov backward equation satisfied by the function

$$\rho(t; b_0, x_0) := \frac{P[T_d(b_0, x_0) \in (t, t + dt]]}{dt}$$

where

$$T_d(b_0, x_0) := \min\{t > 0 \colon X(t) = d > 0 \mid B(0) = b_0, X(0) = x_0 < d\}$$

is the following:

$$\frac{1}{2}\frac{\partial^2 \rho}{\partial b_0^2} + b_0 \frac{\partial \rho}{\partial x_0} = \frac{\partial \rho}{\partial t}$$

The main difficulty is due to the fact that the process $\{X(t), t \geq 0\}$ cannot hit the boundary $X(t) = d$ (for the first time) with $B(t) \leq 0$. Indeed, we deduce from the formula

$$X(t) = x_0 + \int_0^t B(s)\, ds \quad \Longrightarrow \quad \frac{d}{dt} X(t) = B(t)$$

that if $B(t)$ takes on a negative value, then $X(t)$ decreases. Consequently, the process $\{X(t), t \geq 0\}$ can attain a value d (> 0) greater than x_0 at time $T_d(b_0, x_0)$ only if $B[T_d(b_0, x_0)] > 0$, so that the function ρ is not continuous on the boundary.

4.5 Exercises

Section 4.1

Remark. In the following exercises, the process $\{B(t), t \geq 0\}$ is always a standard Brownian motion.

Question no. 1
   We define

$$X(t) = B^2(t) \quad \text{for } t \geq 0$$

(a) Is the stochastic process $\{X(t), t \geq 0\}$ a Wiener process? Justify.

(b) Is $\{X(t), t \geq 0\}$ (i) wide-sense stationary? (ii) mean ergodic? Justify.

Question no. 2
We consider the stochastic process $\{X(t), t \geq 0\}$ defined by

$$X(t) = B(t) + B(t^2) \quad \text{for } t \geq 0$$

(a) Calculate the mean of $X(t)$.

(b) Calculate $\text{Cov}[X(t), X(t + \tau)]$, for $\tau \geq 0$.

(c) Is the stochastic process $\{X(t), t \geq 0\}$ (i) Gaussian? (ii) stationary? (iii) a Brownian motion? Justify.

(d) Calculate the correlation coefficient of $B(t)$ and $B(t^2)$, for $t > 0$.

Question no. 3
Calculate the variance of the random variable $X := B(t) - 2B(\tau)$, for $0 \leq t \leq \tau$.

Question no. 4
Let $X(t) := |B(t)|$, for $t \geq 0$. Is the stochastic process $\{X(t), t \geq 0\}$
(a) Gaussian? (b) stationary? Justify.

Question no. 5
Let $\{X(t), t \geq 0\}$ be the stochastic process defined by

$$X(t) = B(t + 1) - B(1) \quad \text{for } t \geq 0$$

(a) Calculate the autocovariance function of the process $\{X(t), t \geq 0\}$.

(b) Is the process $\{X(t), t \geq 0\}$ (i) Gaussian? (ii) a standard Brownian motion? (iii) stationary? (iv) mean ergodic? Justify.

Question no. 6
Calculate the variance of $X := B(4) - 2B(1)$.

Question no. 7
Is the stochastic process $\{X(t), t \geq 0\}$ defined by

$$X(t) = -B(t) \quad \text{for } t \geq 0$$

Gaussian? Is it a Brownian motion? Justify.

Question no. 8
Let $\{X(t), t \geq 0\}$ be a Gaussian process such that $X(0) = 0$, $E[X(t)] = \mu t$
if $t > 0$, where $\mu \neq 0$, and

$$R_X(t, t + \tau) = 2t + \mu^2 t(t + \tau) \quad \text{for } t, \tau \geq 0$$

Is the stochastic process $\{Y(t), t \geq 0\}$, where $Y(t) := X(t) - \mu t$, a Brownian motion? Justify.

Question no. 9

Is the Wiener process mean ergodic? Justify.

Question no. 10

We set

$$Y(t) = \frac{1}{t}B^2(t) \quad \text{for } t > 0$$

(a) Is the process $\{Y(t), t > 0\}$ Gaussian? Justify.

(b) Calculate the mean of $Y(t)$.

(c) Calculate $\text{Cov}[Y(s), Y(t)]$, for $0 < s \leq t$.

Indication. If $X \sim N(0, \sigma_X^2)$ and $Y \sim N(0, \sigma_Y^2)$, then

$$E[X^2 Y^2] = E[X^2]E[Y^2] + 2\left(E[XY]\right)^2$$

(d) Is $\{Y(t), t > 0\}$ a wide-sense stationary process? Justify.

Question no. 11

Consider the stochastic process $\{X(t), t \geq 0\}$ defined by

$$X(t) = B(t) - B([t])$$

where $[t]$ denotes the integer part of t.

(a) Calculate the mean of $X(t)$.

(b) Calculate $\text{Cov}[X(t_1), X(t_2)]$, for $t_2 > t_1$. Are the random variables $X(t_1)$ and $X(t_2)$ independent? Justify.

(c) Is the stochastic process $\{X(t), t \geq 0\}$ (i) stationary? (ii) a Brownian motion? Justify.

Question no. 12

At each time unit, the standard Brownian motion $\{B(t), t \geq 0\}$ is shifted to $[B(n)]$, where $[\ ]$ denotes the integer part. Let X_n be the position at time n, for $n = 0, 1, 2, \ldots$. Then the process $\{X_n, n = 0, 1, \ldots\}$ is a (discrete-time) Markov chain. Calculate (a) $p_{i,j}$, for $i, j > 0$ and (b) $P[X_1 = 0, X_2 = 0]$.

Question no. 13

We define

$$X(t) = B(\ln(t+1)) \quad \text{for } t \geq 0$$

(a) Is the stochastic process $\{X(t), t \geq 0\}$ a Brownian motion? Justify.

(b) Calculate $E[X^2(t) \mid X(t) > 0]$.

Question no. 14

Let $\{X(t), t \geq 0\}$ be the stochastic process defined by

$$X(t) = \frac{B(t+\epsilon) - B(t)}{\epsilon} \quad \forall\, t \geq 0$$

where ϵ is a positive constant.

(a) Calculate $C_X(t, t+s)$, for $s, t \geq 0$.

(b) Is the process $\{X(t), t \geq 0\}$ (i) Gaussian? (ii) stationary? (iii) a Brownian motion? (iv) mean ergodic? Justify.

Question no. 15
Let X_1, X_2, ... be independent and identically distributed random variables such that $p_{X_1}(x) = 1/2$ if $x = -1$ or 1. We define $Y_0 = 0$ and

$$Y_n = \sum_{k=1}^{n} X_k \quad \text{for } n = 1, 2, \ldots$$

Then the stochastic process $\{Y_n, n = 0, 1, \ldots\}$ is a Markov chain (see p. 85).
We propose to use a standard Brownian motion, $\{B(t), t \geq 0\}$, to approximate the stochastic process $\{Y_n, n = 0, 1, \ldots\}$. Compare the exact value of $P[Y_{30} = 0]$ to $P[-1 < B(30) < 1]$.

Indication. We have that $P[N(0, 1) \leq 0.18] \simeq 0.5714$.

Section 4.2

Question no. 16
Let $\{U(t), t \geq 0\}$ be an Ornstein–Uhlenbeck process defined by

$$U(t) = e^{-t} B(e^{2t})$$

(a) What is the distribution of $U(1) + U(2)$?

(b) We set

$$V(t) = \int_0^t U(s) \, ds$$

(i) Calculate the mean and the variance of $V(t)$.
(ii) Is the process $\{V(t), t \geq 0\}$ Gaussian? Justify.

Question no. 17
Suppose that $\{X(t), t \geq 0\}$ is a Wiener process with drift coefficient M and diffusion coefficient $\sigma^2 = 1$, where M is a random variable having a uniform distribution on the interval $[0, 1]$.

(a) Calculate $E[X(t)]$ and $\text{Cov}[X(s), X(t)]$, for $s, t \geq 0$.

(b) Is the process wide-sense stationary? Justify.

Question no. 18
Let $\{Y(t), t \geq 0\}$ be a geometric Brownian motion.

(a) Show that the density function of the random variable $Y(t)$ is given by the formula (4.52).

(b) Is the process $\{Y(t), t \geq 0\}$ stationary? Justify.

Question no. 19

Let $\{X(t), t \geq 0\}$ be a Wiener process with drift coefficient $\mu > 0$ and diffusion coefficient $\sigma^2 = 1$.

(a) (i) Calculate, as explicitly as possible, $E[X(t) \mid X(t) > 0]$ in terms of $Q(x) := P[N(0,1) > x]$.

(ii) Calculate $E[X(t)X(t+s)]$, for $s, t \geq 0$.

(b) Let $Z(0) = 0$ and

$$Z(t) := \frac{X(t)}{\mu t} - 1 \quad \text{for } t > 0$$

Is the stochastic process $\{Z(t), t \geq 0\}$ a Brownian motion? Justify.

Section 4.3

Question no. 20

(a) Calculate the covariance of X and Y, where

$$X := \int_{-1}^{1} t \, dB(t) \quad \text{and} \quad Y := \int_{-1}^{1} t^2 \, dB(t)$$

(b) Are the random variables X and Y independent? Justify.

Question no. 21

We define the stochastic process $\{Y(t), t \geq 0\}$ by

$$Y(t) = \int_{0}^{t} s \, dB(s) \quad \text{for } t \geq 0$$

(a) Calculate the autocovariance function of the process $\{Y(t), t \geq 0\}$.

(b) Is the process $\{Y(t), t \geq 0\}$ Gaussian? Justify.

(c) Let $\{Z(t), t \geq 0\}$ be the stochastic process defined by $Z(t) = Y^2(t)$. Calculate its mean.

Question no. 22

Calculate the mean and the autocovariance function of the stochastic process $\{Y(t), t \geq 0\}$ defined by

$$Y(t) = e^{ct} \int_{0}^{t} e^{-cs} \, dW(s) \quad \text{for } t \geq 0$$

where c is a real constant.

Section 4.4

Question no. 23

Let $\{W(t), t \geq 0\}$ be a Brownian motion with infinitesimal parameters $\mu = 0$ and σ^2 (> 0). We set

$$Y(t) = W(t^2) \quad \text{for } t \geq 0$$

(a) Is the stochastic process $\{Y(t), t \geq 0\}$ a Brownian motion? Justify.

(b) Calculate $\text{Cov}[Y(s), Y(t)]$, for $0 \leq s < t$.

(c) Let $T_d := \min\{t > 0 : Y(t) = d > 0\}$. Calculate the probability density function of the random variable T_d.

Question no. 24

Let $\{Y(t), t \geq 0\}$ be the stochastic process defined in Question no. 21.

(a) Calculate the distribution of $Y(1) + Y(2)$.

(b) Let $T_1 := \min\{t > 0 : Y(t) = 1\}$. Calculate the probability density function of the random variable T_1.

Question no. 25

We consider the process $\{X(t), t \geq 1\}$ defined by

$$X(t) = e^{-1/t} B(e^{1/t}) \quad \text{for } t \geq 1$$

Suppose that $B(e) = 0$, so that $X(1) = 0$.

(a) Calculate $E[X(t)]$ and $\text{Cov}[X(t), X(t+s)]$, for $t \geq 1, s \geq 0$.

(b) Is the stochastic process $\{X(t), t \geq 1\}$ (i) Gaussian? (ii) an Ornstein–Uhlenbeck process? (iii) stationary? Justify.

(c) Let $T_d := \min\{t > 1 : X(t) = d > 0\}$. Calculate $f_{T_d}(t)$.

Question no. 26

Let $\{Z(t), 0 \leq t \leq 1\}$ be a Brownian bridge. We define

$$Y(t) = \int_0^t Z(\tau) \, d\tau \quad \text{for } 0 \leq t \leq 1$$

(a) Calculate $E[Y(t)]$ and $\text{Cov}[Y(t), Y(t+s)]$, for $0 \leq t \leq 1, s \geq 0, s+t \leq 1$.

(b) Is the stochastic process $\{Y(t), 0 \leq t \leq 1\}$ (i) Gaussian? (ii) stationary? (iii) a Brownian bridge? Justify.

(c) Calculate approximately, if $d > 0$ is small, the probability that the process $\{Y(t), 0 \leq t \leq 1\}$ will reach d in the interval $(0, 1)$.

Question no. 27

Let T_c be the first-passage time to the origin for a standard Brownian motion starting from $c > 0$. We define $S = 1/T_c$.

(a) Calculate the probability density function of S. What is the distribution of the random variable S?

(b) Find a number b (in terms of the constant c) for which we have $P[S \leq b] = 0.5$.

Question no. 28

The nonstationary Ornstein–Uhlenbeck process $\{X(t), t \geq 0\}$ is a Gaussian process such that, if $\sigma = 1$ (see Subsection 4.2.5),

$$E[X(t)] = X(0)e^{-\alpha t} \quad \text{and} \quad V[X(t)] = \frac{1 - e^{-2\alpha t}}{2\alpha}$$

where $\alpha > 0$ is a parameter. Suppose that $X(0) = d > 0$. We define $T_0(d) = \min\{t > 0 : X(t) = 0\}$. Calculate $f_{T_0(d)}(t)$ when $\alpha = 1/2$.

Question no. 29

Let $\{W(t), t \geq 0\}$ be a Brownian motion with drift coefficient μ and diffusion coefficient σ^2. We assume that the flow of a certain river can be modeled by the process $\{X(t), t \geq 0\}$ defined by

$$X(t) = e^{W(t)+k} \quad \forall\, t \geq 0$$

where k is a constant. Next, let d be a value of the flow above which the risk of flooding is high. Suppose that $X(0) = d/3$. Calculate the probability that the flow will reach the critical value d in the interval $(0, 1]$ if $\mu \geq 0$ and $\sigma = 1$.

Question no. 30

Let $\{X(t), t \geq 0\}$ be an Ornstein–Uhlenbeck process for which $\alpha = 1$ and $\sigma^2 = 2$, and let $\{Y(t), t \geq 0\}$ be the process defined by

$$Y(t) = \int_0^t X(s)\, ds$$

Finally, we set
$$Z(t) = X(t) + Y(t) \quad \text{for } t \geq 0$$

(a) Calculate $E[Z(t)]$.

(b) Calculate $\text{Cov}[Z(t), Z(t+s)]$, for $s, t \geq 0$.

Indication. We find that $\text{Cov}[X(t), X(t+s)] = e^{-s}$ and

$$\text{Cov}[Y(t), Y(t+s)] = 2t - 1 + e^{-t} - e^{-s} + e^{-(s+t)} \quad \text{for } s, t \geq 0$$

(c) Is the stochastic process $\{Z(t), t \geq 0\}$ stationary? Justify.

(d) Let $T_d(z) := \min\{t > 0 : Z(t) = d \mid Z(0) = z\ (< d)\}$. Calculate the probability density function of $T_d(z)$.

Question no. 31

Let $\{X(t), t \geq 0\}$ be a Brownian motion with drift coefficient μ and diffusion coefficient σ^2.

(a) Suppose that $\mu = 0$ and that σ^2 is actually a random variable V having a uniform distribution on the interval $(0, 1)$.
   (i) Calculate $E[X(t)]$ and $\text{Cov}[X(s), X(t)]$, for $s, t \geq 0$.
   (ii) Is the stochastic process $\{X(t), t \geq 0\}$ Gaussian and stationary? Justify.

(b) Suppose now that $\mu > 0$ and $\sigma^2 = 1$. Let $T_{-1,1}(x_0)$ be the time the process takes to attain 1 or -1, starting from $x_0 \in [-1, 1]$.
   (i) It can be shown that $m_{-1,1}(x_0) := E[T_{-1,1}(x_0)]$ satisfies the ordinary differential equation

$$\frac{1}{2} \frac{d^2}{dx_0^2} m_{-1,1}(x_0) + \mu \frac{d}{dx_0} m_{-1,1}(x_0) = -1$$

Solve this differential equation, subject to the appropriate boundary conditions, to obtain $m_{-1,1}(x_0)$ explicitly.
   (ii) Similarly, the function $p_{-1,1}(x_0) := P[X(T_{-1,1}(x_0)) = 1]$ is a solution of

$$\frac{1}{2} \frac{d^2}{dx_0^2} p_{-1,1}(x_0) + \mu \frac{d}{dx_0} p_{-1,1}(x_0) = 0$$

Obtain an explicit formula for $p_{-1,1}(x_0)$.

Poisson Processes

5.1 The Poisson process

We already mentioned the Poisson process in Chapters 2 and 3. It is a particular continuous-time Markov chain. In Chapter 4, we asserted that the Wiener process and the Poisson process are the two most important stochastic processes for applications. The Poisson process is notably used in the basic queueing models.

The Poisson process, which will be denoted by $\{N(t), t \geq 0\}$, is also a *pure birth* process (see Subsection 3.3.4). That is, $N(t)$ designates the number of births (or of events, in general) that occurred from 0 up to time t. A process of this type is called a *counting process*.

Definition 5.1.1. *Let $N(t)$ be the number of events that occurred in the interval $[0, t]$. The stochastic process $\{N(t), t \geq 0\}$ is called a* **counting process**.

Counting processes have the following properties, which are deduced directly from their definition.

Properties. i) $N(t)$ is a random variable whose possible values are $0, 1, \ldots$.
ii) The function $N(t)$ is nondecreasing: $N(t_2) - N(t_1) \geq 0$ if $t_2 > t_1 \geq 0$. Moreover, $N(t_2) - N(t_1)$ is the number of events that occurred in the interval $(t_1, t_2]$.

Definition 5.1.2. *A* **Poisson process** *with rate λ (> 0) is a counting process $\{N(t), t \geq 0\}$ having* independent increments *(see p. 50), for which $N(0) = 0$ and*

$$N(\tau + t) - N(\tau) \sim \text{Poi}(\lambda t) \quad \forall \, \tau, t \geq 0 \tag{5.1}$$

Remarks. We deduce from the preceding formula that a Poisson process also has *stationary increments* (see p. 50), because the distribution of $N(\tau + t) - N(\tau)$ does not depend on τ. Moreover, by taking $\tau = 0$, we may write that

$$N(t) = N(0 + t) - N(0) \sim \text{Poi}(\lambda t) \quad \forall \, t \geq 0 \tag{5.2}$$

ii) If we specify in the definition that $\{N(t), t \geq 0\}$ is a process with independent and *stationary* increments, then we may replace the formula (5.1) by the following conditions:

$$P[N(\delta) = 1] = \lambda\delta + o(\delta) \tag{5.3}$$
$$P[N(\delta) = 0] = 1 - \lambda\delta + o(\delta) \tag{5.4}$$

where $o(\delta)$ is such that (see p. 125)

$$\lim_{\delta \downarrow 0} \frac{o(\delta)}{\delta} = 0 \tag{5.5}$$

Thus, the probability that there will be *exactly* one event in an interval of length δ must be *proportional* to the length of the interval plus a term that is negligible if δ is sufficiently small. Furthermore, we have

$$P[N(\delta) \geq 2] = 1 - \{P[N(\delta) = 0] + P[N(\delta) = 1]\} = o(\delta) \tag{5.6}$$

Therefore, it is not impossible that there will be two or more events between an arbitrary t_0 and $t_0 + \delta$. However, it is very unlikely when δ is small.

It is not difficult to show that if $N(\delta) \sim \text{Poi}(\lambda\delta)$, then the conditions (5.3) and (5.4) are satisfied. We have

$$P[\text{Poi}(\lambda\delta) = 0] = e^{-\lambda\delta} = 1 - \lambda\delta + \frac{(\lambda\delta)^2}{2!} + \ldots = 1 - \lambda\delta + o(\delta) \tag{5.7}$$

and

$$P[\text{Poi}(\lambda\delta) = 1] = \lambda\delta e^{-\lambda\delta} = \lambda\delta[1 - \lambda\delta + o(\delta)] = \lambda\delta + o(\delta) \tag{5.8}$$

As will be seen in Section 5.2, in the more general case where $\lambda = \lambda(t)$, it can also be shown that if the conditions (5.3) and (5.4) are satisfied (and if the increments of $\{N(t), t \geq 0\}$ are stationary), then the formula (5.1) is valid. Consequently, we have two ways of determining whether a given stochastic process is a Poisson process.

Since the random variable $N(t)$ has a Poisson distribution with parameter λt, for all $t \geq 0$, we have

$$E[N(t)] = \lambda t \tag{5.9}$$
$$E[N^2(t)] = V[N(t)] + (E[N(t)])^2 = \lambda t + \lambda^2 t^2 \tag{5.10}$$

As we did in the case of the Wiener process, we use the fact that the increments of the Poisson process are independent (and stationary) to calculate its autocorrelation function. We may write, with the help of the formula (5.10), that

$$R_N(t, t+s) := E[N(t)N(t+s)] = E[N(t)\{N(t+s) - N(t)\}] + E[N^2(t)]$$
$$\overset{\text{ind.}}{=} E[N(t)]E[N(t+s) - N(t)] + E[N^2(t)]$$
$$= \lambda t \lambda s + (\lambda t + \lambda^2 t^2) = \lambda^2 t(t+s) + \lambda t \tag{5.11}$$

It follows that

$$C_N(t, t+s) = R_N(t, t+s) - \lambda t[\lambda(t+s)] = \lambda t \tag{5.12}$$

For arbitrary values of t_1 and t_2, the autocovariance function of the Poisson process with rate $\lambda > 0$ is given by

$$C_N(t_1, t_2) = \lambda \min\{t_1, t_2\} \tag{5.13}$$

Remarks. i) Note that the formula above is similar to that obtained for the Wiener process (see p. 178). Actually, the stochastic process $\{N^*(t), t \geq 0\}$ defined by

$$N^*(t) = N(t) - \lambda t \quad \text{for } t \geq 0 \tag{5.14}$$

has zero mean and an autocovariance function identical to that of a Wiener process with diffusion coefficient $\sigma^2 = \lambda$. We can also assert that the Poisson process, $\{N(t), t \geq 0\}$, and the Brownian motion with drift, $\{X(t), t \geq 0\}$, have the same mean and the same autocovariance function if $\mu = \sigma^2 = \lambda$. Moreover, by the *central limit theorem*, we may write that

$$\text{Poi}(\lambda t) \approx \text{N}(\lambda t, \lambda t) \tag{5.15}$$

if λt is sufficiently large.

ii) Since the mean of the Poisson process depends on the variable t, this process is not even wide-sense stationary (see p. 53), even though its increments are stationary. Furthermore, its autocovariance function depends on t_1 *and* on t_2, and not only on $|t_2 - t_1|$.

Example 5.1.1. Suppose that the failures of a certain machine occur according to a Poisson process with rate $\lambda = 2$ per week and that exactly two failures occurred in the interval $[0, 1]$. Let t_0 (> 3) be an arbitrary value of t.

(a) What is the probability that, at time t_0, (at least) two weeks have elapsed since (i) the last failure occurred? (ii) the penultimate failure occurred?

(b) What is the probability that there will be no failures during the two days beginning with t_0 if exactly one failure occurred (in all) over the last two weeks?

Solution. In order to solve a problem on the Poisson process, we must first determine the value of the parameter of the Poisson *distribution* for each question asked. Let $N(t)$ be the number of failures in the interval $[0, t]$, where

t is measured in weeks. Since the average arrival rate of the failures is equal to two per week, we have

$$N(t) \sim \text{Poi}(2t)$$

(a) (i) In this question, we are interested in the number of failures during a two-week interval. As the Poisson process has stationary increments, we can calculate the probability asked for by considering the random variable $N(2) \sim$ Poi(4). We seek

$$P[N(2) = 0] = e^{-4} \simeq 0.0183 \tag{5.16}$$

(ii) Two weeks or more have elapsed since the penultimate failure if and only if the number of failures in the interval $[t_0 - 2, t_0]$ is either 0 or 1. Therefore, we seek

$$P[N(2) \leq 1] = e^{-4}(1 + 4) \simeq 0.0916$$

(b) Since the Poisson process has independent increments, the fact that there has been exactly one failure over the last two weeks does not matter. We are interested in the value of

$$N(t_0 + 2/7) - N(t_0) \overset{\text{d}}{=} N(2/7) \sim \text{Poi}(4/7)$$

We calculate

$$P[N(2/7) = 0] = e^{-4/7} \simeq 0.5647$$

Remarks. i) We could have written instead that $\lambda = 2/7$ per day, and then, in (b), we would have had that $N(2) \sim$ Poi(4/7).

ii) In this problem, we assume that the rate of the Poisson process is the same for every day of the week. In practice, this rate is probably different on Sundays than on Mondays, for instance. It also probably varies at night from its value during the day. If we want to make the problem more realistic, we must use a parameter λ that is not a constant, but rather a function of t, which will be done in Section 5.2.

Proposition 5.1.1. *Let $\{N_1(t), t \geq 0\}$ and $\{N_2(t), t \geq 0\}$ be two independent Poisson processes, with rates λ_1 and λ_2, respectively. The process $\{N(t), t \geq 0\}$ defined by*

$$N(t) = N_1(t) + N_2(t) \quad \forall\, t \geq 0 \tag{5.17}$$

is a Poisson process with rate $\lambda := \lambda_1 + \lambda_2$.

Proof. First, we have

$$N(0) := N_1(0) + N_2(0) = 0 + 0 = 0 \tag{5.18}$$

as required.

Next, given that the increments of the processes $\{N_1(t), t \geq 0\}$ and $\{N_2(t), t \geq 0\}$ are independent, we can check that so are those of the process $\{N(t), t \geq 0\}$. We simply have to write that

$$N(t_k) - N(t_{k-1}) = [N_1(t_k) - N_1(t_{k-1})] + [N_2(t_k) - N_2(t_{k-1})] \qquad (5.19)$$

Finally, the sum of independent Poisson *random variables*, with parameters α_1 and α_2, also has a Poisson distribution, with parameter $\alpha := \alpha_1 + \alpha_2$. Indeed, if $X_i \sim \mathrm{Poi}(\alpha_i)$, then its moment-generating function is given by (see Ex. 1.2.8)

$$M_{X_i}(t) = e^{-\alpha_i} \exp\left\{e^t \alpha_i\right\} \qquad (5.20)$$

It follows that

$$M_{X_1 + X_2}(t) \stackrel{\text{ind.}}{=} M_{X_1}(t) M_{X_2}(t) = e^{-(\alpha_1 + \alpha_2)} \exp\left\{e^t(\alpha_1 + \alpha_2)\right\} \qquad (5.21)$$

Thus, we may write that

$$N(\tau + t) - N(\tau) = [N_1(\tau + t) - N_1(\tau)] + [N_2(\tau + t) - N_2(\tau)]$$
$$\sim \mathrm{Poi}\left((\lambda_1 + \lambda_2)t\right) \quad \forall\, \tau, t \geq 0 \quad \square \qquad (5.22)$$

The preceding proposition can, of course, be generalized to the case when we add j independent Poisson processes, for $j = 2, 3, \ldots$. Conversely, we can decompose a Poisson process $\{N(t), t \geq 0\}$ with rate λ into j independent Poisson processes with rates λ_i ($i = 1, 2, \ldots, j$), where $\lambda_1 + \ldots + \lambda_j = \lambda$, as follows: suppose that each event that occurs is classified, *independently* from the other events, of type i with probability p_i, for $i = 1, \ldots, j$, where $p_1 + \ldots + p_j = 1$. Let $N_i(t)$ be the number of type i events in the interval $[0, t]$. We have the following proposition.

Proposition 5.1.2. *The stochastic processes $\{N_i(t), t \geq 0\}$ defined above are independent Poisson processes, with rates $\lambda_i := \lambda p_i$, for $i = 1, \ldots, j$.*

Proof. Since $N(t) = \sum_{i=1}^{j} N_i(t)$, we may write that

$$P\left[N_1(t) = n_1, \ldots, N_j(t) = n_j\right]$$
$$= \sum_{k=0}^{\infty} P\left[N_1(t) = n_1, \ldots, N_j(t) = n_j \mid N(t) = k\right] P[N(t) = k]$$
$$= P\left[N_1(t) = n_1, \ldots, N_j(t) = n_j \mid N(t) = n_1 + \ldots + n_j\right]$$
$$\times P[N(t) = n_1 + \ldots + n_j] \qquad (5.23)$$

Let $n := n_1 + \ldots + n_j$. The conditional probability above is given by

$$P\left[N_1(t) = n_1, \ldots, N_j(t) = n_j \mid N(t) = n\right] = \frac{n!}{n_1! \times \ldots \times n_j!} \prod_{i=1}^{j} p_i^{n_i} \qquad (5.24)$$

This is an application of the *multinomial distribution* (see p. 4), which generalizes the binomial distribution. It follows that

$$P\left[N_1(t) = n_1, \ldots, N_j(t) = n_j\right] = \left(\frac{n!}{n_1! \times \ldots \times n_j!} \prod_{i=1}^{j} p_i^{n_i}\right) e^{-\lambda t} \frac{(\lambda t)^n}{n!}$$

$$= \prod_{i=1}^{j} e^{-\lambda p_i t} \frac{(\lambda p_i t)^{n_i}}{n_i!} \tag{5.25}$$

from which we find that

$$P\left[N_i(t) = n_i\right] = e^{-\lambda p_i t} \frac{(\lambda p_i t)^{n_i}}{n_i!} \quad \text{for } i = 1, \ldots, j \tag{5.26}$$

and

$$P\left[N_1(t) = n_1, \ldots, N_j(t) = n_j\right] = \prod_{i=1}^{j} P\left[N_i(t) = n_i\right] \tag{5.27}$$

Therefore, we can assert that the random variables $N_i(t)$ have Poisson distributions with parameters $\lambda p_i t$, for $i = 1, \ldots, j$, and are independent.

Now, given that $\{N(t), t \geq 0\}$ has independent and stationary increments, the processes $\{N_i(t), t \geq 0\}$ have independent and stationary increments $\forall i$ as well. It follows that $N_i(\tau + t) - N_i(\tau) \sim \text{Poi}(\lambda p_i t)$, for $i = 1, \ldots, j$.

Finally, since $N(0) = 0$, we have that $N_1(0) = \ldots = N_j(0) = 0$. Hence, we may conclude that the processes $\{N_i(t), t \geq 0\}$ satisfy all the conditions in Definition 5.1.2 and are independent. □

Remarks. i) We can use the other way of characterizing a Poisson process, namely the one when we calculate the probability of the number of events in an interval of length δ. We have

$$P\left[N_i(\delta) = 1\right] = P\left[N_i(\delta) = 1 \mid N(\delta) = 1\right] P\left[N(\delta) = 1\right]$$
$$+ P\left[N_i(\delta) = 1 \mid N(\delta) > 1\right] P\left[N(\delta) > 1\right]$$
$$= p_i e^{-\lambda \delta} \lambda \delta + o(\delta) = p_i[\lambda \delta + o(\delta)] + o(\delta)$$
$$= \lambda p_i \delta + o(\delta)$$

Moreover, we may write that

$$P\left[N_i(\delta) = 0\right] = 1 - \lambda p_i \delta + o(\delta) \tag{5.28}$$

because $N_i(\delta) \leq N(\delta)$, for all i, which implies that

$$P\left[N_i(\delta) \geq 2\right] \leq P\left[N(\delta) \geq 2\right] = o(\delta) \tag{5.29}$$

We could complete the proof and show that the processes $\{N_i(t), t \geq 0\}$ are indeed *independent* Poisson processes, with rates λp_i, for $i = 1, \ldots, j$.

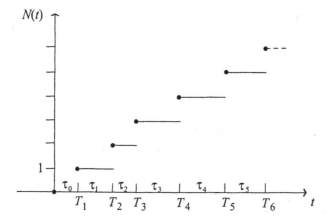

Fig. 5.1. Example of a trajectory of a Poisson process.

ii) The fact than an arbitrary event is classified of type i must not depend on the *time* at which this event occurred. We will return to this case further on in this section.

Since a Poisson process is a particular continuous-time Markov chain, we can assert that the time τ_i that the process spends in state $i \in \{0, 1, \ldots\}$ has an *exponential* distribution with parameter $\nu_i > 0$ (see p. 121) and that the random variables τ_0, τ_1, ... are *independent* (by the Markov property). Furthermore, because the Poisson process has *independent* and *stationary* increments, the process starts anew, from a probabilistic point of view, from any time instant. It follows that the τ_i's, for $i = 0, 1, \ldots$, are *identically distributed*. All that remains to do is thus to find the common parameter of the random variables τ_i. We have

$$P[\tau_0 > t] = P[N(t) = 0] = P[\text{Poi}(\lambda t) = 0] = e^{-\lambda t} \tag{5.30}$$

$$\implies \quad f_{\tau_0}(t) = \frac{d}{dt}\left(1 - e^{-\lambda t}\right) = \lambda e^{-\lambda t} \quad \text{for } t \geq 0 \tag{5.31}$$

That is, $\tau_0 \sim \text{Exp}(\lambda)$. Therefore, we can state the following proposition.

Proposition 5.1.3. *Let $\{N(t), t \geq 0\}$ be a Poisson process with rate λ, and let τ_i be the time that the process spends in state i, for $i = 0, 1, \ldots$. The random variables τ_0, τ_1, ... are* independent *and τ_i has an exponential $Exp(\lambda)$ distribution, for all i.*

Notation. We designate by T_1, T_2, ... the *arrival times* of the events of the Poisson process $\{N(t), t \geq 0\}$ (see Fig. 5.1).

Corollary 5.1.1. *In a Poisson process with rate λ, the time needed to obtain a total of n events, from any time instant, has a gamma distribution with parameters $\alpha = n$ and λ.*

Proof. The arrival time T_n of the nth event, from the initial time 0, can be represented as follows (because $\{N(t), t \geq 0\}$ is a continuous-time stochastic process):

$$T_n = \sum_{i=0}^{n-1} \tau_i \quad \text{for } n = 1, 2, \ldots \tag{5.32}$$

Since the τ_i's are independent random variables, all of which having an exponential distribution with parameter λ, we can indeed assert that $T_n \sim G(\alpha = n, \lambda)$ (see Prop. 3.3.6). Moreover, we deduce from the fact that the Poisson process has *independent* and *stationary* increments that

$$P[T_{n+k} \leq t + t_k \mid N(t_k) = k] = P[T_n \leq t] \quad \text{for all } k = 0, 1, \ldots \tag{5.33}$$

That is, if we know that at a given time t_k there have been exactly k events since the initial time, then the time needed for n additional events to occur, from t_k, has the same distribution as T_n. $\square$

The preceding results and the following relation:

$$N(t) \geq n \quad \Longleftrightarrow \quad T_n \leq t \tag{5.34}$$

provide us with yet another way of defining a Poisson process. This alternative definition may be easier to check in some cases.

Proposition 5.1.4. *Let $X_i \sim Exp(\lambda)$, for $i = 1, 2, \ldots$, be independent random variables, and let $T_0 := 0$ and*

$$T_n := \sum_{i=1}^{n} X_i \quad \text{for } n = 1, 2, \ldots \tag{5.35}$$

We set

$$N(t) = \max\{n \geq 0 : T_n \leq t\} \tag{5.36}$$

Then, $\{N(t), t \geq 0\}$ is a Poisson process with rate λ.

Remarks. i) Since $P[T_1 > 0] = 1$, we indeed have

$$N(0) = \max\{n \geq 0 : T_n \leq 0\} = 0 \tag{5.37}$$

ii) We also have

$$N(t) = n \quad \Longleftrightarrow \quad \{T_n \leq t\} \cap \{T_{n+1} > t\} \tag{5.38}$$

from which we obtain the definition of $N(t)$ in terms of T_n above.

iii) To calculate the probability of the event $\{N(t) = n\}$, it suffices to notice that

$$P[N(t) = n] = P[N(t) \geq n] - P[N(t) \geq n+1] = P[T_n \leq t] - P[T_{n+1} \leq t]$$
$$(5.39)$$

and to use the generalization of the formula (3.173) to the case when Y has a $G(n, \lambda)$ distribution:

$$P[Y \leq y] = 1 - P[W \leq n-1] = P[W \geq n], \quad \text{where } W \sim \text{Poi}(\lambda y) \quad (5.40)$$

We indeed obtain that

$$P[N(t) = n] = P[T_n \leq t] - P[T_{n+1} \leq t] \quad (5.41)$$
$$= P[\text{Poi}(\lambda t) \geq n] - P[\text{Poi}(\lambda t) \geq n+1] = P[\text{Poi}(\lambda t) = n]$$

iv) Let

$$T_d^* := \min\{t > 0 \colon N(t) \geq d \, (> 0)\} \quad (5.42)$$

If $d \in \{1, 2, \dots\}$, then we deduce from the relation (5.34) that $T_d^* \sim G(d, \lambda)$. When $d \notin \{1, 2, \dots\}$, we only have to replace d by $[d] + 1$, that is, the smallest integer larger than d. Note that, since $\lim_{t \to \infty} N(t) = \infty$, we may write that $P[T_d^* < \infty] = 1$, for any real number $d \in (0, \infty)$.

A more difficult problem consists in finding the distribution of the random variable

$$T_{c,d}^* := \min\{t > 0 \colon N(t) \geq ct + d\} \quad (5.43)$$

where $c > 0$ and $d \geq 0$.

Example 5.1.2. Suppose that the random variables X_i have a uniform distribution on the interval $(0, 1]$, rather than an exponential distribution, in the preceding proposition. To obtain a Poisson process by proceeding as above, it suffices to define

$$Y_i = -\frac{1}{\lambda} \ln X_i \quad \text{for } i = 1, 2, \dots$$

Indeed, we then have, for $y \geq 0$:

$$P[Y_i \leq y] = P[\ln X_i \geq -\lambda y] = P\left[X_i \geq e^{-\lambda y}\right] = 1 - e^{-\lambda y}$$

so that

$$f_{Y_i}(y) = \lambda e^{-\lambda y} \quad \text{for } y \geq 0$$

Thus, the stochastic process $\{N(t), t \geq 0\}$ defined by $N(t) = 0$, for $t < Y_1$, and

$$N(t) = \max\left\{n \geq 1 \colon \sum_{i=1}^{n} Y_i \leq t\right\} \quad \text{for } t \geq Y_1 \quad (5.44)$$

is a Poisson process with rate λ.

Remark. If we define $Y_0 = 0$, then we may write that

$$N(t) = \max\left\{n \geq 0: \sum_{i=0}^{n} Y_i \leq t\right\} \quad \forall\, t \geq 0 \qquad (5.45)$$

Proposition 5.1.5. *Let $\{N(t), t \geq 0\}$ be a Poisson process with rate λ. We have that $T_1 \mid \{N(t) = 1\} \sim U(0, t]$, where T_1 is the arrival time of the first event of the process.*

Proof. For $0 < s \leq t$, we have

$$
\begin{aligned}
P[T_1 \leq s \mid N(t) = 1] &= \frac{P[T_1 \leq s, N(t) = 1]}{P[N(t) = 1]} \\
&= \frac{P[N(s) = 1, N(t) - N(s) = 0]}{P[N(t) = 1]} \\
&\stackrel{\text{ind.}}{=} \frac{(\lambda s e^{-\lambda s}) e^{-\lambda(t-s)}}{\lambda t e^{-\lambda t}} = \frac{s}{t} \quad \square
\end{aligned}
$$

Remarks. i) The result actually follows from the fact that the increments of the Poisson process are independent and stationary. This indeed implies that the probability that an event occurs in an arbitrary interval must depend only on the length of this interval.

ii) More generally, if T^* denotes the arrival time of the *only* event in the interval $(t_1, t_2]$, where $0 \leq t_1 < t_2$, then T^* is uniformly distributed on this interval.

iii) The random variable $T_1 \mid \{N(t) = 1\}$ is different from $T_1^* := T_1 \mid \{T_1 \leq t\}$. The variable T_1^* has a *truncated* exponential distribution:

$$P[T_1^* \leq s] = P[T_1 \leq s \mid T_1 \leq t] = \frac{P[T_1 \leq s]}{P[T_1 \leq t]} = \frac{1 - e^{-\lambda s}}{1 - e^{-\lambda t}} \quad \text{for } 0 < s \leq t \qquad (5.46)$$

Note that the occurrence of the event $\{N(t) = 1\}$ implies the occurrence of $\{T_1 \leq t\}$. However, $\{T_1 \leq t\} \implies \{N(t) \geq 1\}$. On the other hand, we may write that

$$T_1 \mid \{N(t) = 1\} \equiv T_1 \mid \{T_1 \leq t, T_2 > t\} \qquad (5.47)$$

We would like to generalize the preceding proposition by calculating the distribution of $(T_1, \ldots, T_n)$, given that exactly n events occurred in the interval $(0, t]$. Let us first consider the case when $n = 2$. Let $0 < t_1 < t_2 \leq t$.

To obtain the conditional distribution function of the random vector (T_1, T_2), given that $N(t) = 2$, we begin by calculating

$$P[T_1 \leq t_1, T_2 \leq t_2, N(t) = 2]$$

$$= P[N(t_1) = 1, N(t_2) - N(t_1) = 1, N(t) - N(t_2) = 0]$$
$$+ P[N(t_1) = 2, N(t) - N(t_1) = 0]$$

$$= \left(\lambda t_1 e^{-\lambda t_1} \lambda (t_2 - t_1) e^{-\lambda(t_2 - t_1)} e^{-\lambda(t - t_2)} \right) + \left(\frac{(\lambda t_1)^2}{2!} e^{-\lambda t_1} e^{-\lambda(t - t_1)} \right)$$

$$= \left(\lambda^2 t_1 (t_2 - t_1) + \frac{(\lambda t_1)^2}{2!} \right) e^{-\lambda t} \tag{5.48}$$

where we used the fact that the increments of the process $\{N(t), t \geq 0\}$ are independent and stationary, from which we find that

$$P[T_1 \leq t_1, T_2 \leq t_2 \mid N(t) = 2] = \frac{P[T_1 \leq t_1, T_2 \leq t_2, N(t) = 2]}{P[N(t) = 2]}$$

$$= \frac{\left(\lambda^2 t_1 (t_2 - t_1) + \frac{1}{2}(\lambda t_1)^2 \right) e^{-\lambda t}}{\frac{1}{2}(\lambda t)^2 e^{-\lambda t}}$$

$$= \frac{2 t_1 (t_2 - t_1) + t_1^2}{t^2} \tag{5.49}$$

so that

$$f_{T_1, T_2 \mid N(t)}(t_1, t_2 \mid 2) = \frac{\partial^2}{\partial t_1 \partial t_2} \left\{ \frac{2 t_1 (t_2 - t_1)}{t^2} + \frac{t_1^2}{t^2} \right\}$$

$$= \frac{2}{t^2} + 0 = \frac{2}{t^2} \quad \text{for } 0 < t_1 < t_2 \leq t \tag{5.50}$$

Remark. From the preceding formula, we calculate

$$f_{T_1 \mid N(t)}(t_1 \mid 2) = \int_{t_1}^{t} \frac{2}{t^2} \, dt_2 = \frac{2(t - t_1)}{t^2} \quad \text{for } 0 < t_1 \leq t \tag{5.51}$$

Note that the distribution of $T_1 \mid \{N(t) = 2\}$ is *not* uniform on the interval $(0, t]$, contrary to that of $T_1 \mid \{N(t) = 1\}$. We also have

$$f_{T_2 \mid N(t)}(t_2 \mid 2) = \int_{0}^{t_2} \frac{2}{t^2} \, dt_1 = \frac{2 t_2}{t^2} \quad \text{for } 0 < t_2 \leq t \tag{5.52}$$

Finally, we may write that

$$f_{T_2 \mid T_1, N(t)}(t_2 \mid t_1, 2) = \frac{f_{T_1, T_2 \mid N(t)}(t_1, t_2 \mid 2)}{f_{T_1 \mid N(t)}(t_1 \mid 2)} = \frac{2/t^2}{2(t - t_1)/t^2}$$

$$= \frac{1}{t - t_1} \quad \text{for } 0 < t_1 < t_2 \leq t \tag{5.53}$$

That is, $T_2 \mid \{T_1 = t_1, N(t) = 2\} \sim U(t_1, t]$, from which we deduce that $(T_2 - T_1) \mid \{T_1 = t_1, N(t) = 2\} \sim U(0, t - t_1]$.

To obtain the formula in the general case, we can consider all the possible values of the random variables $N(t_1)$, $N(t_2) - N(t_1)$, $\ldots$, $N(t_n) - N(t_{n-1})$, given that $T_1 \leq t_1, T_2 \leq t_2, \ldots, T_n \leq t_n, N(t) = n$. The total number of ways to place the n events in the intervals $(0, t_1]$, $(t_1, t_2]$, $\ldots$, $(t_{n-1}, t_n]$ is given by the multinomial coefficients (see p. 4). Moreover, $N(t_1)$ must be greater than or equal to 1, $N(t_2)$ must be greater than or equal to 2, etc. However, it suffices to notice that the only nonzero term, after having differentiated the conditional distribution function with respect to each of the variables $t_1, \ldots, t_n$, will be the one for which the event

$$F = \{N(t_1) = 1\} \cap \{N(t_2) - N(t_1) = 1\} \cap \ldots \cap \{N(t_n) - N(t_{n-1}) = 1\}$$
(5.54)

occurs. Let

$$G = \{N(t_1) > 1\} \cap \{N(t_2) \geq 2\} \cap \ldots \cap \{N(t_n) \geq n\} \quad (5.55)$$

We have

$$P[T_1 \leq t_1, T_2 \leq t_2, \ldots, T_n \leq t_n, N(t) = n]$$

$$= P[F \cap \{N(t) = n\}] + P[G \cap \{N(t) = n\}]$$
$$= P[F \cap \{N(t) - N(t_n) = 0\}] + P[G \cap \{N(t) = n\}]$$
$$= \lambda t_1 e^{-\lambda t_1} \times \prod_{k=2}^{n} \lambda(t_k - t_{k-1})e^{-\lambda(t_k - t_{k-1})} \times e^{-\lambda(t - t_n)}$$
$$+ P[G \cap \{N(t) = n\}]$$
$$= \lambda^n t_1 e^{-\lambda t} \prod_{k=2}^{n}(t_k - t_{k-1}) + P[G \cap \{N(t) = n\}] \quad (5.56)$$

It follows that

$$P[T_1 \leq t_1, T_2 \leq t_2, \ldots, T_n \leq t_n \mid N(t) = n]$$
$$= \frac{P[T_1 \leq t_1, T_2 \leq t_2, \ldots, T_n \leq t_n, N(t) = n]}{P[N(t) = n]}$$
$$= \frac{\lambda^n t_1 e^{-\lambda t} \prod_{k=2}^{n}(t_k - t_{k-1})}{(\lambda t)^n e^{-\lambda t}/n!} + P[G \mid N(t) = n] \quad (5.57)$$

Since at least one t_i is not present in the term $P[G \mid N(t) = n]$, we may write that

$$f_{T_1, T_2, \ldots, T_n \mid N(t)}(t_1, t_2, \ldots, t_n \mid n)$$
$$= \frac{\partial^n}{\partial t_1 \partial t_2 \cdots \partial t_n} \frac{t_1 \prod_{k=2}^{n}(t_k - t_{k-1})}{t^n/n!}$$

$$= \frac{n!}{t^n} \quad \text{for } 0 < t_1 < t_2 < \ldots < t_n \leq t \qquad (5.58)$$

We have proved the following proposition.

Proposition 5.1.6. *Let $\{N(t), t \geq 0\}$ be a Poisson process with rate λ. Given that $N(t) = n$, the n arrival times of the events, $T_1, \ldots, T_n$, have the joint probability density function given by the formula (5.58).*

Remarks. i) Another (intuitive) way of obtaining the joint density function of the random vector $(T_1, \ldots, T_n)$, given that $N(t) = n$, is to use the fact that, for an arbitrary event that occurred in the interval $(0, t]$, the arrival time of this event has a uniform distribution on this interval. We can consider the n arrival times of the events as being the values taken by n *independent* $U(0, t]$ random variables U_i placed in increasing order (since $T_1 < T_2 < \ldots < T_n$). There are $n!$ ways of putting the variables $U_1, \ldots, U_n$ in increasing order. Moreover, by independence, we may write, for $0 < t_1 < t_2 < \ldots < t_n \leq t$, that

$$f_{U_1,\ldots,U_n}(t_1, \ldots, t_n) = \prod_{i=1}^{n} f_{U_i}(t_i) = \prod_{i=1}^{n} \frac{1}{t} = \frac{1}{t^n} \qquad (5.59)$$

Thus, we retrieve the formula (5.58):

$$f_{T_1,\ldots,T_n \mid N(t)}(t_1, \ldots, t_n \mid n) = n! f_{U_1,\ldots,U_n}(t_1, \ldots, t_n) = \frac{n!}{t^n} \qquad (5.60)$$

if $0 < t_1 < t_2 < \ldots < t_n \leq t$.

ii) When we place some random variables $X_1, \ldots, X_n$ in increasing order, we generally use the notation $X_{(1)}, \ldots, X_{(n)}$, where $X_{(i)} \leq X_{(j)}$ if $i < j$. In the continuous case, we always have that $X_{(i)} < X_{(j)}$ if $i < j$. The variables placed in increasing order are called the *order statistics* of $X_1, \ldots, X_n$. We have that $X_{(1)} = \min\{X_1, \ldots, X_n\}$ and $X_{(n)} = \max\{X_1, \ldots, X_n\}$.

Let us now return to the problem of decomposing the events of a Poisson process, $\{N(t), t \geq 0\}$, into two or more types.

Proposition 5.1.7. *Suppose that an event of a Poisson process with rate λ, $\{N(t), t \geq 0\}$, that occurs at time s is classified, independently from the other events, of type i with probability $p_i(s)$, where $i = 1, \ldots, j$ and $\sum_{i=1}^{j} p_i(s) = 1$. Let $N_i(t)$ be the number of type i events in the interval $[0, t]$. The $N_i(t)$'s are independent random variables having Poisson distributions with parameters*

$$\lambda_i(t) := \lambda \int_0^t p_i(s) \, ds \quad for \; i = 1, 2, \ldots, j \qquad (5.61)$$

Proof. We know that the arrival time of an arbitrary event that occurred in the interval $(0, t]$ has a uniform distribution on this interval. Consequently, we may write that the probability p_i that this event is of type i is given by

$$p_i = \int_0^t p_i(s) \frac{1}{t} \, ds \quad \text{for } i = 1, 2, \ldots, j \tag{5.62}$$

Proceeding as in the proof of Proposition 5.1.2, we find that the random variable $N_i(t)$ has a Poisson distribution with parameter

$$\lambda_i(t) = \lambda p_i t = \lambda \int_0^t p_i(s) \, ds \quad \text{for } i = 1, 2, \ldots, j \tag{5.63}$$

and that the variables $N_1(t), N_2(t), \ldots, N_j(t)$ are independent. $\square$

Remark. The random variables $N_i(t)$ have Poisson distributions. However, the stochastic processes $\{N_i(t), t \geq 0\}$ are *not* Poisson processes, unless the functions $p_i(s)$ do not depend on s, for all i.

Example 5.1.3. Customers arriving at a car dealer (open from 9 a.m. to 9 p.m.) can be classified in two categories: those who intend to buy a car (type I) and those who are just looking at the cars or want to ask some information (type II). Suppose that

$$P[\text{Customer is of type I}] = \begin{cases} 1/2 \text{ from 9 a.m. to 6 p.m.} \\ 1/4 \text{ from 6 p.m. to 9 p.m.} \end{cases}$$

independently from one customer to another and that the arrivals constitute a Poisson process with rate λ per day.

(a) Calculate the variance of the number of type I customers arriving in one day if $\lambda = 50$.

(b) Suppose that the average profit per car sold is equal to \$1000 and that $\lambda = 10$. What is the average profit for the dealer from 9 a.m. to 6 p.m. on a given day, knowing that at least two cars were sold during this time period?

Solution. (a) Let $N_I(t)$ be the number of type I customers in the interval $[0, t]$, where t is in (opening) hours. We can write that

$$N_I(12) \sim \text{Poi}\left(\frac{50}{12} \int_9^{21} p_I(s) \, ds\right) \equiv \text{Poi}(21.875)$$

because

$$\int_9^{21} p_I(s) \, ds = \int_9^{18} \frac{1}{2} \, ds + \int_{18}^{21} \frac{1}{4} \, ds = 21/4$$

We seek $V[N_I(12)] = 21.875$.

(b) Let $N_1(t)$ be the number of sales in t days, from 9 a.m. to 6 p.m. The process $\{N_1(t), t \geq 0\}$ is a Poisson process with rate

$$\lambda = 10 \times \frac{9}{12} \times \frac{1}{2} = 3.75$$

We are looking for $1000E[N_1(1) \mid N_1(1) \geq 2] := 1000x$. We have

$$E[N_1(1)] = 0 + E[N_1(1) \mid N_1(1) = 1]P[N_1(1) = 1] + xP[N_1(1) \geq 2]$$

It follows that

$$3.75 = 1 \cdot 3.75e^{-3.75} + x\left[1 - e^{-3.75}(1 + 3.75)\right]$$

We find that $x \simeq 4.12231$, so that the average profit is given by (approximately) \$4122.31.

Proposition 5.1.8. *Let $\{N_1(t), t \geq 0\}$ and $\{N_2(t), t \geq 0\}$ be independent Poisson processes, with rates λ_1 and λ_2, respectively. We define the event $F_{n_1,n_2} = n_1$ events of the process $\{N_1(t), t \geq 0\}$ occur before n_2 events of the process $\{N_2(t), t \geq 0\}$ have occurred. We have*

$$P[F_{n_1,n_2}] = P[X \geq n_1], \quad \text{where } X \sim B\left(n = n_1 + n_2 - 1, p := \frac{\lambda_1}{\lambda_1 + \lambda_2}\right) \tag{5.64}$$

That is,

$$P[F_{n_1,n_2}] = \sum_{i=n_1}^{n_1+n_2-1} \binom{n_1 + n_2 - 1}{i} \left(\frac{\lambda_1}{\lambda_1 + \lambda_2}\right)^i \left(\frac{\lambda_2}{\lambda_1 + \lambda_2}\right)^{n_1+n_2-1-i} \tag{5.65}$$

Proof. Let $N(t) := N_1(t) + N_2(t)$, and let E_j be the random experiment that consists in observing whether the jth event of the process $\{N(t), t \geq 0\}$ is an event of the process $\{N_1(t), t \geq 0\}$ or not. Given that the Poisson process has independent and stationary increments, the E_j's are *Bernoulli trials*, that is, *independent* trials for which the probability that the jth trial is a *success* is the same for all j. Then the random variable X that counts the number of successes in n trials has, by definition, a binomial distribution with parameters n and (see Prop. 3.3.4)

$$p := P[T_{1,1} < T_{2,1}] = \frac{\lambda_1}{\lambda_1 + \lambda_2} \tag{5.66}$$

where $T_{1,1} \sim \text{Exp}(\lambda_1)$ and $T_{2,1} \sim \text{Exp}(\lambda_2)$ are independent random variables. Since F_{n_1,n_2} occurs if and only if there are *at least* n_1 events of the process $\{N_1(t), t \geq 0\}$ among the first $n_1 + n_2 - 1$ events of the process $\{N(t), t \geq 0\}$, we obtain the formula (5.65). $\square$

Example 5.1.4. Let $\{X(t), 0 \leq t \leq 1\}$ be the stochastic process defined from a Poisson process $\{N(t), t \geq 0\}$ with rate λ as follows:

$$X(t) = N(t) - tN(1) \quad \text{for } 0 \leq t \leq 1$$

Note that, like the *Brownian bridge* (see p. 196), this process is such that $X(0) = X(1) = 0$. We calculate

$$E[X(t)] = E[N(t) - tN(1)] = E[N(t)] - tE[N(1)] = \lambda t - t(\lambda \cdot 1) = 0$$

for all $t \in [0, 1]$. It follows that if $0 \leq t_i \leq 1$, for $i = 1, 2$, then

$$\begin{aligned}
C_X(t_1, t_2) &= E\left[X(t_1)X(t_2)\right] - 0^2 \\
&= E\left\{[N(t_1) - t_1N(1)][N(t_2) - t_2N(1)]\right\} \\
&= E\left[N(t_1)N(t_2)\right] - t_1 E\left[N(1)N(t_2)\right] - t_2 E\left[N(t_1)N(1)\right] \\
&\quad + t_1 t_2 E[N^2(1)] \\
&= \lambda^2 t_1 t_2 + \lambda \min\{t_1, t_2\} - t_1\left(\lambda^2 t_2 + \lambda t_2\right) - t_2\left(\lambda^2 t_1 + \lambda t_1\right) \\
&\quad + t_1 t_2\left(\lambda^2 + \lambda\right) \\
&= \lambda \min\{t_1, t_2\} - \lambda t_1 t_2
\end{aligned}$$

where we used the formula

$$R_N(t_1, t_2) = C_N(t_1, t_2) + E[N(t_1)]E[N(t_2)] = \lambda \min\{t_1, t_2\} + \lambda^2 t_1 t_2$$

We can generalize the process above by defining $\{X(t), 0 \leq t \leq c\}$ by

$$X(t) = N(t) - \frac{t}{c}N(c) \quad \text{for } 0 \leq t \leq c$$

where c is a positive constant.

Example 5.1.5. We know that the Poisson process is not wide-sense stationary (see p. 232). On the other hand, its increments are stationary. Consider the process $\{X(t), t \geq 0\}$ defined by

$$X(t) = N(t + c) - N(t) \quad \text{for } t \geq 0$$

where c is a positive constant and $\{N(t), t \geq 0\}$ is a Poisson process with rate λ. Note that $X(0) = N(c) \sim \text{Poi}(\lambda c)$. Thus, the initial value of the process is random. We have

$$N(t + c) - N(t) \sim \text{Poi}(\lambda c) \quad \Longrightarrow \quad E[X(t)] = \lambda c \quad \text{for all } t \geq 0$$

Next, by using the formula for $R_N(t_1, t_2)$ in the preceding example, we calculate (for $s, t \geq 0$)

$$R_X(t, t+s) := E\left[\{N(t+c) - N(t)\}\{N(t+s+c) - N(t+s)\}\right]$$

$$= E[N(t+c)N(t+s+c)] - E[N(t+c)N(t+s)]$$
$$- E[N(t)N(t+s+c)] + E[N(t)N(t+s)]$$
$$= \lambda\left[(t+c) - t - \min\{c,s\} - t + t\right]$$
$$+ \lambda^2\left[(t+c)(t+s+c) - (t+c)(t+s) - t(t+s+c) + t(t+s)\right]$$
$$= \lambda\left(c - \min\{c,s\}\right) + \lambda^2 c^2$$

It follows that

$$C_X(t,t+s) := R_X(t,t+s) - E[X(t)]E[X(t+s)]$$
$$= \lambda\left(c - \min\{c,s\}\right) + \lambda^2 c^2 - (\lambda c)^2$$
$$= \lambda\left(c - \min\{c,s\}\right) = \begin{cases} 0 & \text{if } c \leq s \\ \lambda(c-s) & \text{if } c > s \end{cases}$$

We can therefore write that $C_X(t,t+s) = C_X(s)$. Thus, the stochastic process $\{X(t), t \geq 0\}$ is wide-sense stationary (because $E[X(t)] \equiv \lambda c$), which actually follows from the stationary increments of the Poisson process. Moreover, the fact that the autocovariance function $C_X(s)$ is equal to zero, for $c \leq s$, is a consequence of the independent increments of $\{N(t), t \geq 0\}$.

Example 5.1.6. We suppose that the number $N(t)$ of visitors of a certain Web site in the interval $[0,t]$ is such that $\{N(t), t \geq 0\}$ is a Poisson process with rate λ per hour. Calculate the probability that there have been more visitors from 8 a.m. to 9 a.m. than from 9 a.m. to 10 a.m., given that there have been 10 visitors from 8 a.m. to 10 a.m.

Solution. Let N_1 (respectively, N_2) be the number of visitors from 8 a.m. to 9 a.m. (resp., from 9 a.m. to 10 a.m.). We can assert that the random variables $N_1 \sim \text{Poi}(5)$ and $N_2 \sim \text{Poi}(5)$ are independent. Furthermore, we have $N_i \mid \{N_1 + N_2 = 10\} \sim \text{B}(n = 10, p = 1/2)$, for $i = 1,2$. We seek the probability $x := P[N_1 > N_2 \mid N_1 + N_2 = 10]$. By symmetry, we can write that

$$1 = x + P[N_1 = N_2 \mid N_1 + N_2 = 10] + x = 2x + \binom{10}{5}(1/2)^{10} \simeq 2x + 0.2461$$

Then $x \simeq 0.3770$.

Example 5.1.7. We define $M(t) = N(t) - (t^2/2)$, for $t \geq 0$, where $\{N(t), t \geq 0\}$ is a Poisson process with rate λ. Calculate $P[1 < S_1 \leq \sqrt{2}]$, where $S_1 := \min\{t \geq 0 : M(t) \geq 1\}$.

Solution. The increments of the Poisson process being independent and stationary, we can write that

$$P[1 < S_1 \leq \sqrt{2}]$$

$$= P[N(1) = 0, N(\sqrt{2}) - N(1) \geq 2] + P[N(1) = 1, N(\sqrt{2}) - N(1) \geq 1]$$
$$= e^{-\lambda}\left[1 - e^{-(\sqrt{2}-1)\lambda}\left(1 + (\sqrt{2}-1)\lambda\right)\right] + \lambda e^{-\lambda}\left[1 - e^{-(\sqrt{2}-1)\lambda}\right]$$

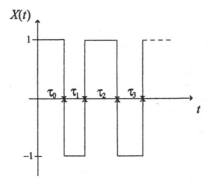

Fig. 5.2. Example of a trajectory of a telegraph signal.

5.1.1 The telegraph signal

As we did with the Brownian motion and in Examples 5.1.4 and 5.1.5, we can define stochastic processes from a Poisson process. An interesting particular transformation of the Poisson process is the *telegraph signal* $\{X(t), t \geq 0\}$, defined as follows:

$$X(t) = (-1)^{N(t)} = \begin{cases} 1 \text{ if } N(t) = 0, 2, 4, \ldots \\ -1 \text{ if } N(t) = 1, 3, 5, \ldots \end{cases} \tag{5.67}$$

An example of a trajectory of a telegraph signal is shown in Fig. 5.2.

Remark. Note that $X(0) = 1$, because $N(0) = 0$. Thus, the initial value of the process is *deterministic*. To make the starting point of the process random, we can simply multiply $X(t)$ by a random variable Z that is independent of $X(t)$, for all t, and that takes on the value 1 or -1 with probability $1/2$. It is as if we tossed a fair coin at time $t \geq 0$ to determine whether $X(t) = 1$ or -1.

The process $\{Y(t), t \geq 0\}$, where $Y(t) := Z \cdot X(t)$, for all $t \geq 0$, is called a *random telegraph signal.* We may write that $Z = Y(0)$. Moreover, to be precise, we then use the expression *semirandom* telegraph signal to designate the process $\{X(t), t \geq 0\}$. We already encountered the random telegraph signal in Example 2.3.2.

To obtain the distribution of the random variable $X(t)$, it suffices to calculate

$$P[X(t) = 1] = \sum_{k=0}^{\infty} P[N(t) = 2k] = \sum_{k=0}^{\infty} e^{-\lambda t} \frac{(\lambda t)^{2k}}{(2k)!}$$

$$= e^{-\lambda t} \frac{e^{\lambda t} + e^{-\lambda t}}{2} = \frac{1 + e^{-2\lambda t}}{2} \quad \forall\, t \geq 0 \tag{5.68}$$

because

$$\cosh \lambda t := \frac{e^{\lambda t} + e^{-\lambda t}}{2} = \sum_{k=0}^{\infty} \frac{(\lambda t)^{2k}}{(2k)!} \tag{5.69}$$

where "cosh" denotes the *hyperbolic cosine*. It follows that

$$P[X(t) = -1] = 1 - \frac{1 + e^{-2\lambda t}}{2} = \frac{1 - e^{-2\lambda t}}{2} \qquad (5.70)$$

In the case of the process $\{Y(t), t \geq 0\}$, we have

$$
\begin{aligned}
P[Y(t) = 1] &= P[Z \cdot X(t) = 1] \\
&= P[X(t) = 1 \mid Z = 1]P[Z = 1] \\
&\quad + P[X(t) = -1 \mid Z = -1]P[Z = -1] \\
&\overset{\text{ind.}}{=} \frac{1}{2}\{P[X(t) = 1] + P[X(t) = -1]\} = \frac{1}{2} \qquad (5.71)
\end{aligned}
$$

Thus, $P[Y(t) = 1] = P[Y(t) = -1] = 1/2$, for all $t \geq 0$, so that $E[Y(t)] \equiv 0$, whereas

$$E[X(t)] = 1 \cdot \frac{1 + e^{-2\lambda t}}{2} + (-1) \cdot \frac{1 - e^{-2\lambda t}}{2} = e^{-2\lambda t} \quad \forall\, t \geq 0 \qquad (5.72)$$

To obtain the autocorrelation function of the (semirandom) telegraph signal, we make use of the definition $X(t) = (-1)^{N(t)}$. If $s > 0$, we may write that

$$
\begin{aligned}
R_X(t, t + s) &:= E[X(t)X(t + s)] = E\left[(-1)^{N(t)}(-1)^{N(t+s)}\right] \\
&= E\left[(-1)^{2N(t)}(-1)^{N(t+s)-N(t)}\right] = E\left[(-1)^{N(t+s)-N(t)}\right] \\
&= E\left[(-1)^{N(s)}\right] = E[X(s)] = e^{-2\lambda s} \qquad (5.73)
\end{aligned}
$$

where we used the fact that the increments of the Poisson process are stationary. We then have

$$C_X(t, t + s) = e^{-2\lambda s} - e^{-2\lambda t}e^{-2\lambda(t+s)} = e^{-2\lambda s}\left(1 - e^{-4\lambda t}\right) \quad \forall\, s, t \geq 0 \qquad (5.74)$$

Finally, since

$$
\begin{aligned}
Y(t)Y(t + s) &:= [Z \cdot X(t)][Z \cdot X(t + s)] = Z^2 \cdot X(t)X(t + s) \\
&= 1 \cdot X(t)X(t + s) = X(t)X(t + s) \qquad (5.75)
\end{aligned}
$$

we find that

$$C_Y(t, t + s) = R_Y(t, t + s) = R_X(t, t + s) = e^{-2\lambda s} \quad \forall\, s, t \geq 0 \qquad (5.76)$$

We deduce from what precedes that the process $\{Y(t), t \geq 0\}$ is *wide-sense stationary*. As we mentioned in Example 2.3.2, it can even be shown that it is *strict*-sense stationary. On the other hand, $\{X(t), t \geq 0\}$ is not WSS, because its mean depends on t (and $C_X(t, t + s) \neq C_X(s)$). Furthermore, in Example 2.3.2, we showed that the *random* telegraph signal is a *mean ergodic* process.

5.2 Nonhomogeneous Poisson processes

In many applications, it is not realistic to assume that the average arrival rate of events of a counting process, $\{N(t), t \geq 0\}$, is constant. In practice, this rate generally depends on the variable t. For example, the average arrival rate of customers into a store is not the same during the entire day. Similarly, the average arrival rate of cars on a highway fluctuates between its maximum during rush hours and its minimum during slack hours. We will generalize the definition of a Poisson process to take this fact into account.

Definition 5.2.1. *Let* $\{N(t), t \geq 0\}$ *be a counting process with* independent increments. *This process is called a* **nonhomogeneous** *(or* **nonstationary***)* **Poisson process** *with* **intensity function** $\lambda(t) \geq 0$, *for* $t \geq 0$, *if* $N(0) = 0$ *and*

i) $P[N(t + \delta) - N(t) = 1] = \lambda(t)\delta + o(\delta)$,

ii) $P[N(t + \delta) - N(t) \geq 2] = o(\delta)$.

Remark. The condition i) implies that the process $\{N(t), t \geq 0\}$ does *not* have *stationary increments* unless $\lambda(t) \equiv \lambda > 0$. In this case, $\{N(t), t \geq 0\}$ becomes a *homogeneous* Poisson process, with rate λ.

As in the particular case when the average arrival rate of events is constant, we find that the number of events that occur in a given interval has a Poisson distribution.

Proposition 5.2.1. *Let* $\{N(t), t \geq 0\}$ *be a nonhomogeneous Poisson process with intensity function* $\lambda(t)$. *We have*

$$N(s + t) - N(s) \sim Poi(m(s + t) - m(s)) \quad \forall \ s, t \geq 0 \qquad (5.77)$$

where

$$m(s) := \int_0^s \lambda(\tau) \, d\tau \qquad (5.78)$$

Proof. Let

$$p_n(s, t) := P[N(s + t) - N(s) = n] \quad \text{for } n = 0, 1, 2, \ldots \qquad (5.79)$$

Using the fact that the increments of the process $\{N(t), t \geq 0\}$ are independent, and using the two conditions in the definition above, we may write, for $n = 1, 2, \ldots$, that

$$
\begin{aligned}
p_n&(s, t + \delta) \\
&= P[N(s + t) - N(s) = n, N(s + t + \delta) - N(s + t) = 0] \\
&\quad + P[N(s + t) - N(s) = n - 1, N(s + t + \delta) - N(s + t) = 1] + o(\delta)
\end{aligned}
$$

$$= p_n(s,t)[1 - \lambda(s+t)\delta + o(\delta)] + p_{n-1}(s,t)[\lambda(s+t)\delta + o(\delta)] + o(\delta)$$

from which we find that

$$p_n(s,t+\delta) - p_n(s,t) = \lambda(s+t)\delta[p_{n-1}(s,t) - p_n(s,t)] + o(\delta) \qquad (5.80)$$

Dividing both sides by δ and taking the limit as δ decreases to zero, we obtain

$$\frac{\partial}{\partial t}p_n(s,t) = \lambda(s+t)[p_{n-1}(s,t) - p_n(s,t)] \qquad (5.81)$$

When $n = 0$, the equation above becomes

$$\frac{\partial}{\partial t}p_0(s,t) = -\lambda(s+t)p_0(s,t) \qquad (5.82)$$

The variable s may be considered as a constant. Therefore, this equation is a first-order, linear, homogeneous ordinary differential equation. Its general solution is given by

$$p_0(s,t) = c_0 \exp\left\{-\int_s^{s+t} \lambda(\tau)\,d\tau\right\} \qquad (5.83)$$

where c_0 is a constant. Making use of the boundary condition $p_0(s,0) = 1$, we obtain $c_0 = 1$, so that

$$p_0(s,t) = \exp\left\{-\int_s^{s+t} \lambda(\tau)\,d\tau\right\} = e^{m(s)-m(s+t)} \quad \text{for } s,t \geq 0 \qquad (5.84)$$

Substituting this solution into Eq. (5.81), we find that

$$\frac{\partial}{\partial t}p_1(s,t) = \lambda(s+t)\left[e^{m(s)-m(s+t)} - p_1(s,t)\right] \qquad (5.85)$$

We may rewrite this equation as follows:

$$\frac{\partial}{\partial t}p_1(s,t) = \left[e^{m(s)-m(s+t)} - p_1(s,t)\right]\frac{\partial}{\partial t}[m(s+t) - m(s)] \qquad (5.86)$$

We easily check that the solution of this nonhomogeneous differential equation, which satisfies the boundary condition $p_1(s,0) = 0$, is

$$p_1(s,t) = e^{m(s)-m(s+t)}[m(s+t) - m(s)] \quad \text{for } s,t \geq 0 \qquad (5.87)$$

Finally, we can show, by mathematical induction, that

$$p_n(s,t) = e^{m(s)-m(s+t)}\frac{[m(s+t) - m(s)]^n}{n!} \quad \forall\, s,t \geq 0 \text{ and } n = 0,1,\dots \ \square$$

$$(5.88)$$

Remarks. i) The function $m(t)$ is called the *mean-value function* of the process.

ii) Suppose that $\lambda(t)$ is a twice-differentiable function such that $\lambda(t) \leq \lambda$, for all $t \geq 0$. We can then obtain a nonhomogeneous Poisson process with intensity function $\lambda(t)$ from a homogeneous Poisson process $\{N_1(t), t \geq 0\}$, with rate λ, by supposing that an event that occurs at time t is counted with probability $\lambda(t)/\lambda$. Let

$F = $ exactly one of the events that occur in the interval $(t, t+\delta]$ is counted,

and let

$N(t)=$ the number of events counted in the interval $[0, t]$.

We have (see Prop. 5.1.7)

$$P[N(t + \delta) - N(t) = 1] = \sum_{k=1}^{\infty} P\left[\{N_1(t + \delta) - N_1(t) = k\} \cap F\right]$$

$$= [\lambda\delta + o(\delta)] \int_{t}^{t+\delta} \frac{\lambda(u)}{\lambda} \frac{1}{\delta} \, du + o(\delta)$$

$$= [\lambda\delta + o(\delta)] \frac{\lambda(t + c\delta)}{\lambda} + o(\delta) \qquad (5.89)$$

for some $c \in (0, 1)$, by the *mean-value theorem for integrals*. Since

$$\lambda(t + c\delta) = \lambda(t) + c\delta\lambda'(t) + o(\delta) \qquad (5.90)$$

we may write that

$$P[N(t + \delta) - N(t) = 1] = [\lambda\delta + o(\delta)] \frac{\lambda(t) + c\delta\lambda'(t) + o(\delta)}{\lambda} + o(\delta)$$

$$= \lambda(t)\delta + o(\delta) \qquad (5.91)$$

iii) If we assume that

$$\int_{t_0}^{\infty} \lambda(t) \, dt = \infty \quad \text{for all } t_0 \in [0, \infty) \qquad (5.92)$$

then the probability that at least one event will occur after t_0 is equal to 1:

$$\lim_{s \to \infty} P[N(t_0 + s) - N(t_0) \geq 1] = 1 - \lim_{s \to \infty} P[N(t_0 + s) - N(t_0) = 0]$$

$$= 1 - \lim_{s \to \infty} \exp\left\{-\int_{t_0}^{t_0+s} \lambda(t) \, dt\right\}$$

$$= 1 - 0 = 1 \qquad (5.93)$$

Note that the formula (5.92) is valid when $\lambda(t) \equiv \lambda > 0$.

Let T_1 be the random variable that denotes the arrival time of the first event of the process $\{N(t), t \geq 0\}$. We will now calculate the distribution of T_1, given that $N(t) = 1$, as we did in the case of the homogeneous Poisson process (see Prop. 5.1.5).

Proposition 5.2.2. *Let $\{N(t), t \geq 0\}$ be a nonhomogeneous Poisson process with intensity function $\lambda(t)$. The probability density function of the random variable $S := T_1 \mid \{N(t) = 1\}$ is given by*

$$f_S(s) = \frac{\lambda(s)}{m(t)} \quad \text{for } 0 < s \leq t \tag{5.94}$$

Proof. If $0 < s \leq t$, we may write that

$$P[T_1 \leq s \mid N(t) = 1] = \frac{P[N(s) = 1, N(t) - N(s) = 0]}{P[N(t) = 1]}$$

$$\stackrel{\text{ind.}}{=} \frac{\left(m(s)e^{-m(s)}\right) e^{-m(t)+m(s)}}{m(t)e^{-m(t)}} = \frac{m(s)}{m(t)}$$

Since

$$\frac{d}{ds} m(s) = \frac{d}{ds} \int_0^s \lambda(u) \, du = \lambda(s) \tag{5.95}$$

we obtain the formula (5.94). □

Remark. The formula for the distribution function of the arrival time S of the *single* event that occurred in an arbitrary interval $(\tau, \tau + t]$ is

$$F_S(s) = \frac{m(s) - m(\tau)}{m(\tau + t) - m(\tau)} \quad \text{for } 0 \leq \tau < s \leq \tau + t \tag{5.96}$$

so that

$$f_S(s) = \frac{\lambda(s)}{m(\tau + t) - m(\tau)} \quad \text{for } 0 \leq \tau < s \leq \tau + t \tag{5.97}$$

where $S := T \mid \{N(\tau + t) - N(\tau) = 1\}$ and T is the arrival time of the first event of the nonhomogeneous Poisson process in the interval $(\tau, \tau + t]$.

Example 5.2.1. Let $\{N(t), t \geq 0\}$ be a nonhomogeneous Poisson process with intensity function $\lambda(t) > 0$, for $t \geq 0$. We set

$$M(t) = N\left(m^{-1}(t)\right) \quad \text{for all } t \geq 0$$

where $m^{-1}(t)$ is the *inverse* function of the mean-value function of the process.

Remark. This inverse function exists, because we assumed (in this example) that the intensity function $\lambda(t)$ is *strictly* positive, for all t, so that $m(t)$ is a *strictly increasing* function.

We then have

$$M(0) = N\left(m^{-1}(0)\right) = N(0) = 0$$

Moreover, we calculate, for all $\tau, t \geq 0$ and $n = 0, 1, \ldots,$

$$
\begin{aligned}
P[M(\tau + t) - M(\tau) = n] &= P\left[N\left(m^{-1}(\tau + t)\right) - N\left(m^{-1}(\tau)\right) = n\right] \\
&= P\left[\mathrm{Poi}\left(m\left(m^{-1}(\tau + t)\right) - m\left(m^{-1}(\tau)\right)\right) = n\right] \\
&= P\left[\mathrm{Poi}(\tau + t - \tau) = n\right] = P\left[\mathrm{Poi}(t) = n\right]
\end{aligned}
$$

Finally, since the function $m^{-1}(t)$ is strictly increasing, we may assert that the stochastic process $\{M(t), t \geq 0\}$, like $\{N(t), t \geq 0\}$, has independent increments. We then deduce, from what precedes, that $\{M(t), t \geq 0\}$ is actually a homogeneous Poisson process, with rate $\lambda = 1$.

5.3 Compound Poisson processes

Definition 5.3.1. *Let $X_1, X_2, \ldots$ be independent and identically distributed random variables, and let N be a random variable whose possible values are all positive integers and that is independent of the X_k's. The variable*

$$
S_N := \sum_{k=1}^{N} X_k \tag{5.98}
$$

is called a **compound random variable.**

We already gave the formulas for the mean and the variance of S_N [see Eqs. (1.89) and (1.90)]. We will now prove these formulas.

Proposition 5.3.1. *The mean and the variance of the random variable S_N defined above are given, respectively, by*

$$
E\left[\sum_{k=1}^{N} X_k\right] = E[N]E[X_1] \tag{5.99}
$$

and

$$
V\left[\sum_{k=1}^{N} X_k\right] = E[N]V[X_1] + V[N](E[X_1])^2 \tag{5.100}
$$

Proof. First, we have

$$
E\left[\sum_{k=1}^{n} X_k\right] = \sum_{k=1}^{n} E[X_k] \overset{\text{i.d.}}{=} nE[X_1] \tag{5.101}
$$

Then, since N is independent of the X_k's, we may write that

$$
E\left[\sum_{k=1}^{N} X_k \,\middle|\, N = n\right] = E\left[\sum_{k=1}^{n} X_k\right] = nE[X_1] \tag{5.102}
$$

so that

$$E\left[\sum_{k=1}^{N} X_k \Big| N\right] = NE[X_1] \tag{5.103}$$

It follows that

$$E\left[\sum_{k=1}^{N} X_k\right] = E\left[E\left[\sum_{k=1}^{N} X_k \Big| N\right]\right] = E[NE[X_1]] = E[N]E[X_1] \tag{5.104}$$

Remark. It is not necessary that the random variables X_k be independent among themselves for the formula (5.99) to be valid.

Next, proceeding as above, we find that

$$V\left[\sum_{k=1}^{N} X_k \Big| N = n\right] \stackrel{\text{i.i.d.}}{=} nV[X_1] \quad \Longrightarrow \quad V\left[\sum_{k=1}^{N} X_k \Big| N\right] = NV[X_1] \tag{5.105}$$

With the help of the formula (see p. 28)

$$V[S_N] = E[V[S_N \mid N]] + V[E[S_N \mid N]] \tag{5.106}$$

we may then write that

$$V\left[\sum_{k=1}^{N} X_k\right] = E[NV[X_1]] + V[NE[X_1]]$$

$$= E[N]V[X_1] + (E[X_1])^2 V[N] \quad \square \tag{5.107}$$

Definition 5.3.2. *Let $\{N(t), t \geq 0\}$ be a Poisson process with rate λ, and let $X_1, X_2, \ldots$ be random variables that are i.i.d. and independent of the process $\{N(t), t \geq 0\}$. The stochastic process $\{Y(t), t \geq 0\}$ defined by*

$$Y(t) = \sum_{k=1}^{N(t)} X_k \quad \forall \, t \geq 0 \quad (and \ Y(t) = 0 \ if \ N(t) = 0) \tag{5.108}$$

is called a **compound Poisson process.**

Remarks. i) This is another way of generalizing the Poisson process, since if the random variables X_k are actually the constant 1, then the processes $\{Y(t), t \geq 0\}$ and $\{N(t), t \geq 0\}$ are identical.

ii) A Poisson process, $\{N(t), t \geq 0\}$, only counts the number of events that occurred in the interval $[0, t]$, while the process $\{Y(t), t \geq 0\}$ gives, for example, the sum of the lengths of telephone calls that happened in $[0, t]$, or the total number of persons who were involved in car accidents in this interval, etc. Note that we must assume that the lengths of the calls or the numbers of persons involved in distinct accidents are independent and identically distributed random variables. We could consider the two-dimensional process $\{(N(t), Y(t)), t \geq 0\}$ to retain all the information of interest.

Using Proposition 5.3.1 with $S_N = Y(t)$ and $N = N(t)$, we obtain

$$E[Y(t)] = E[N(t)]E[X_1] = \lambda t E[X_1] \tag{5.109}$$

and

$$V[Y(t)] = E[N(t)]V[X_1] + V[N(t)](E[X_1])^2$$
$$= \lambda t \left(V[X_1] + (E[X_1])^2\right) = \lambda t E[X_1^2] \tag{5.110}$$

We can easily calculate the moment-generating function of the random variable $Y(t)$. Let

$$M_1(s) \equiv M_{X_1}(s) := E\left[e^{sX_1}\right] \tag{5.111}$$

We have

$$M_{Y(t)}(s) := E\left[e^{sY(t)}\right] = E\left[e^{s(X_1+\dots+X_{N(t)})}\right] = E\left[E\left[e^{s(X_1+\dots+X_{N(t)})}\big|N(t)\right]\right]$$
$$\stackrel{\text{ind.}}{=} \sum_{n=0}^{\infty} E\left[e^{s(X_1+\dots+X_n)}\right] e^{-\lambda t}\frac{(\lambda t)^n}{n!} \stackrel{\text{i.i.d.}}{=} \sum_{n=0}^{\infty} [M_1(s)]^n e^{-\lambda t}\frac{(\lambda t)^n}{n!}$$
$$= e^{-\lambda t}e^{M_1(s)\lambda t} = \exp\{\lambda t[M_1(s) - 1]\} \tag{5.112}$$

This formula enables us to check the results obtained above. In particular, we have

$$E[Y(t)] = \frac{d}{ds}M_{Y(t)}(s)\Big|_{s=0} = M_{Y(t)}(s)\lambda t M_1'(s)\Big|_{s=0}$$
$$= M_{Y(t)}(0)\lambda t M_1'(0) = 1\lambda t E[X_1] = \lambda t E[X_1] \tag{5.113}$$

When X_1 is a *discrete* random variable, whose possible values are $1, 2, \dots, j$, we may write that

$$Y(t) = \sum_{i=1}^{j} iN_i(t) \tag{5.114}$$

where $N_i(t)$ is the number of random variables X_k (associated with some random events) that took on the value i in the interval $[0, t]$. By Proposition 5.1.2, the processes $\{N_i(t), t \geq 0\}$ are independent Poisson processes with rates $\lambda p_{X_1}(i)$, for $i = 1, \dots, j$.

Remarks. i) This representation of the process $\{Y(t), t \geq 0\}$ can be generalized to the case when X_1 is an arbitrary discrete random variable.

ii) The moment-generating function of the random variable $Y(t)$ in the case above becomes

$$M_{Y(t)}(s) = \exp\left\{\lambda t \sum_{i=1}^{j} \left(e^{si} - 1\right) p_{X_1}(i)\right\} \tag{5.115}$$

Since $\lim_{t\to\infty} N(t) = \infty$, we deduce from the central limit theorem the following proposition.

Proposition 5.3.2. *For t sufficiently large, we may write that*

$$Y(t) \approx N\left(\lambda t E[X_1], \lambda t E[X_1^2]\right) \tag{5.116}$$

Remark. For the approximation to be good, the number of variables in the sum must be approximately equal to 30 or more (or maybe less), depending on the degree of asymmetry of (the distribution of) the random variable X_1 with respect to its mean.

Finally, let $\{Y_1(t), t \geq 0\}$ and $\{Y_2(t), t \geq 0\}$ be *independent* compound Poisson processes, defined by

$$Y_i(t) = \sum_{k=1}^{N_i(t)} X_{i,k} \quad \forall\, t \geq 0 \quad (\text{and } Y_i(t) = 0 \text{ if } N_i(t) = 0) \tag{5.117}$$

where $\{N_i(t), t \geq 0\}$ is a Poisson process with rate λ_i, for $i = 1, 2$. We know that the process $\{N(t), t \geq 0\}$, where $N(t) := N_1(t) + N_2(t) \,\forall\, t \geq 0$, is a Poisson process with rate $\lambda := \lambda_1 + \lambda_2$ (because the two Poisson processes are independent; see Prop. 5.1.1). Let X_k be the random variable associated with the kth event of the process $\{N(t), t \geq 0\}$. We may write that

$$X_k \stackrel{\mathrm{d}}{=} \begin{cases} X_{1,k} \text{ with probability } p := \dfrac{\lambda_1}{\lambda_1 + \lambda_2} \\[2mm] X_{2,k} \text{ with probability } 1 - p \end{cases} \tag{5.118}$$

(That is, X_k has the same *distribution* as $X_{1,k}$ (respectively, $X_{2,k}$) with probability p (resp., $1 - p$).) Thus, we have

$$P[X_k \leq x] = P[X_{1,k} \leq x]p + P[X_{2,k} \leq x](1 - p) \tag{5.119}$$

Since the random variables $X_1, X_2, \ldots$ are i.i.d., and are independent of the Poisson process $\{N(t), t \geq 0\}$, we may assert that the process $\{Y(t), t \geq 0\}$ defined by

$$Y(t) = Y_1(t) + Y_2(t) \quad \text{for } t \geq 0 \tag{5.120}$$

is a *compound Poisson process* as well.

Example 5.3.1. Suppose that the customers of a certain insurance company pay premiums to the company at the constant rate α (per time unit) and that the company pays indemnities to its customers according to a Poisson process with rate λ. If the amount paid per claim is a random variable, and if the variables corresponding to distinct claims are independent and identically distributed, then the capital $C(t)$ at the company's disposal at time t is given by

$$C(t) = C(0) + \alpha t - Y(t)$$

where $\{Y(t), t \geq 0\}$ is a compound Poisson process (if, in addition, the indemnity amounts do not depend on the *number* of indemnities paid). An important question is to be able to determine the risk that the random variable $C(t)$ will take on a value smaller than or equal to zero, that is, the risk that the company will go bankrupt. If $\alpha < \lambda\mu$, where $\mu := E[X_1]$ is the average indemnity paid by the company, then we find that the probability that it will eventually go bankrupt is equal to 1, which is logical.

5.4 Doubly stochastic Poisson processes

In Section 5.2, we generalized the definition of a Poisson process by allowing the average arrival rate of events of the process to be a *deterministic* function $\lambda(t)$. We will now generalize further the basic Poisson process by supposing that the function $\lambda(t)$ is a random variable $\Lambda(t)$. Thus, the set $\{\Lambda(t), t \geq 0\}$ is a stochastic process. For this reason, the process $\{N(t), t \geq 0\}$ is called a *doubly stochastic Poisson process*. First, we consider the case when the random variable $\Lambda(t)$ does not depend on t.

Definition 5.4.1. *Let Λ be a positive random variable. If the counting process $\{N(t), t \geq 0\}$, given that $\Lambda = \lambda$, is a Poisson process with rate λ, then the stochastic process $\{N(t), t \geq 0\}$ is called a* **conditional** (or **mixed**) **Poisson process**.

Proposition 5.4.1. *The conditional Poisson process $\{N(t), t \geq 0\}$ has stationary, but not independent increments.*

Proof. Consider the case when Λ is a discrete random variable whose possible values are in the set $\{1, 2, \ldots\}$. We have

$$P[N(\tau + t) - N(\tau) = n] = \sum_{k=1}^{\infty} P[N(\tau + t) - N(\tau) = n \mid \Lambda = k] = p_\Lambda(k)$$

$$= \sum_{k=1}^{\infty} e^{-kt} \frac{(kt)^n}{n!} p_\Lambda(k) \qquad (5.121)$$

Since $N(0) = 0$, for any value of Λ, we conclude that

$$P[N(\tau + t) - N(\tau) = n] = P[N(t) = n] \quad \text{for all } \tau, t \geq 0 \text{ and } n = 0, 1, \ldots$$
$$(5.122)$$

Thus, the increments of the process $\{N(t), t \geq 0\}$ are stationary.

Next, using the formula (5.121) above, we may write that

$$P[\Lambda = j \mid N(t) = n] = \frac{P[N(t) = n \mid \Lambda = j]P[\Lambda = j]}{P[N(t) = n]}$$

$$= \frac{e^{-jt}[(jt)^n/n!]p_\Lambda(j)}{\sum_{k=1}^{\infty} e^{-kt}[(kt)^n/n!]p_\Lambda(k)}$$

$$= \frac{e^{-jt}j^n p_\Lambda(j)}{\sum_{k=1}^{\infty} e^{-kt}k^n p_\Lambda(k)} \qquad (5.123)$$

Therefore, the number of events that occurred in the interval $[0, t]$ gives us some information about the probability that the random variable Λ took on the value j. Now, the larger j is, the larger is the expected number of events that will occur in the interval $(t, t + \tau]$, where $\tau > 0$. Consequently, we must conclude that the increments of the process $\{N(t), t \geq 0\}$ are not independent. □

Remarks. i) In the case of the homogeneous Poisson process, with rate $\lambda > 0$, we have that $p_\Lambda(\lambda) = 1$, so that

$$P[\Lambda = \lambda \mid N(t) = n] = \frac{e^{-\lambda t}[(\lambda t)^n/n!] \times 1}{e^{-\lambda t}[(\lambda t)^n/n!]} = 1 \quad \text{for } n = 0, 1, \ldots \quad (5.124)$$

ii) The proposition above implies that a conditional Poisson process is *not* a Poisson process unless Λ is a constant.

iii) If the parameter λ of the homogeneous Poisson process $\{N(t), t \geq 0\}$ is unknown, we can estimate it by taking *observations* of the random variable $N(t)$, for an arbitrary $t > 0$. Suppose that we collected n (independent) observations of $N(t)$. Let $\bar{X}$ be the average number of events that occurred in the interval $[0, t]$. That is,

$$\bar{X} := \frac{\sum_{k=1}^{n} X_k}{n} \qquad (5.125)$$

where X_k is the kth observation of $N(t)$. The best *estimator* of the parameter λ is given by

$$\hat{\lambda} := \frac{\bar{X}}{t} \qquad (5.126)$$

Thus, the larger the number of events that occurred in the interval $[0, t]$ is, the larger the *estimated* value of λ is. However, once this parameter has been estimated, we start anew and we *assume* that the increments are independent.

iv) When Λ is a continuous random variable, we find that

$$f_{\Lambda|N(t)}(\lambda \mid n) = \frac{e^{-\lambda t}\lambda^n f_\Lambda(\lambda)}{\int_0^\infty e^{-\mu t}\mu^n f_\Lambda(\mu)\, d\mu} \quad \forall\, \lambda > 0 \qquad (5.127)$$

Given that

$$E[N(t) \mid \Lambda] = \Lambda t \quad \text{and} \quad V[N(t) \mid \Lambda] = \Lambda t \qquad (5.128)$$

we calculate

$$E[N(t)] = E[E[N(t) \mid \Lambda]] = tE[\Lambda] \qquad (5.129)$$

and

$$\begin{aligned} V[N(t)] &= E[V[N(t) \mid \Lambda]] + V[E[N(t) \mid \Lambda]] \\ &= E[\Lambda t] + V[\Lambda t] = tE[\Lambda] + t^2 V[\Lambda] \end{aligned} \qquad (5.130)$$

Example 5.4.1. Suppose that the rate of a Poisson process is a continuous random variable Λ such that

$$f_\Lambda(\lambda) = \frac{2}{\lambda^3} \quad \text{if } \lambda \geq 1$$

It can be shown that

$$P[N(t) > n] = \int_0^\infty P[\Lambda > \lambda] t e^{-\lambda t} \frac{(\lambda t)^n}{n!}\, d\lambda$$

Use this formula to calculate the probability that there will be more than two events during an arbitrary time unit.

Solution. First, we calculate

$$P[\Lambda > \lambda] = \int_\lambda^\infty \frac{2}{x^3}\, dx = \frac{1}{\lambda^2} \quad \text{for } \lambda \geq 1$$

We seek

$$P[N(1) > 2] = \int_0^1 1 \cdot e^{-\lambda}\frac{\lambda^2}{2}\, d\lambda + \int_1^\infty \frac{1}{\lambda^2}e^{-\lambda}\frac{\lambda^2}{2}\, d\lambda = \frac{1}{2}\int_0^1 \lambda^2 e^{-\lambda}\, d\lambda + \frac{1}{2}e^{-1}$$

Doing the integral above by parts, we find that

$$P[N(1) > 2] = \frac{1}{2}\left(2 - 5e^{-1} + e^{-1}\right) = 1 - 2e^{-1} \simeq 0.2642$$

Example 5.4.2. If Λ has a geometric distribution, with parameter $p \in (0, 1)$, then we deduce from the formula (5.121) that

$$P[N(t) = n] = \sum_{k=1}^{\infty} e^{-kt} \frac{(kt)^n}{n!} q^{k-1} p = \frac{p}{q} \frac{t^n}{n!} \sum_{k=1}^{\infty} \left(e^{-t} q\right)^k k^n$$

Thus, we have

$$P[N(t) = 0] = \frac{p}{q} \sum_{k=1}^{\infty} \left(e^{-t} q\right)^k = \frac{p}{q} \left(\frac{e^{-t} q}{1 - e^{-t} q}\right)$$

Moreover, let $r := e^{-t} q$. We may write that

$$P[N(t) = 1] = \frac{p}{q} t \sum_{k=1}^{\infty} k r^k = \frac{p}{q} tr \sum_{k=1}^{\infty} \frac{d}{dr} r^k = \frac{p}{q} tr \frac{d}{dr} \left(\frac{r}{1-r}\right) = \frac{p}{q} \frac{tr}{(1-r)^2}$$

Example 5.4.3 (**The Pólya[1] process**). Suppose that Λ has a gamma distribution, with parameters $\alpha = k \in \mathbb{N}$ and $\beta > 0$. We have

$$P[N(t) = n]$$

$$= \int_0^{\infty} P[N(t) = n \mid \Lambda = \lambda] f_\Lambda(\lambda) \, d\lambda$$

$$= \int_0^{\infty} e^{-\lambda t} \frac{(\lambda t)^n}{n!} (\beta e^{-\beta \lambda}) \frac{(\beta \lambda)^{k-1}}{(k-1)!} \, d\lambda$$

$$= \frac{t^n}{n!} \frac{\beta^k}{(k-1)!} \int_0^{\infty} e^{-(t+\beta)\lambda} \lambda^{n+k-1} \, d\lambda$$

$$= \frac{t^n}{n!} \frac{\beta^k}{(k-1)!} \frac{1}{(t+\beta)^{n+k}} \int_0^{\infty} e^{-x} x^{n+k-1} \, dx \quad (\text{with } x = (t+\beta)\lambda)$$

$$= \frac{t^n}{n!} \frac{\beta^k}{(k-1)!} \frac{1}{(t+\beta)^{n+k}} \Gamma(n+k)$$

$$= \frac{t^n}{n!} \frac{\beta^k}{(k-1)!} \frac{1}{(t+\beta)^{n+k}} (n+k-1)!$$

$$= \binom{n+k-1}{n} p^k (1-p)^n \quad \text{for } n = 0, 1, \ldots$$

where

$$p := \frac{\beta}{t + \beta}$$

---

[1] George Pólya, 1887–1985, was born in Hungary and died in the United States. He contributed to several domains of mathematics, including probability theory, number theory, mathematical physics, and complex analysis.

We say that $N(t)$ has a *negative binomial*, or *Pascal*[2] *distribution*, with parameters k and p. Furthermore, the process $\{N(t), t \geq 0\}$ is called a *negative binomial process*, or a *Pólya process*.

Remark. There exist various ways of defining a negative binomial distribution. The version above generalizes the geometric distribution when we define it as being a random variable that counts the number of *failures*, in Bernoulli trials, *before* the first *success*, which can be checked by setting $k = 1$. Thus, here $N(t)$ would correspond to the variable that counts the number of failures before the kth success. Note also that if $k = 1$, then Λ has an exponential distribution with parameter β.

Definition 5.4.2. *Let* $\{\Lambda(t), t \geq 0\}$ *be a stochastic process for which* $\Lambda(t)$ *is nonnegative, for all* $t \geq 0$. *If, given that* $\Lambda(t) = \lambda(t)$, *for all* $t \geq 0$, *the counting process* $\{N(t), t \geq 0\}$ *is a nonhomogeneous Poisson process with intensity function* $\lambda(t)$, *then* $\{N(t), t \geq 0\}$ *is called a* **doubly stochastic Poisson process** *or a* **Cox**[3] **process**.

Remarks. i) If the random variable $\Lambda(t)$ does not depend on t, we retrieve the conditional Poisson process. Moreover, if $\Lambda(t)$ is a deterministic function $\lambda(t)$, then $\{N(t), t \geq 0\}$ is a nonhomogeneous Poisson process, with intensity function $\lambda(t)$. The doubly stochastic Poisson process thus includes the homogeneous, nonhomogeneous, and conditional Poisson processes as particular cases.

ii) The doubly stochastic Poisson process is not a Poisson process unless the random variable $\Lambda(t)$ is equal to the constant $\lambda > 0$, for all $t \geq 0$.

iii) The process $\{\Lambda(t), t \geq 0\}$ is called the *intensity process*.

We can write, for all $t_2 \geq t_1 \geq 0$, that

$$N(t_2) - N(t_1) \mid \{\Lambda(t) = \lambda(t) \; \forall \, t \geq 0\} \sim \text{Poi}(m(t_2) - m(t_1)) \qquad (5.131)$$

where

$$m(t) := \int_0^t \lambda(s) \, ds \qquad (5.132)$$

It can also be shown that

$$P[N(t_2) - N(t_1) = k \mid \{\Lambda(t), 0 \leq t_1 \leq t \leq t_2\}] = e^{\int_{t_1}^{t_2} \Lambda(t) \, dt} \frac{\left(\int_{t_1}^{t_2} \Lambda(t) \, dt\right)^k}{k!}$$
$$(5.133)$$

for $k = 0, 1, \ldots$.

---

[2] See p. 135.
[3] Sir David Cox, professor at the University of Oxford, in England.

To simplify the writing, set

$$N(t) \mid \{\Lambda(s), 0 \leq s \leq t\} = N(t) \mid \Lambda(0, t) \qquad (5.134)$$

We then have

$$E[N(t) \mid \Lambda(0, t)] = \int_0^t \Lambda(s) \, ds \qquad (5.135)$$

so that

$$E[N(t)] = E[E[N(t) \mid \Lambda(0, t)]] = E\left[\int_0^t \Lambda(s) \, ds\right] = \int_0^t E[\Lambda(s)] \, ds \quad (5.136)$$

Similarly, we have

$$E[N^2(t) \mid \Lambda(0, t)] = \int_0^t \Lambda(s) \, ds + \left(\int_0^t \Lambda(s) \, ds\right)^2 \qquad (5.137)$$

Since

$$E\left[\left(\int_0^t \Lambda(s) \, ds\right)^2\right] = E\left[\int_0^t \int_0^t \Lambda(s)\Lambda(u) \, ds \, du\right]$$

$$= \int_0^t \int_0^t E[\Lambda(s)\Lambda(u)] \, ds \, du \qquad (5.138)$$

we may write that

$$E[N^2(t)] = E[E[N^2(t) \mid \Lambda(0, t)]] = \int_0^t E[\Lambda(s)] \, ds + \int_0^t \int_0^t R_\Lambda(s, u) \, ds \, du$$

$$(5.139)$$

Example 5.4.4. Suppose that $\{\Lambda(t), t \geq 0\}$ is a homogeneous Poisson process, with rate $\lambda > 0$. Then

$$E[\Lambda(t)] = \lambda t \quad \text{and} \quad R_\Lambda(s, u) = \lambda \min\{s, u\} + \lambda^2 su$$

We calculate

$$E[N(t)] = \int_0^t \lambda s \, ds = \lambda \frac{t^2}{2}$$

and

$$E[N^2(t)] = \lambda \frac{t^2}{2} + \int_0^t \int_0^t [\lambda \min\{s, u\} + \lambda^2 su] \, ds \, du$$

$$= \lambda \frac{t^2}{2} + \int_0^t \int_0^t \lambda \min\{s, u\} \, ds \, du + \lambda^2 \frac{t^4}{4}$$

$$= \lambda \frac{t^2}{2} + \lambda \int_0^t \left[\int_0^u s \, ds + \int_u^t u \, ds\right] du + \lambda^2 \frac{t^4}{4}$$

$$= \lambda \frac{t^2}{2} + \lambda \frac{t^3}{3} + \lambda^2 \frac{t^4}{4}$$

It follows that

$$V[N(t)] = \lambda \frac{t^2}{2} + \lambda \frac{t^3}{3} + \lambda^2 \frac{t^4}{4} - \left(\lambda \frac{t^2}{2}\right)^2 = \lambda \left(\frac{t^2}{2} + \frac{t^3}{3}\right)$$

Example 5.4.5. If $\{\Lambda(t), t \geq 0\}$ is a geometric Brownian motion, then $\Lambda(t)$ has a lognormal distribution (see p. 186), with parameters μt and $\sigma^2 t$. Its mean is given by

$$E[\Lambda(t)] = \exp\{\mu t + \frac{1}{2}\sigma^2 t\}$$

We then have

$$E[N(t)] = \int_0^t \exp\{\mu s + \frac{1}{2}\sigma^2 s\} \, ds = \frac{1}{\mu + \frac{1}{2}\sigma^2}[\exp\{\mu t + \frac{1}{2}\sigma^2 t\} - 1]$$

if $\mu + \frac{1}{2}\sigma^2 \neq 0$ (and $E[N(t)] = t$ if $\mu + \frac{1}{2}\sigma^2 = 0$).

5.5 Filtered Poisson processes

After having generalized the homogeneous Poisson process in various ways, we now generalize the compound Poisson process.

Definition 5.5.1. *Let $X_1, X_2, \ldots$ be random variables that are i.i.d. and independent of the Poisson process $\{N(t), t \geq 0\}$, with rate $\lambda > 0$. We say that the stochastic process $\{Y(t), t \geq 0\}$ defined by*

$$Y(t) = \sum_{k=1}^{N(t)} w(t, T_k, X_k) \quad \forall \, t \geq 0 \quad (Y(t) = 0 \text{ if } N(t) = 0) \qquad (5.140)$$

*where the T_k's are the arrival times of the events of the Poisson process and the function $w(\cdot, \cdot, \cdot)$ is called the **response function**, is a **filtered Poisson** process.*

Remarks. i) The compound Poisson process is the particular case obtained by setting $w(t, T_k, X_k) = X_k$.

ii) We can say that the response function gives the value at time t of a *signal* that occurred at time T_k and for which the quantity X_k has been added to the filtered Poisson process. The random variable $Y(t)$ is then the sum of the value at time t of every signal that occurred since the initial time 0.

iii) An application of this type of process is the following: the variable $Y(t)$ represents the flow of a certain river at time t (since the beginning of the period considered, for example, since the beginning of the thawing period), the T_k's are the arrival times of the precipitation events (snow or rain), and X_k is the quantity of precipitation observed at time T_k and measured in inches of water or water equivalent of snow. In practice, the precipitation does not fall to the ground instantly. However, we must *discretize* the time variable, because the flow of rivers is not measured in a continuous way, but rather, in many cases, only once per day. Then, the index k represents the kth day since the initial time and X_k is the quantity of precipitation observed on this kth day. A classic response function in this example is given by

$$w(t, T_k, X_k) = X_k e^{-(t-T_k)/c} \quad \text{for } t \geq T_k \tag{5.141}$$

where c is a positive constant that depends on each river and must be estimated.

Suppose that we replace the random variable T_k by the deterministic variable s in the function $w(t, T_k, X_k)$. Let $C_w(\theta)$ be the characteristic function of the random variable $w(t, s, X_k)$:

$$C_w(\theta) \ (= C_w(\theta, t, s)) \ := E\left[e^{j\theta w(t,s,X_k)}\right] \tag{5.142}$$

Making use of Proposition 5.1.6, we can show that the characteristic function of $Y(t)$ is given by

$$C_{Y(t)}(\theta) = \exp\left\{-\lambda t + \lambda \int_0^t C_w(\theta) \, ds\right\} \tag{5.143}$$

from which we obtain the following proposition.

Proposition 5.5.1. *If $E[w^2(t, s, X_k)] < \infty$, then the mean and the variance of $Y(t)$ are given by*

$$E[Y(t)] = \lambda \int_0^t E[w(t, s, X_k)] \, ds \tag{5.144}$$

and

$$V[Y(t)] = \lambda \int_0^t E[w^2(t, s, X_k)] \, ds \tag{5.145}$$

Remarks. i) When X_k is a continuous random variable, the mathematical expectations $E[w^n(t, s, X_k)]$, for $n = 1, 2$, are calculated as follows:

$$E[w^n(t, s, X_k)] = \int_{-\infty}^{\infty} w^n(t, s, x) f_{X_k}(x) \, dx \tag{5.146}$$

ii) It can also be shown that

$$\text{Cov}[Y(t_1), Y(t_2)] = \lambda \int_0^{\min\{t_1, t_2\}} E[w(t_1, s, X_k) w(t_2, s, X_k)] \, ds \quad \forall \, t_1, t_2 \geq 0$$

$$(5.147)$$

Example 5.5.1. Suppose that the random variables X_k have an exponential distribution with parameter μ in the formula (5.141). We can then write that

$$C_w(\theta) := E\left[e^{j\theta w(t,s,X_k)}\right] = E\left[e^{j\theta X_k e^{-(t-s)/c}}\right] = E\left[e^{j\theta g(t,s,c)X_k}\right]$$

where

$$g(t, s, c) := e^{-(t-s)/c}$$

We have

$$E\left[e^{j\theta g(t,s,c)X_k}\right] = \int_0^\infty e^{j\theta g(t,s,c)x} \mu e^{-\mu x} \, dx = \frac{\mu}{\mu - j\theta g(t,s,c)}$$

Next, we calculate

$$\int_0^t C_w(\theta) \, ds = \int_0^t \frac{\mu}{\mu - j\theta g(t,s,c)} \, ds = \int_0^t \frac{\mu - j\theta g(t,s,c) + j\theta g(t,s,c)}{\mu - j\theta g(t,s,c)} \, ds$$

$$= t + \int_0^t \frac{j\theta g(t,s,c)}{\mu - j\theta g(t,s,c)} \, ds$$

Given that

$$\frac{\partial}{\partial s} g(t, s, c) = e^{-(t-s)/c} \frac{1}{c} = \frac{1}{c} g(t, s, c)$$

we may write that

$$\int_0^t \frac{j\theta g(t,s,c)}{\mu - j\theta g(t,s,c)} \, ds = -c \ln\left[\mu - j\theta g(t,s,c)\right]\big|_{s=0}^{s=t}$$

$$= -c\ln(\mu - j\theta) + c\ln(\mu - j\theta e^{-t/c})$$

It follows that

$$C_{Y(t)}(\theta) = \exp\left\{-\lambda t + \lambda[t - c\ln(\mu - j\theta) + c\ln(\mu - j\theta e^{-t/c})]\right\}$$

$$= \left(\frac{\mu - j\theta e^{-t/c}}{\mu - j\theta}\right)^{\lambda c}$$

Using this formula, we find that

$$E[Y(t)] = \frac{\lambda c}{\mu}\left(1 - e^{-t/c}\right) \quad \text{and} \quad V[Y(t)] = \frac{\lambda c}{\mu^2}\left(1 - e^{-2t/c}\right)$$

Remarks. i) In this example, we have

$$E[w^n(t,s,X_k)] = E\left[X_k^n e^{-n(t-s)/c}\right] = e^{-n(t-s)/c} E\left[X_k^n\right]$$

$$= e^{-n(t-s)/c} \frac{\Gamma(n+1)}{\mu^n} < \infty \quad \text{for } n = 0,1,\ldots$$

We then deduce from the formulas (5.144) and (5.145) that

$$E[Y(t)] = \lambda \int_0^t E[w(t,s,X_k)]\, ds = \lambda \int_0^t e^{-(t-s)/c} \frac{1}{\mu}\, ds = \frac{\lambda c}{\mu}\left(1 - e^{-t/c}\right)$$

and

$$V[Y(t)] = \lambda \int_0^t E[w^2(t,s,X_k)]\, ds = \lambda \int_0^t e^{-2(t-s)/c} \frac{2}{\mu^2}\, ds$$

$$= \frac{\lambda c}{\mu^2}\left(1 - e^{-2t/c}\right)$$

as above.

ii) We can check that the covariance of the random variables $Y(t)$ and $Y(t+\tau)$ is given by

$$\text{Cov}[Y(t), Y(t+\tau)] = \frac{\lambda c}{\mu^2} e^{\tau/c}\left(1 - e^{-2t/c}\right) \quad \forall\, t, \tau \geq 0$$

Finally, we can generalize the notion of filtered Poisson process by setting

$$Y(t) = \sum_{k=1}^{N(t)} W_k(t, T_k) \quad \forall\, t \geq 0 \quad \text{(and } Y(t) = 0 \text{ if } N(t) = 0) \qquad (5.148)$$

where $\{W_k(t,s), t \geq 0, s \geq 0\}$ is a stochastic process (with two time parameters). We assume that the processes $\{W_k(t,s), t \geq 0, s \geq 0\}$ are independent and identically distributed and are also independent of the Poisson process $\{N(t), t \geq 0\}$ (having rate $\lambda > 0$). We say that $\{Y(t), t \geq 0\}$ is a *generalized filtered Poisson process*. It can be shown that

$$E[Y(t)] = \lambda \int_0^t E[W_1(t,s)]\, ds \quad \text{and} \quad V[Y(t)] = \lambda \int_0^t E[W_1^2(t,s)]\, ds$$

$$(5.149)$$

5.6 Renewal processes

An essential characteristic of the Poisson process, $\{N(t), t \geq 0\}$, is that the time between consecutive events is a random variable having an exponential distribution with parameter λ, regardless of the state in which the process is. Moreover, the random variables τ_i denoting the time that $\{N(t), t \geq 0\}$ spends in state i, for $i = 0, 1, \ldots$, are independent.

We also know that the Poisson process is a particular continuous-time Markov chain. Another way of generalizing the Poisson process is now to suppose that the nonnegative variables τ_i are independent and identically distributed, so that they can have any distribution, whether discrete or continuous.

Definition 5.6.1. *Let $\{N(t), t \geq 0\}$ be a counting process, and let τ_i be the random variable denoting the time that the process spends in state i, for $i = 0, 1, \ldots$. The process $\{N(t), t \geq 0\}$ is called a* **renewal process** *if the nonnegative variables $\tau_0, \tau_1, \ldots$ are independent and identically distributed.*

Remarks. i) It is sometimes possible (see Ex. 5.1.2) to transform into a Poisson process a renewal process for which the time spent in the states $0, 1, \ldots$ does not have an exponential distribution.

ii) We say that a *renewal* has occurred every time an event of the counting process takes place.

iii) Some authors suppose that the random variables $\tau_0, \tau_1, \ldots$ are *strictly* positive.

iv) We can generalize the definition above by supposing that the random variable τ_0 is independent of the other variables, $\tau_1, \tau_2, \ldots$ but does not necessarily have the same distribution as these variables. In this case, $\{N(t), t \geq 0\}$ is called a *modified* or *delayed* renewal process.

The time T_n of the nth renewal of the (continuous-time) stochastic process $\{N(t), t \geq 0\}$, defined by [see Eq. (5.32)]

$$T_n = \sum_{i=0}^{n-1} \tau_i \quad \text{for } n = 1, 2, \ldots \tag{5.150}$$

satisfies the relation $T_n \leq t \Leftrightarrow N(t) \geq n$. By setting $T_0 = 0$, we may write, as in Proposition 5.1.4, that

$$N(t) = \max\{n \geq 0 : T_n \leq t\} \tag{5.151}$$

Since

$$P[N(t) = n] = P[N(t) \geq n] - P[N(t) \geq n + 1] \tag{5.152}$$

we can state the following proposition.

Proposition 5.6.1. *The probability mass function of the random variable $N(t)$ can be obtained from the formula*

$$P[N(t) = n] = P[T_n \leq t] - P[T_{n+1} \leq t] \tag{5.153}$$

for $t \geq 0$ and for $n = 0, 1, \ldots$.

In general, it is difficult to find the exact distribution function of the variable T_n. We have

$$C_{T_n}(\omega) := E\left[e^{j\omega T_n}\right] = E\left[e^{j\omega \sum_{i=0}^{n-1} \tau_i}\right] = E\left[\prod_{i=0}^{n-1} e^{j\omega \tau_i}\right]$$

$$\stackrel{\text{ind.}}{=} \prod_{i=0}^{n-1} E\left[e^{j\omega \tau_i}\right] = \prod_{i=0}^{n-1} C_{\tau_i}(\omega) \tag{5.154}$$

When τ_i is a continuous random variable, the characteristic function $C_{T_n}(\omega)$ is the Fourier transform of the probability density function $f_{T_n}(t)$ of T_n. We can then write that

$$f_{T_n}(t) = f_{\tau_0}(t_0) * f_{\tau_1}(t_1) * \ldots * f_{\tau_{n-1}}(t_{n-1}) \tag{5.155}$$

where $t_0 + t_1 + \ldots + t_{n-1} = t$. That is, the density function of T_n is the *convolution product* of the density functions of $\tau_0, \ldots, \tau_{n-1}$.

In some cases, we know the exact distribution of the random variable T_n. For example, if $\tau_i \sim \text{Exp}(\lambda)$, then

$$T_n \sim G(n, \lambda) \tag{5.156}$$

Similarly, if $\tau_i \sim \text{Poi}(\lambda)$, we have

$$T_n \sim \text{Poi}(n\lambda) \tag{5.157}$$

If n is large enough, we deduce from the central limit theorem that

$$T_n \approx N(n\mu, n\sigma^2) \tag{5.158}$$

where $\mu := E[\tau_i]$ and $\sigma^2 := V[\tau_i]\ \forall\ i$.

Example 5.6.1. We can use the formula (5.153) to check that, for a Poisson process with rate λ, we have

$$P[N(t) = n] = e^{-\lambda t}\frac{(\lambda t)^n}{n!} \quad \text{for } t \geq 0 \text{ and } n = 0, 1, \ldots$$

In this case,

$$T_n \sim G(n, \lambda) \quad \Longrightarrow \quad P[T_n \leq t] = \int_0^t \lambda e^{-\lambda x}\frac{(\lambda x)^{n-1}}{(n-1)!}\ dx$$

Let

$$I_n := \int_0^t e^{-\lambda x} x^n\ dx \quad \text{for } n = 0, 1, \ldots$$

Integrating by parts, we obtain

$$I_n = -x^n \frac{e^{-\lambda x}}{\lambda}\Big|_0^t + \frac{n}{\lambda} I_{n-1}$$

$$= \cdots = -\sum_{k=1}^{n} \frac{n!}{(n-k+1)!} x^{n-k+1} \frac{e^{-\lambda x}}{\lambda^k}\Big|_0^t + \frac{n!}{\lambda^n} I_0$$

$$= -\sum_{k=1}^{n} \frac{n!}{(n-k+1)!} t^{n-k+1} \frac{e^{-\lambda t}}{\lambda^k} + \frac{n!}{\lambda^n} I_0$$

where

$$I_0 := \int_0^t e^{-\lambda x}\, dx = \frac{1 - e^{-\lambda t}}{\lambda}$$

It follows that

$$P[T_{n+1} \le t] = \frac{\lambda^{n+1}}{n!}\left[-\sum_{k=1}^{n} \frac{n!}{(n-k+1)!} t^{n-k+1} \frac{e^{-\lambda t}}{\lambda^k} + \frac{n!}{\lambda^n}\left(\frac{1 - e^{-\lambda t}}{\lambda} \right) \right]$$

$$= 1 - e^{-\lambda t} - \sum_{k=1}^{n} \frac{(\lambda t)^{n-k+1}}{(n-k+1)!} e^{-\lambda t} = 1 - \sum_{i=0}^{n} \frac{(\lambda t)^i}{i!} e^{-\lambda t}$$

Similarly,

$$P[T_n \le t] = 1 - \sum_{i=0}^{n-1} \frac{(\lambda t)^i}{i!} e^{-\lambda t}$$

Thus, we indeed have

$$P[N(t) = n] = \left(1 - \sum_{i=0}^{n-1} \frac{(\lambda t)^i}{i!} e^{-\lambda t} \right) - \left(1 - \sum_{i=0}^{n} \frac{(\lambda t)^i}{i!} e^{-\lambda t} \right) = \frac{(\lambda t)^n}{n!} e^{-\lambda t}$$

When τ_i is a discrete random variable, the probability $P[\tau_i = 0]$ may be *strictly* positive. That is, the time needed for a renewal to occur may be equal to zero. For example, if $\tau_i \sim \text{Poi}(\lambda)$, then $P[\tau_i = 0] = e^{-\lambda} > 0$. However, if the nonnegative random variable τ_i is not the constant 0, we may write that $\mu > 0$. If we assume that $\mu < \infty$ and $\sigma^2 < \infty$, the *strong law of large numbers* (see p. 32) then implies that

$$1 = P\left[\lim_{n \to \infty} \frac{\sum_{i=0}^{n-1} \tau_i}{n} = \mu \right] = P\left[\lim_{n \to \infty} \frac{T_n}{n} = \mu \right] \implies P\left[\lim_{n \to \infty} T_n = \infty \right] = 1 \tag{5.159}$$

from which we deduce from the formula (5.151) that we can write

$$P[N(t) = \infty] = 0 \quad \text{for all } t < \infty \tag{5.160}$$

because the random variable T_n will eventually be larger than any finite t. That is, there *cannot* be an *infinite* number of renewals in a finite-time interval. However, given that $P[\tau_i = \infty] = 0$, we may conclude that

$$P\left[\lim_{t \to \infty} N(t) = \infty\right] = 1 \tag{5.161}$$

Since, in most cases, it is very difficult to explicitly calculate the probability of the event $\{N(t) = n\}$, we must generally content ourselves with finding the mean of the random variable $N(t)$.

Definition 5.6.2. *The function $m_N(t) := E[N(t)]$ is called the* **renewal function** *(or* **mean-value function***) of the renewal process* $\{N(t), t \geq 0\}$.

Proposition 5.6.2. *The renewal function $m_N(t)$ can be calculated as follows:*

$$m_N(t) = \sum_{n=1}^{\infty} P[T_n \leq t] \tag{5.162}$$

Proof. As $N(t)$ is a random variable taking its values in the set $\mathbb{N}^0 = \{0, 1, \dots\}$, we may write (using the relation $T_n \leq t \Leftrightarrow N(t) \geq n$) that

$$E[N(t)] := \sum_{i=0}^{\infty} iP[N(t) = i] = \sum_{i=1}^{\infty} iP[N(t) = i] = \sum_{i=1}^{\infty} \sum_{n=1}^{i} P[N(t) = i]$$

$$= \sum_{n=1}^{\infty} \sum_{i=n}^{\infty} P[N(t) = i] = \sum_{n=1}^{\infty} P[N(t) \geq n]$$

$$= \sum_{n=1}^{\infty} P[T_n \leq t] \quad \square \tag{5.163}$$

Remarks. i) Since it is also difficult to calculate the probability $P[T_n \leq t]$, in practice the proposition does not enable us very often to obtain the function $m_N(t)$. Moreover, if we managed to calculate $P[N(t) = n]$, for all n, then it is perhaps simpler to find the mean of $N(t)$ by directly using the definition of $E[N(t)]$.

ii) We will further discuss another technique that enables us, when the τ_i's are continuous random variables, to calculate $m_N(t)$, namely by solving an *integral equation*.

Example 5.6.2. Suppose that τ_i has a Bernoulli distribution, with parameter $p \in (0, 1)$. We then have

$$T_n := \sum_{i=0}^{n-1} \tau_i \sim B(n, p)$$

so that

$$P[T_n \leq t] = \sum_{k=0}^{[t]} \binom{n}{k} p^k (1-p)^{n-k}$$

where $[t]$ denotes the *integer part* of t. Theoretically, we can calculate $m_N(t)$ by finding the value of the sum

$$S(t,p) := \sum_{n=1}^{\infty} \sum_{k=0}^{[t]} \binom{n}{k} p^k (1-p)^{n-k}$$

where $\binom{n}{k} = 0$ if $k > n$.

Consider the particular case when $p = 1/2$. We have

$$S(t,1/2) = \sum_{n=1}^{\infty} \left\{ (1/2)^n \sum_{k=0}^{[t]} \binom{n}{k} \right\}$$

We find, making use of a mathematical software package, that $S(0,1/2) = 1$, $S(1,1/2) = 3$, $S(2,1/2) = 5$, $S(3,1/2) = 7$, etc., from which we deduce the formula

$$S(t,1/2) = 2[t] + 1 \quad \text{for all } t \geq 0$$

Now, let $t = r \in \{0, 1, \dots\}$. Given that $\tau_i = 0$ or 1, for all i, we may write that $N(r) \geq r$. We find that

$$N(r) = \begin{cases} r & \text{with probability } p^{r+1} \\ \\ r+1 & \text{with probability } \binom{r+1}{1} p^{r+1}(1-p) \\ \\ r+2 & \text{with probability } \binom{r+2}{2} p^{r+1}(1-p)^2 \\ \\ \cdots \quad \cdots \end{cases}$$

In general, we have

$$P[N(r) = r+k] = \binom{r+k}{k} p^{r+1}(1-p)^k \quad \text{for } k = 0, 1, \dots$$

Remark. We have that $P[N(r) = r] = p^{r+1}$, and not p^r, because $\{N(r) = r\}$ if and only if (iff) the first $r+1$ renewals each take one time unit. In particular, $N(0) = 0$ iff $\tau_0 = 1$. Similarly, $N(1) = 1$ iff $\tau_0 = 1$ *and* $\tau_1 = 1$, etc.

We may write that $X := N(r) + 1$ has a negative binomial distribution (see p. 135) with parameters $r + 1$ and p. Now, the mean of this distribution is given by $(r+1)/p$, from which we deduce that

$$E[N(r)] = \frac{r+1}{p} - 1 \quad \Longrightarrow \quad E[N(t)] = \frac{[t]+1}{p} - 1$$

Note that if $p = 1/2$, then we retrieve the formula $S(t, 1/2) = 2[t]+1$ obtained previously.

Suppose that the τ_i's are *continuous* random variables. We then deduce from Eq. (5.162) that

$$\frac{d}{dt}m_N(t) = \sum_{n=1}^{\infty} f_{T_n}(t) \tag{5.164}$$

from which, taking the Laplace transform of both members of the preceding equation, we obtain

$$m_N^*(\alpha) := \int_0^{\infty} e^{\alpha t} dm_N(t) = \sum_{n=1}^{\infty} \int_0^{\infty} e^{\alpha t} f_{T_n}(t) \, dt \tag{5.165}$$

$$\Longleftrightarrow \quad m_N^*(\alpha) = \sum_{n=1}^{\infty} M_{T_n}(\alpha) \tag{5.166}$$

where M_{T_n} is the moment-generating function of the random variable T_n (see p. 19) and α is a negative constant. Since the τ_i's are independent and identically distributed random variables, we may write that

$$M_{T_n}(\alpha) \stackrel{\text{ind.}}{=} \prod_{i=0}^{n-1} M_{\tau_i}(\alpha) \stackrel{\text{i.d.}}{=} [M_{\tau_0}(\alpha)]^n \tag{5.167}$$

The parameter α being negative, we have that $M_{\tau_0}(\alpha) \in (0,1)$. It follows that

$$m_N^*(\alpha) = \sum_{n=1}^{\infty} [M_{\tau_0}(\alpha)]^n = \frac{M_{\tau_0}(\alpha)}{1 - M_{\tau_0}(\alpha)} \tag{5.168}$$

so that

$$M_{\tau_0}(\alpha) = \frac{m_N^*(\alpha)}{1 + m_N^*(\alpha)} \tag{5.169}$$

By the uniqueness of the Laplace transform (and of the inverse Laplace transform), we deduce from the last equation that to each renewal function $m_N(t)$ corresponds a different density function $f_{\tau_0}(t)$. Similarly, Eq. (5.168) implies that the density function $f_{\tau_0}(t)$ uniquely determines the function $m_N(t)$. Moreover, it can be shown that these results hold true, whether the random variables τ_i are discrete or continuous. We can thus state the following proposition.

Proposition 5.6.3. *There is a* one-to-one *correspondence* between the distri-bution function of the random variables τ_i and the renewal function $m_N(t)$.

Since $m_N(t) = \lambda t$ in the case of the Poisson process, we have the following corollary.

Corollary 5.6.1. *The Poisson process is the* only *renewal process having a* linear *renewal function.*

Remark. The Poisson process is also the *only Markovian* renewal process.

In the case when the random variables τ_i are continuous, we have

$$m_N(t) := E[N(t)] = E[E[N(t) \mid \tau_0]] = \int_0^\infty E[N(t) \mid \tau_0 = \tau] f_{\tau_0}(\tau) \, d\tau$$
$$(5.170)$$

Moreover, the relation

$$\tau_0 > t \quad \Longleftrightarrow \quad N(t) = 0 \tag{5.171}$$

implies that

$$m_N(t) = \int_0^t E[N(t) \mid \tau_0 = \tau] f_{\tau_0}(\tau) \, d\tau \tag{5.172}$$

Finally, we may write (the τ_i's being i.i.d. random variables) that

$$E[N(t) \mid \tau_0 = \tau] = 1 + E[N(t - \tau)] = 1 + m_N(t - \tau) \quad \text{for } 0 \le \tau \le t \tag{5.173}$$

It follows that

$$m_N(t) = \int_0^t [1 + m_N(t - \tau)] f_{\tau_0}(\tau) \, d\tau \tag{5.174}$$

$$\Longleftrightarrow \quad m_N(t) = F_{\tau_0}(t) + \int_0^t m_N(t - \tau) f_{\tau_0}(\tau) \, d\tau \quad \text{for } t \ge 0 \tag{5.175}$$

Definition 5.6.3. *Equation (5.175) is called the* **renewal equation** *(for the* renewal function) of the process $\{N(t), t \ge 0\}$.

To explicitly calculate the renewal function, it is often easier to solve the integral equation (5.175). Setting $s = t - \tau$, we can rewrite it as follows:

$$m_N(t) = F_{\tau_0}(t) + \int_0^t m_N(s) f_{\tau_0}(t - s) \, ds \quad \text{for } t \ge 0 \tag{5.176}$$

Next, differentiating both members of the preceding equation, we obtain

$$m_N'(t) = f_{\tau_0}(t) + m_N(t) f_{\tau_0}(0) + \int_0^t m_N(s) f_{\tau_0}'(t - s) \, ds \tag{5.177}$$

When $f_{\tau_0}(t - s)$ is a polynomial function, we can differentiate repeatedly the members of this equation, until we obtain a *differential equation* for the function $m_N(t)$. We have, in particular, the boundary condition $m_N(0) = 0$.

Remark. We can also sometimes obtain a differential equation for $m_N(t)$ even if $f_{\tau_0}(t - s)$ is not a polynomial function.

Example 5.6.3. Let

$$f_{\tau_0}(t) = 4t^3 \quad \text{if } 0 \le t \le 1$$

We have [see Eq. (5.177)]:

$$m_N'(t) = 4t^3 + m_N(t) \times 0 + \int_0^t m_N(s)12(t - s)^2 \, ds$$

$$= 4t^3 + \int_0^t m_N(s)12(t - s)^2 \, ds \quad \text{for } 0 \le t \le 1$$

Note that this equation implies that $m_N'(0) = 0$. We differentiate once more:

$$m_N''(t) = 12t^2 + \int_0^t m_N(s)24(t - s) \, ds$$

from which we deduce that $m_N''(0) = 0$. Next, we have

$$m_N'''(t) = 24t + 24 \int_0^t m_N(s) \, ds$$

so that $m_N'''(0) = 0$. Finally, we obtain the ordinary differential equation

$$m_N^{(4)}(t) = 24 + 24 m_N(t)$$

The general solution of this equation is given by

$$m_N(t) = -1 + c_1 \cos kt + c_2 \sin kt + c_3 e^{kt} + c_4 e^{-kt}$$

where $k := (24)^{1/4}$ and $c_1, \ldots, c_4$ are constants. The particular solution that satisfies the conditions $m_N(0) = m_N'(0) = m_N''(0) = m_N'''(0) = 0$ is

$$m_N(t) = -1 + \frac{1}{2} \cos kt + \frac{1}{4} \left(e^{kt} + e^{-kt} \right) \quad \text{for } 0 \le t \le 1$$

Note that the solution above is only valid for $t \in [0, 1]$. Moreover, since the random variable τ_0 takes its values in the interval $[0, 1]$, the mean $m_N(1)$ must be greater than 1. We calculate

$$m_N(1) = -1 + \frac{1}{2} \cos k + \frac{1}{4} \left(e^k + e^{-k} \right) \simeq 1.014$$

Equation (5.175) is a particular case of the *general renewal equation*

$$g(t) = h(t) + \int_0^\infty g(t - \tau) dF_{\tau_0}(\tau) \quad \text{for } t \geq 0 \qquad (5.178)$$

where $h(t)$ and $F_{\tau_0}(t)$ are functions defined for all $t \geq 0$. The following proposition can be proved.

Proposition 5.6.4. *The function $g(t)$ defined by*

$$g(t) = h(t) + \int_0^t h(t - \tau) \, dm_N(\tau) \quad \text{for } t \geq 0 \qquad (5.179)$$

is a solution of the renewal equation (5.178).

Remarks. i) Equation (5.175) is obtained from (5.178) by taking $g(t) = m_N(t)$ and $h(t) = F_{\tau_0}(t)$ and by noticing that $m_N(t) = 0$ if $t < 0$. If $h(t)$ is a function bounded on finite intervals, then $g(t)$ is bounded on finite intervals as well. Moreover, $g(t)$ is the *unique* solution of (5.179) with this property. Note that the distribution function $F_{\tau_0}(t)$ ($= h(t)$) is evidently a bounded function, because $0 \leq F_{\tau_0}(t) \leq 1 \ \forall \, t \geq 0$.

ii) We can replace $dm_N(\tau)$ by $m_N'(\tau) \, d\tau$ in the integral above.

Corollary 5.6.2. *If τ_0 is a continuous random variable, then the second moment of $N(t)$ about the origin, $E[N^2(t)]$, is given by*

$$E[N^2(t)] = m_N(t) + 2 \int_0^t m_N(t - \tau) \, dm_N(\tau) \quad \text{for } t \geq 0 \qquad (5.180)$$

Proof. We may write that

$$E[N^2(t)] = \int_0^\infty E[N^2(t) \mid \tau_0 = \tau] f_{\tau_0}(\tau) \, d\tau \qquad (5.181)$$

$$= \int_0^t E[N^2(t) \mid \tau_0 = \tau] f_{\tau_0}(\tau) \, d\tau = \int_0^t E\left[(1 + N(t - \tau))^2 \right] f_{\tau_0}(\tau) \, d\tau$$

$$= \int_0^t \left\{ 1 + 2E[N(t - \tau)] + E[N^2(t - \tau)] \right\} f_{\tau_0}(\tau) \, d\tau$$

$$= F_{\tau_0}(t) + 2 \int_0^t m_N(t - \tau) f_{\tau_0}(\tau) \, d\tau + \int_0^t E[N^2(t - \tau)] f_{\tau_0}(\tau) \, d\tau$$

Making use of Eq. (5.175), we obtain

$$E[N^2(t)] = 2m_N(t) - F_{\tau_0}(t) + \int_0^t E[N^2(t - \tau)] f_{\tau_0}(\tau) \, d\tau \qquad (5.182)$$

This equation is of the form of that in (5.178), with $g(t) = E[N^2(t)]$ and $h(t) = 2m_N(t) - F_{\tau_0}(t)$ (since $g(t) = 0$ if $t < 0$). It follows, by Proposition 5.6.4, that

$$E[N^2(t)] = 2m_N(t) - F_{\tau_0}(t) + \int_0^t [2m_N(t-\tau) - F_{\tau_0}(t-\tau)]\, dm_N(\tau)$$
(5.183)

We have

$$
\begin{aligned}
\int_0^t F_{\tau_0}(t-\tau)\, dm_N(\tau) &= F_{\tau_0}(t-\tau)m_N(\tau)\big|_0^t + \int_0^t f_{\tau_0}(t-\tau)m_N(\tau)\, d\tau \\
&= 0 + \int_0^t f_{\tau_0}(\tau)m_N(t-\tau)\, d\tau \\
&\overset{(5.175)}{=} m_N(t) - F_{\tau_0}(t)
\end{aligned}
$$
(5.184)

from which we obtain Eq. (5.180). □

Example 5.6.4. When $\{N(t), t \geq 0\}$ is a Poisson process with rate λ, we have

$$E[N^2(t)] = V[N(t)] + (E[N(t)])^2 = \lambda t + (\lambda t)^2$$

which is indeed equal to

$$
\begin{aligned}
m_N(t) + 2\int_0^t m_N(t-\tau)\, dm_N(\tau) &= \lambda t + 2\int_0^t \lambda(t-\tau)\lambda\, d\tau \\
&= \lambda t + 2\lambda^2\left(t^2 - \frac{t^2}{2}\right)
\end{aligned}
$$

Example 5.6.5. When

$$f_{\tau_0}(t) = 1 \quad \text{if } 0 \leq t \leq 1$$

we find that

$$E[N(t)] = e^t - 1 \quad \text{for } 0 \leq t \leq 1$$

It follows that

$$
\begin{aligned}
E[N^2(t)] &= e^t - 1 + 2\int_0^t \left(e^{t-\tau} - 1\right)e^\tau\, d\tau \\
&= e^t - 1 + 2\int_0^t \left(e^t - e^\tau\right)\, d\tau = e^t - 1 + 2\left[te^t - (e^t - 1)\right] \\
&= e^t(2t - 1) + 1 \quad \text{for } 0 \leq t \leq 1
\end{aligned}
$$

As in Example 5.6.3, the fact that $0 \leq \tau_0 \leq 1$ implies that $E[N^k(1)] > 1$, for $k = 1, 2, \ldots$. We have

$$E[N(1)] = e - 1 \quad \text{and} \quad E[N^2(1)] = e + 1$$

We know that the sum of two *independent* Poisson processes, with rates λ_1 and λ_2, is a Poisson process with rate $\lambda := \lambda_1 + \lambda_2$ (see p. 234). The following proposition can be proved.

Proposition 5.6.5. *Let* $\{N_1(t), t \geq 0\}$ *and* $\{N_2(t), t \geq 0\}$ *be independent renewal processes. The process* $\{N(t), t \geq 0\}$ *defined by*

$$N(t) = N_1(t) + N_2(t) \quad \text{for } t \geq 0 \tag{5.185}$$

is a renewal process if and only if $\{N_i(t), t \geq 0\}$ *is a Poisson process, for* $i = 1, 2.$

Finally, the following result can be proved as well.

Proposition 5.6.6. *Let* $\{N(t), t \geq 0\}$ *be a renewal process. The kth moment of* $N(t)$ *about the origin (exists and) is finite, for all* $k \in \{1, 2, 3, \ldots\}$:

$$E[N^k(t)] < \infty \quad \text{for all } t < \infty \tag{5.186}$$

Remark. We deduce from the proposition that $m_N(t) < \infty \; \forall \, t < \infty$. Thus, not only can the number of renewals in an interval of finite length not be infinite (see p. 270), but the mathematical expectation of the number of renewals in this interval cannot be infinite either.

5.6.1 Limit theorems

Note first that $T_{N(t)}$ designates the time instant of the last renewal that occurred before or at time t. Similarly, $T_{N(t)+1}$ denotes the time instant of the first renewal after time t.

We will show that the average number of renewals per time unit, namely $N(t)/t$, tends to $1/\mu$ as $t \to \infty$, where $\mu := E[\tau_i] \; \forall \, i$ is assumed to be finite. The constant $\lambda := 1/\mu$ is called the *rate* of the process.

Proposition 5.6.7. *We have, with probability 1:*

$$\frac{N(t)}{t} \to \lambda \quad \text{as } t \to \infty \tag{5.187}$$

where $\lambda = 1/\mu > 0.$

Proof. We may write that

$$T_{N(t)} \leq t < T_{N(t)+1} \tag{5.188}$$

$$\Longleftrightarrow \quad \frac{T_{N(t)}}{N(t)} \leq \frac{t}{N(t)} < \frac{T_{N(t)+1}}{N(t)} \quad \text{(if } N(t) > 0) \tag{5.189}$$

Now, by the strong law of large numbers (since $E[|\tau_i|] = E[\tau_i] < \infty$), we have

$$\frac{T_{N(t)}}{N(t)} := \sum_{i=0}^{N(t)-1} \frac{\tau_i}{N(t)} \to \mu \quad \text{as } N(t) \to \infty \tag{5.190}$$

with probability 1. Furthermore, because $N(t) \to \infty$ as $t \to \infty$, we may write that (with probability 1)

$$\lim_{t \to \infty} \frac{T_{N(t)}}{N(t)} = \mu \tag{5.191}$$

Finally, given that

$$\frac{T_{N(t)+1}}{N(t)} = \frac{T_{N(t)+1}}{N(t)+1}\left(1 + \frac{1}{N(t)}\right) \tag{5.192}$$

we deduce that

$$\mu \le \lim_{t \to \infty} \frac{t}{N(t)} \le \mu \quad \Longleftrightarrow \quad \frac{1}{\mu} \le \lim_{t \to \infty} \frac{N(t)}{t} \le \frac{1}{\mu} \tag{5.193}$$

with probability 1. $\square$

The next theorem, which states that the average *expected* number of renewals per time unit also converges to λ, is not a direct consequence of the preceding proposition, because the fact that a sequence $\{X_n\}_{n=1}^{\infty}$ of random variables converges to a constant c does not imply that the sequence $\{E[X_n]\}_{n=1}^{\infty}$ converges to c.

Theorem 5.6.1 (Elementary renewal theorem). *If the mean $E[\tau_0]$ is finite, then we have*

$$\lim_{t \to \infty} \frac{E[N(t)]}{t} = \lambda \tag{5.194}$$

Remark. i) Actually, if $\mu = \infty$, the preceding result is still valid (by setting $\lambda := 1/\mu = 0$). Similarly, Proposition 5.6.7 is valid when $\mu = \infty$ as well.

ii) There is no mention of probability in the theorem, because $E[N(t)]$ is a deterministic function of t (while $N(t)$ is a random variable).

The following two theorems can also be shown.

Theorem 5.6.2. *If $\mu := E[\tau_0] < \infty$ and $\sigma^2 := V[\tau_0] < \infty$, then*

$$\lim_{t \to \infty} \frac{V[N(t)]}{t} = \sigma^2 \lambda^3 \tag{5.195}$$

Theorem 5.6.3 (Central limit theorem for renewal processes). *If $E[\tau_0^2]$ is finite, then we may write that*

$$N(t) \approx N\left(\frac{t}{\mu}, \frac{t\sigma^2}{\mu^3}\right) \equiv N\left(\lambda t, \sigma^2 \lambda^3 t\right) \quad \text{if t is large enough} \tag{5.196}$$

Example 5.6.6. If $\{N(t), t \geq 0\}$ is a Poisson process with rate λ_0, then $\tau_0 \sim$ $\mathrm{Exp}(\lambda_0)$, so that $\mu := E[\tau_0] = 1/\lambda_0$ and $\sigma^2 := V[\tau_0] = 1/\lambda_0^2$. We have indeed

$$N(t) \sim \mathrm{Poi}(\lambda_0 t) \approx \mathrm{N}(\lambda_0 t, \lambda_0 t) \equiv \mathrm{N}(\lambda t, \lambda_0^{-2}\lambda^3 t)$$

where $\lambda := 1/\mu = \lambda_0$.

The proposition that follows can be used to show Theorem 5.6.1.

Proposition 5.6.8. *We have*

$$E[T_{N(t)+1}] = E[\tau_0]\{E[N(t)] + 1\} \tag{5.197}$$

Proof. By conditioning on the value taken by the random variable τ_0, which is assumed to be continuous, we obtain

$$\begin{aligned}
E[T_{N(t)+1}] &= \int_0^\infty E[T_{N(t)+1} \mid \tau_0 = \tau] f_{\tau_0}(\tau)\, d\tau \\
&= \int_0^t \{\tau + E[T_{N(t-\tau)+1}]\} f_{\tau_0}(\tau)\, d\tau + \int_t^\infty \tau f_{\tau_0}(\tau)\, d\tau \\
&= \int_0^\infty \tau f_{\tau_0}(\tau)\, d\tau + \int_0^t E[T_{N(t-\tau)+1}] f_{\tau_0}(\tau)\, d\tau \\
&= E[\tau_0] + \int_0^t E[T_{N(t-\tau)+1}] f_{\tau_0}(\tau)\, d\tau \tag{5.198}
\end{aligned}$$

This equation is the particular case of the renewal equation (5.178), which is obtained with $g(t) = E[T_{N(t)+1}]$ and $h(t) = E[\tau_0]$. Dividing both members of the equation by $E[\tau_0]$ and subtracting the constant 1, we find that

$$\begin{aligned}
g^*(t) &:= \frac{E[T_{N(t)+1}]}{E[\tau_0]} - 1 = \int_0^t \frac{E[T_{N(t-\tau)+1}]}{E[\tau_0]} f_{\tau_0}(\tau)\, d\tau \tag{5.199} \\
&= \int_0^{t'} [g^*(t-\tau) + 1] f_{\tau_0}(\tau)\, d\tau = F_{\tau_0}(t) + \int_0^t g^*(t-\tau) f_{\tau_0}(\tau)\, d\tau
\end{aligned}$$

Now, this equation is the same as Eq. (5.175). Therefore, by the uniqueness of the solution, we may write that

$$m_N(t) = \frac{E[T_{N(t)+1}]}{E[\tau_0]} - 1 \quad \Longleftrightarrow \quad E[T_{N(t)+1}] = E[\tau_0][m_N(t) + 1] \quad \square \tag{5.200}$$

Definition 5.6.4. *Let t be a fixed time instant. The random variable*

$$A(t) := t - T_{N(t)} \tag{5.201}$$

*is called the **age** of the renewal process at time t, while*

$$D(t) := T_{N(t)+1} - t \tag{5.202}$$

*is the **remaining** or **excess lifetime** of the process at time t (see Fig. 5.3).*

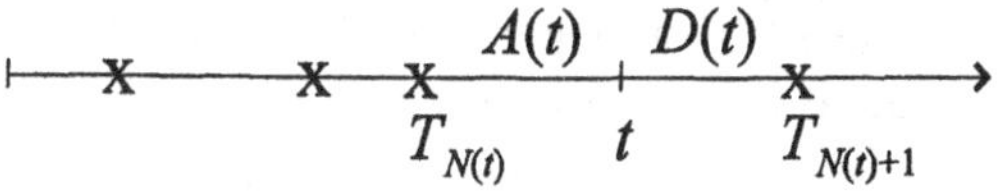

Fig. 5.3. Age and remaining lifetime of a renewal process.

By proceeding as above, we can prove the following proposition.

Proposition 5.6.9. *The distribution function of $D(t)$ is given by*

$$F_{D(t)}(r) = F_{\tau_0}(t + r) - \int_0^t [1 - F_{\tau_0}(t + r - \tau)] \, dm_N(\tau) \qquad (5.203)$$

for $r \geq 0$. In the case of $A(t)$, we find that

$$P[A(t) \geq a] = 1 - F_{\tau_0}(t) + \int_0^{t-a} [1 - F_{\tau_0}(t - \tau)] \, dm_N(\tau) \quad \text{if } 0 \leq a \leq t$$
$$(5.204)$$

and $P[A(t) \geq a] = 0$ if $a > t$.

When t is large enough and τ_0 is a continuous random variable, we find that

$$f_{D(t)}(r) \simeq \frac{1}{E[\tau_0]}[1 - F_{\tau_0}(r)] \quad \text{for } r \geq 0 \qquad (5.205)$$

In fact, this formula for the probability density function of $D(t)$ is *exact* if the renewal process $\{N(t), t \geq 0\}$ is *stationary*.

Suppose now that at the moment of the nth renewal of the process $\{N(t), t \geq 0\}$, we receive, or we have accumulated, a *reward R_n* (which may, actually, be a *cost*). Suppose also that the rewards $\{R_n\}_{n=1}^{\infty}$ are independent and identically distributed random variables. However, in general, R_n will depend on τ_{n-1}, that is, the length of the nth renewal period, called a *cycle*. Let $R(t)$ be the total reward received in the interval $[0, t]$. That is,

$$R(t) = \sum_{n=1}^{N(t)} R_n \qquad (R(t) = 0 \text{ if } N(t) = 0) \qquad (5.206)$$

We will show that the average reward received by time unit, *in the limit*, is equal to the average reward received during a cycle, divided by the average length of a cycle.

Proposition 5.6.10. *If $E[R_1] < \infty$ and $E[\tau_0] < \infty$, then we have*

$$P\left[\lim_{t \to \infty} \frac{R(t)}{t} = \frac{E[R_1]}{E[\tau_0]}\right] = 1 \qquad (5.207)$$

Proof. It suffices to use the strong law of large numbers and Proposition 5.6.7. Since $N(t) \to \infty$ as $t \to \infty$, we may write

$$P\left[\lim_{t\to\infty} \sum_{n=1}^{N(t)} \frac{R_n}{N(t)} = E[R_1]\right] = 1 \tag{5.208}$$

from which we have

$$\lim_{t\to\infty} \frac{R(t)}{t} := \lim_{t\to\infty} \frac{1}{t} \sum_{n=1}^{N(t)} R_n = \lim_{t\to\infty}\left[\frac{\sum_{n=1}^{N(t)} R_n}{N(t)}\right] \lim_{t\to\infty}\left[\frac{N(t)}{t}\right]$$

$$= E[R_1] \times \frac{1}{E[\tau_0]} \tag{5.209}$$

with probability 1. $\square$

Remarks. i) It can also be shown that

$$\lim_{t\to\infty} \frac{E[R(t)]}{t} = \frac{E[R_1]}{E[\tau_0]} \tag{5.210}$$

ii) If the *reward* R_n may take positive *and* negative values, then we must assume that $E[|R_1|]$ is finite.

With the help of the preceding proposition, we will prove the following result.

Proposition 5.6.11. *If τ_0 is a continuous random variable, then we have, with probability 1:*

$$\lim_{t\to\infty} \frac{\int_0^t A(\tau)\, d\tau}{t} = \lim_{t\to\infty} \frac{\int_0^t D(\tau)\, d\tau}{t} = \frac{E[\tau_0^2]}{2E[\tau_0]} \tag{5.211}$$

Proof. Suppose that the reward received at time t is given by $A(t)$. Then, we may write that

$$E[R_1] = E\left[\int_0^{\tau_0} \tau\, d\tau\right] = E\left[\frac{\tau_0^2}{2}\right] = \frac{E[\tau_0^2]}{2} \tag{5.212}$$

To obtain the other result, we simply have to suppose that the reward received at time t is rather given by $D(t)$, and then the variable τ is replaced by $\tau_0 - \tau$ in the integral. $\square$

Remark. Note that we calculated the *average value* of the age (or of the remaining lifetime) of the process over a long period, which is a *temporal mean*, and not a mathematical expectation. If $E[\tau_0^2] < \infty$, then it can also be shown that

$$\lim_{t\to\infty} E[A(t)] = \lim_{t\to\infty} E[D(t)] = \frac{E[\tau_0^2]}{2E[\tau_0]} \tag{5.213}$$

Example 5.6.7. If $\{N(t), t \geq 0\}$ is a Poisson process, then, making use of the fact that it has independent and stationary increments, we can assert that $D(t)$ has an exponential distribution with parameter λ. Thus, we have

$$E[D(t)] = \frac{1}{\lambda}$$

Moreover, $\tau_0 \sim \text{Exp}(\lambda)$, too, so that

$$\lim_{t \to \infty} \frac{\int_0^t D(\tau) \, d\tau}{t} = \frac{\frac{1}{\lambda^2} + \left(\frac{1}{\lambda}\right)^2}{2/\lambda} = \frac{1}{\lambda}$$

So, in this case, the temporal mean and the mathematical expectation of $D(t)$ are equal, for all $t \geq 0$.

Example 5.6.8. (a) Let $\{N(t), t \geq 0\}$ be a counting process for which the time between two events has a $U[0, S]$ distribution, where

$$S = \begin{cases} 1 \text{ with probability } 1/2 \\ 2 \text{ with probability } 1/2 \end{cases}$$

Explain why $\{N(t), t \geq 0\}$ is *not* a renewal process.

(b) In part (a), suppose that S is rather a random variable whose value is determined at time $t = 0$ and after each event by tossing a fair coin. More precisely, we have

$$S = \begin{cases} 1 \text{ if "tails" is obtained} \\ 2 \text{ if "heads" is obtained} \end{cases}$$

In this case, the stochastic process $\{N(t), t \geq 0\}$ *is* a renewal process.
 (i) Calculate $m_N(t)$, for $0 \leq t \leq 1$.
 (ii) Suppose that we receive a reward equal to \$1 when the length of a cycle is greater than 1 (and \$0 otherwise). Calculate the average reward per time unit over a long period.

Solution. (a) $\{N(t), t \geq 0\}$ is not a renewal process, because the times between the successive events are not independent random variables. Indeed, the smaller τ_0 is, the larger the probability that $S = 1$ is. So, the τ_k's depend on τ_0, for all $k \geq 1$.

(b) (i) Let τ be the length of a cycle. We have

$$F(t) \equiv P[\tau \leq t] = P[\tau \leq t \mid S = 1] \underbrace{P[S = 1]}_{1/2} + P[\tau \leq t \mid S = 2] P[S = 2]$$

That is,

$$F(t) = \frac{1}{2}\left(t + \frac{t}{2}\right) = \frac{3t}{4} \quad \text{for } 0 \leq t \leq 1$$

It follows that the renewal equation is

$$m_N(t) = \frac{3t}{4} + \int_0^t m(t-x)\frac{3}{4}\,dx \overset{y=t-x}{=} \frac{3t}{4} + \frac{3}{4}\int_0^t m_N(y)\,dy$$

$$\implies \quad m_N'(t) = \frac{3}{4} + \frac{3}{4}m_N(t)$$

We deduce from the initial condition $m_N(0) = 0$ and from the formula (3.233), p. 130, that

$$m_N(t) = e^{3t/4}\left(0 + \int_0^t e^{-3y/4}\frac{3}{4}\,dy\right) = e^{3t/4} - 1 \quad \text{for } 0 \le t \le 1$$

(ii) We have

$$E[\tau_0] = E[\tau_0 \mid S=1]\underbrace{P[S=1]}_{1/2} + E[\tau_0 \mid S=2]P[S=2] = \frac{1}{2}\left(\frac{1}{2}+1\right) = \frac{3}{4}$$

and

$$E[R_1] = \underbrace{E[R_1 \mid S=1]}_{0}\,P[S=1] + \underbrace{E[R_1 \mid S=2]}_{1/2}\,P[S=2] = \frac{1}{4}$$

Therefore, the average reward per time unit (over a long period) is given by $(1/4)/(3/4) = 1/3$.

5.6.2 Regenerative processes

We know that the Poisson process, $\{N(t), t \ge 0\}$, starts anew, probabilistically, from any time instant, because it has independent and stationary increments. However, since it is a counting process, once the first event occurred, $N(t)$ will never be equal to 0 again. We are interested, in this subsection, in processes that are certain to eventually return to their initial state and that, at the moment of this return, start afresh probabilistically. This type of stochastic process is said to be *regenerative*.

Definition 5.6.5. *Let τ_0 be the time that the discrete-state stochastic process $\{X(t), t \ge 0\}$ spends in the initial state $X(0)$. Suppose that*
i) $P[\exists\, t > \tau_0 : X(t) = X(0)] = 1$,
ii) the processes $\{X(t) - X(0), t \ge 0\}$ and $\{Y(t), t \ge 0\}$, where

$$Y(t) := X(t+T_1) - X(T_1) \tag{5.214}$$

and T_1 is the time of first return to the initial state, are identically distributed,
iii) the stochastic process $\{Y(t), t \ge 0\}$ is independent of the process $\{X(t), 0 \le t \le T_1\}$.
The process $\{X(t), t \ge 0\}$ is then called a **regenerative process**.

Remarks. i) Let T_n be the time of the nth return to the initial state, for $n = 1, 2, \dots$. We assume that $P[T_1 < \infty] = 1$. It then follows that the time instants $T_2, T_3, \dots$ must also (exist and) be finite with probability 1.

ii) A regenerative process may be a discrete-time process $\{X_n, n = 0, 1, \dots\}$ as well. In this case, τ_0 is the time that the process takes to make a transition from the initial state to an arbitrary state, which may be the initial state (for example, if $\{X_n, n = 0, 1, \dots\}$ is a Markov chain).

iii) The *cycles* of the regenerative process are the processes

$$\{C_0(t), 0 \le t \le T_1\} \quad \text{and} \quad \{C_n(t), T_n < t \le T_{n+1}\} \quad \text{for } n = 1, 2, \dots$$
(5.215)

These stochastic processes are independent and identically distributed, for $n = 0, 1, \dots$. Similarly, the random variables $X_0 := T_1$ and $X_n := T_{n+1} - T_n$, for $n = 1, 2, \dots$, are i.i.d.

iv) According to the definition above, a renewal process is not regenerative, because it is a particular counting process. However, we can associate a renewal process $\{N(t), t \ge 0\}$ with a regenerative process $\{X(t), t \ge 0\}$, by setting that $N(t)$ is the number of times that $\{X(t), t \ge 0\}$ came back to the initial state and started anew in the interval $[0, t]$. That is, $N(t)$ is the number of cycles completed in this interval. Conversely, we can define a regenerative process $\{X(t), t \ge 0\}$ from a renewal process $\{N(t), t \ge 0\}$ (for which $N(0) = 0$) as follows:

$$X(t) := \begin{cases} N(t) & \text{if } 0 \le t < T_k^* \\ N(t) - k & \text{if } T_k^* \le t < T_{2k}^* \\ N(t) - 2k & \text{if } T_{2k}^* \le t < T_{3k}^* \\ \cdots & \cdots \end{cases}$$
(5.216)

for some $k \in \{2, 3, \dots\}$, where $T_1^*, T_2^*, \dots$ are the arrival times of the renewals. Note that with $k = 1$, we would have that $X(t) \equiv 0$. Moreover, here $T_1 = T_k^*$.

Example 5.6.9. The queueing model $M/M/1$, described in Example 3.1.6 (and which will be studied in detail in Chapter 6), is an example of a regenerative process. In this model, $X(t)$ designates the number of persons in a system at time t. Thus, if $X(0) = 0$, then we are certain that the system will eventually be empty again and that, from that point on, everything will start anew.

Example 5.6.10. Let $\{X_n, n = 0, 1, \dots\}$ be a discrete-time Markov chain, whose state space is the set $\{0, 1\}$ and whose one-step transition probability matrix is given by

$$\mathbf{P} = \begin{bmatrix} 1/2 & 1/2 \\ 1 & 0 \end{bmatrix}$$

Suppose that $X_0 = 0$. Since the chain is irreducible and has a finite number of states, we may assert that it is recurrent. Therefore, the probability that the process will visit the initial state 0 again is equal to 1. It follows that this Markov chain is a regenerative process. Suppose that every transition of the chain takes *one* time unit. Then, we may write that $P[T_1 = 1] = P[T_1 = 2] = 1/2$. Finally, let $N(0) = 0$ and $N(t)$ be the number of visits to state 0 in the interval $(0, t]$, for $t > 0$. The process $\{N(t), t \geq 0\}$ is then a renewal process.

Proposition 5.6.12. *If $E[T_1] < \infty$, then the proportion π_k of time that a regenerative process spends in state k is given, in the limit, by*

$$\pi_k = \frac{E[\text{time spent in state } k \text{ in the interval } [0, T_1]]}{E[T_1]} \tag{5.217}$$

Proof. Suppose that we receive an instantaneous reward of \$1 per time unit when the process $\{X(t), t \geq 0\}$ is in state k, and \$0 otherwise. Then the total reward received in the interval $[0, t]$ is given by

$$R(t) = \int_0^t I_{\{X(\tau)=k\}} \, d\tau \tag{5.218}$$

where $I_{\{X(\tau)=k\}}$ is the *indicator variable* of the event $\{X(\tau) = k\}$ [see Eq. (3.43)]. If $E[T_1] < \infty$, Proposition 5.6.10 implies that

$$\lim_{t \to \infty} \frac{R(t)}{t} = \frac{E[R_1]}{E[T_1]} \tag{5.219}$$

(with probability 1), where R_1 is the time spent by the process in state k in the interval $[0, T_1]$. Now, $\lim_{t \to \infty} R(t)/t$ is the proportion of time that the process spends in state k. $\square$

Remark. The proposition is valid for any type of random variable T_1. When T_1 is a *continuous* variable, it can also be shown that

$$\lim_{t \to \infty} P[X(t) = k] = \pi_k \tag{5.220}$$

That is, π_k is the *limiting probability* that the process will be in state k at time t as well.

Example 5.6.11. In the case of the process $\{X(t), t \geq 0\}$ defined by Eq. (5.216), we can directly write that $\pi_0 = \pi_1 = \ldots = \pi_{k-1} = 1/k$, because the random variables $T_1^*, T_2^*, \ldots$ are independent and identically distributed.

Example 5.6.12. In Example 5.6.10, the reward R_1 is equal to 1, because the Markov chain spends exactly one time unit in the initial state 0 per cycle. It follows that

$$\pi_0 = \frac{1}{E[T_1]} = \frac{1}{3/2} = \frac{2}{3}$$

We can check, using Theorem 3.2.1, that the proportion π_0 is indeed equal to 2/3. Note that we may apply this theorem to obtain the proportion π_0, whether the chain is periodic or not. Here, the chain is ergodic, since $p_{0,0} = 1/2 > 0$. Then, π_0 is also the limiting probability that the process will be in state 0 after a very large number of transitions.

Example 5.6.13. We consider a discrete-time Markov chain $\{X_n, n = 0, 1, \ldots\}$ with state space $\{0, 1, 2\}$ and transition matrix $\mathbf{P}$ given by

$$\mathbf{P} = \begin{bmatrix} 0 & 1/2 & 1/2 \\ 1 & 0 & 0 \\ 0 & 1/2 & 1/2 \end{bmatrix}$$

Suppose that the initial state is 0 and that when the process enters state i, it remains there during a random time having mean μ_i, for $i = 0, 1, 2$, independently from one visit to another. Calculate the proportion of time that the process spends in state 0 over a long period.

Solution. The Markov chain is recurrent, because it is irreducible and has a finite number of states. Then the process $\{X_n, n = 0, 1, \ldots\}$ is regenerative. Let T_1 be the time of first return to state 0 and N be the number of transitions needed to return to 0. We have

$$E[T_1 \mid N = n] = \mu_0 + \mu_1 + (n - 2)\mu_2 \quad \text{for } n = 2, 3, \ldots$$

Since $P[N = n] = (1/2)^{n-1}$, for $n = 2, 3, \ldots$, it follows that

$$E[T_1] = E[E[T_1 \mid N]] = \sum_{n=2}^{\infty} [\mu_0 + \mu_1 + (n - 2)\mu_2](1/2)^{n-1}$$

Finally, we can write that

$$\sum_{n=2}^{\infty} (n - 2)(1/2)^{n-1} = \sum_{n=1}^{\infty} (n - 1)(1/2)^n = E[\text{Geom}(1/2)] - 1 = 1$$

so that

$$E[T_1] = \mu_0 + \mu_1 + \mu_2$$

The proportion of time that the process spends in state 0 over a long period is thus given by $\mu_0/(\mu_0 + \mu_1 + \mu_2)$.

Remark. Since the matrix $\mathbf{P}$ is finite and doubly stochastic, we deduce from Proposition 3.2.6 (the Markov chain being irreducible and aperiodic) that the limiting probabilities π_k exist and are given by $1/3$, for $k = 0, 1, 2$. Hence, we deduce that the proportion of time requested is indeed equal to $\mu_0/(\mu_0 + \mu_1 + \mu_2)$.

An important particular case of a regenerative process is the one for which the state space of the process $\{X(t), t \geq 0\}$ contains only two elements, which will be denoted by 0 and 1. For example, $X(t) = 0$ if some machine is down, and $X(t) = 1$ if it is operating at time t. Suppose that the machine is brand-new at the initial time 0, so that $X(0) = 1$, and that the time during which the machine operates, before breaking down, is a random variable S_1. Next, the repairman takes a random time U_1 to set the machine going again, and then the system starts afresh. Let

$$T_n := S_n + U_n \quad \text{for } n = 1, 2, \dots \tag{5.221}$$

where S_n is the operating time of the machine between the $(n-1)$st and the nth breakdown, and U_n is the repair time of the nth breakdown, for $n = 2, 3, \dots$. We assume that the random variables $\{S_n\}_{n=1}^{\infty}$ are i.i.d. and that so are the r.v.s $\{U_n\}_{n=1}^{\infty}$. However, the r.v.s S_n and U_n may be *dependent*. This type of stochastic process is often called an *alternating renewal process*.

Remark. Let $N(t)$ be the number of times that the machine has been repaired in the interval $[0, t]$. The process $\{N(t), t \geq 0\}$ is a renewal process and the T_n's are the arrival times of the renewals. However, the process $\{X(t), t \geq 0\}$ defined above is not a renewal process (according to our definition), because it is not a counting process.

By using the renewal equation, we can show the following proposition.

Proposition 5.6.13. *Let $\pi_1(t)$ be the probability that $X(t) = 1$. We have*

$$\pi_1(t) = 1 - F_{S_1}(t) + \int_0^t [1 - F_{S_1}(t-\tau)] \, dm_N(\tau) \tag{5.222}$$

Next, we deduce directly from Proposition 5.6.12 that

$$\pi_1 = 1 - \pi_0 = \frac{E[S_1]}{E[S_1] + E[U_1]} = \frac{E[S_1]}{E[T_1]} \tag{5.223}$$

Moreover, if T_1 is a continuous random variable, then

$$\lim_{t \to \infty} \pi_1(t) = \pi_1 \tag{5.224}$$

Finally, let $\{N(t), t \geq 0\}$ be a renewal process and

$$X(t) := \begin{cases} 1 \text{ if } A(t) < a \\ 0 \text{ if } A(t) \geq a \end{cases} \tag{5.225}$$

where $a > 0$ is a constant and $A(t)$ is defined in Eq. (5.201). The process $\{X(t), t \geq 0\}$ is then an alternating renewal process, and we deduce from the formula (5.223) that

$$\pi_1 = \frac{E[S_1]}{E[T_1]} = \frac{E[\min\{T_1^*, a\}]}{E[T_1^*]} \tag{5.226}$$

where T_1^* is the arrival time of the first renewal of the stochastic process $\{N(t), t \geq 0\}$.

Remark. If $T_1^* < a$, then we set $U_1 = 0$.

In the continuous case, we have

$$E[\min\{T_1^*, a\}] = \int_0^\infty \min\{t, a\} f_{T_1^*}(t) dt = \int_0^a t f_{T_1^*}(t) dt + \int_a^\infty a f_{T_1^*}(t) dt \tag{5.227}$$

We calculate

$$\int_0^a t f_{T_1^*}(t) dt = t F_{T_1^*}(t)\Big|_0^a - \int_0^a F_{T_1^*}(t) dt = a F_{T_1^*}(a) - \int_0^a F_{T_1^*}(t) dt \tag{5.228}$$

from which we obtain the formula

$$E[\min\{T_1^*, a\}] = a F_{T_1^*}(a) - \int_0^a F_{T_1^*}(t) dt + a P[T_1^* > a] = a - \int_0^a F_{T_1^*}(t) dt \tag{5.229}$$

We can thus write that

$$\pi_1 = \frac{\int_0^a P[T_1^* > t] dt}{E[T_1^*]} \tag{5.230}$$

Remark. We obtain exactly the same formula as above for π_1 if we replace $A(t)$ by $D(t)$ (defined in (5.202)) in (5.225).

Example 5.6.14. If $T_1^* \sim \text{Exp}(\lambda)$, then we have that $P[T_1^* > t] = e^{-\lambda t}$. It follows that

$$\pi_1 = \frac{\int_0^a e^{-\lambda t} dt}{1/\lambda} = 1 - e^{-\lambda a}$$

5.7 Exercises

Section 5.1

Question no. 1
Let $\{N(t), t \geq 0\}$ be a Poisson process with rate λ. We define the stochastic process $\{X(t), 0 \leq t \leq c\}$ by

$$X(t) = N(t) - \frac{t}{c} N(c) \quad \text{for } 0 \leq t \leq c$$

where $c > 0$ is a constant.
(a) Calculate the mean of $X(t)$.
(b) Calculate the autocovariance function of the process $\{X(t), 0 \leq t \leq c\}$.
(c) Is the process $\{X(t), 0 \leq t \leq c\}$ wide-sense stationary? Justify.

Question no. 2

We suppose that customers arrive at a bank counter according to a Poisson process with rate $\lambda = 10$ per hour. Let $N(t)$ be the number of customers in the interval $[0, t]$.

(a) What is the probability that no customers arrive over a 15-minute time period?

(b) Knowing that eight customers arrive during a given hour, what is the probability that at most two customers will arrive over the following hour?

(c) Given that a customer arrived during a certain 15-minute time period, what is the probability that he arrived during the first 5 minutes of the period considered?

(d) Let $X(t) := N(t)/t$, for $t > 0$. Calculate the autocovariance function, $C_X(t_1, t_2)$, of the stochastic process $\{X(t), t > 0\}$, for $t_1, t_2 > 0$.

Question no. 3

The stochastic process $\{X(t), t \geq 0\}$ is defined by

$$X(t) = \frac{N\left(t + \delta^2\right) - N(t)}{\delta} \quad \text{for } t \geq 0$$

where $\{N(t), t \geq 0\}$ is a Poisson process with rate $\lambda > 0$, and $\delta > 0$ is a constant.

(a) Is the process $\{X(t), t \geq 0\}$ a Poisson process? Justify.

(b) Calculate the mean of $X(t)$.

(c) Calculate the autocovariance function of the process $\{X(t), t \geq 0\}$, for $t_1 = 1$, $t_2 = 2$, and $\delta = 1$.

(d) Let $Z_n := N(n)$, for $n = 0, 1, 2, \ldots$.

(i) The stochastic process $\{Z_n, n = 0, 1, \ldots\}$ *is a Markov chain. Justify* this assertion.

(ii) Calculate $p_{i,j}$, for $i, j \in \{0, 1, 2, \ldots\}$.

Question no. 4

Let $N(t)$ be the number of failures of a computer system in the interval $[0, t]$. We suppose that $\{N(t), t \geq 0\}$ is a Poisson process with rate $\lambda = 1$ per week.

(a) Calculate the probability that

(i) the system operates without failure during two consecutive weeks,

(ii) the system will have exactly two failures during a given week, knowing that it operated without failure during the previous two weeks,

(iii) less than two weeks elapse before the third failure occurs.

(b) Let

$$Z(t) := e^{-N(t)} \quad \text{for } t \geq 0$$

Is the stochastic process $\{Z(t), t \geq 0\}$ wide-sense stationary? Justify.

Indication. We have that $E[e^{-sX}] = \exp\left[\alpha\left(e^{-s} - 1\right)\right]$ if $X \sim \text{Poi}(\alpha)$.

Question no. 5

A man plays independent repetitions of the following game: at each repetition, he throws a dart onto a circular target. Suppose that the distance D between the impact point of the dart and the center of the target has a U[0, 30] distribution. If $D \leq 5$, the player wins \$1; if $5 < D < 25$, the player neither wins nor loses anything; if $D \geq 25$, the player loses \$1. The player's initial fortune is equal to \$1, and he will stop playing when either he is ruined or his fortune reaches \$3. Let X_n be the fortune of the player after n repetitions. Then the stochastic process $\{X_n, n = 0, 1, \dots\}$ is a Markov chain.

(a) Find the one-step transition probability matrix of the chain.

(b) Calculate $E[X_2^2]$.

Suppose now that the man never stops playing, so that the state space of the Markov chain is the set $\{0, \pm 1, \pm 2, \dots\}$. Suppose also that the duration T (in seconds) of a repetition of the game has an exponential distribution with mean 30. Then the stochastic process $\{N(t), t \geq 0\}$, where $N(t)$ denotes the number of repetitions completed in the interval $[0, t]$, is a Poisson process with rate $\lambda = 2$ per minute.

(c) Calculate the probability that the player will have completed at least three repetitions in less than two minutes (from the initial time).

(d) Calculate (approximately) the probability $P[N(25) \leq 50]$.

Question no. 6

Let $N(t)$ be the number of telephone calls received at an exchange in the interval $[0, t]$. We suppose that $\{N(t), t \geq 0\}$ is a Poisson process with rate $\lambda = 10$ per hour. Calculate the probability that no calls will be received during each of two consecutive 15-minute periods.

Question no. 7

The customers of a newspaper salesperson arrive according to a Poisson process with rate $\lambda = 2$ per minute. Calculate the probability that at least one customer will arrive in the interval $(t_0, t_0 + 2]$, given that there has been exactly one customer in the interval $(t_0 - 1, t_0 + 1]$, where $t_0 \geq 1$.

Question no. 8

The stochastic process $\{X(t), t \geq 0\}$ is defined by

$$X(t) = N(t + 1) - N(1) \quad \text{for } t \geq 0$$

where $\{N(t), t \geq 0\}$ is a Poisson process with rate $\lambda > 0$. Calculate $C_X(s, t)$, for $0 \leq s \leq t$.

Question no. 9 We define

$$Y(t) = \begin{cases} 1 \text{ if } N(t) = 0, 2, 4, \ldots \\ 0 \text{ if } N(t) = 1, 3, 5, \ldots \end{cases}$$

where $\{N(t), t \geq 0\}$ is a Poisson process with rate $\lambda = 1$. It can be shown that $P[Y(t) = 1] = (1 + e^{-2t})/2$, for $t \geq 0$. Next, let $X_n := Y(n)$ for $n = 0, 1, \ldots$. Then $\{X_n, n = 0, 1, \ldots\}$ *is* a Markov chain. Calculate

(a) its one-step transition probability matrix,

(b) the limiting probabilities of the chain, if they exist.

Question no. 10
   The failures of a certain machine occur according to a Poisson process with rate $\lambda = 1$ per week.

(a) What is the probability that the machine will have at least one failure during each of the first two weeks considered?

(b) Suppose that exactly five failures have occurred during the first four weeks considered. Let M be the number of failures during the fourth of these four weeks. Calculate $E[M \mid M > 0]$.

Question no. 11
   Let $\{N(t), t \geq 0\}$ be a Poisson process with rate $\lambda > 0$. We define

$$N_1(t) = N(\sqrt{t}), \quad N_2(t) = N(2t), \quad \text{and} \quad N_3(t) = N(t + 2) - N(2)$$

and we consider the processes $\{N_k(t), t \geq 0\}$, for $k = 1, 2, 3$. Which of these stochastic processes is (or are) also a Poisson process? Justify.

Question no. 12
   The power failures in a certain region occur according to a Poisson process with rate $\lambda_1 = 1/5$ per week. Moreover, the duration X (in hours) of a given power failure has an exponential distribution with parameter $\lambda_2 = 1/2$. Finally, we assume that the durations of the various power failures are independent random variables.

(a) What is the probability that the longest failure, among the first three power failures observed, lasts more than four hours?

(b) Suppose that there has been exactly one power failure during the first week considered. What is the probability that the failure had still not been repaired at the end of the week in question?

Question no. 13
   A machine is made up of two components that operate independently. The lifetime X_i (in days) of component i has an exponential distribution with parameter λ_i, for $i = 1, 2$.
   Suppose that the two components are placed in series and that as soon as a component fails, it is replaced by a new one. Let $N(t)$ be the number of

replacements in the interval $(0, t]$. We can show that the stochastic process $\{N(t), t \geq 0\}$ is a Poisson process. Give its rate λ.

Indication. If X_0 is the time between two successive replacements, then we can write that $X_0 = \min\{X_1, X_2\}$.

Question no. 14

We consider the process $\{X(t), 0 \leq t \leq 1\}$ defined by

$$X(t) = N(t^2) - t^2 N(1) \quad \text{for } 0 \leq t \leq 1$$

where $\{N(t), t \geq 0\}$ is a Poisson process with rate $\lambda > 0$.

(a) Calculate the autocorrelation function, $R_X(t_1, t_2)$, of the stochastic process $\{X(t), 0 \leq t \leq 1\}$ at $t_1 = 1/4$ and $t_2 = 1/2$.

(b) Calculate $P[X(t) > 0 \mid N(1) = 1]$, for $0 < t < 1$.

Question no. 15

Let $\{N(t), t \geq 0\}$ be a Poisson process with rate λ. Show that, given that $N(t) = n$, $N(s)$ has a binomial distribution with parameters n and s/t, where $0 < s < t$, for all $\lambda \, (> 0)$.

Question no. 16

Travelers arrive at a bus station from 6 a.m., according to a Poisson process with rate $\lambda = 1$ per minute. The first bus leaves T minutes after 6 a.m.

(a) Calculate the mean and the variance of the number of travelers ready to board this bus if (i) T has an exponential distribution with mean equal to 15 minutes and (ii) T is uniformly distributed between 0 and 20 minutes.

(b) Calculate the average number of passengers on the bus if it leaves at 6:15 and if its capacity is 20 passengers.

Question no. 17

We suppose that every visitor to a museum, independently from the others, moves around the museum for T minutes, where T is a uniform random variable between 30 and 90 minutes. Moreover, the visitors arrive according to a Poisson process with rate $\lambda = 2$ per minute. If the museum opens its doors at 9 a.m. and closes at 6 p.m., find the mean and the variance of the number of visitors in the museum (i) at 10 a.m. and (ii) at time t_0, where t_0 is comprised between 10:30 a.m. and 6 p.m.

Question no. 18

Suppose that events occur at random in the plane, in such a way that the number of events in a region R is a random variable having a Poisson distribution with parameter λA, where λ is a positive constant and A denotes the area of the region R. A point is taken at random in the plane. Let D be the distance between this point and the nearest event. Calculate the mean of the random variable D.

Question no. 19
 The various types of traffic accidents that occur in a certain tunnel, over a given time period, constitute independent stochastic processes. We suppose that accidents stopping the traffic in the northbound (respectively, southbound) direction in the tunnel occur according to a Poisson process with rate α (resp., β). Moreover, accidents causing the complete closure of the tunnel occur at rate γ, also according to a Poisson process. Let T_N (resp., T_S) be the time during which the traffic is *not* stopped in the northbound (resp., southbound) direction, from the moment when the tunnel has just reopened after its complete closure.

(a) Calculate the probability density function of the random variable T_N.

(b) Show that

$$P[T_N > t, T_S > \tau] = e^{-\alpha t - \beta \tau - \gamma \max\{t, \tau\}} \quad \text{for } t, \tau \geq 0$$

(c) Check your answer in (a) with the help of the formula in (b).

Question no. 20
 Let $\{N^*(t), t \geq 0\}$ be the stochastic process that counts only the even events (that is, events nos. $2, 4, \ldots$) of the Poisson process $\{N(t), t \geq 0\}$ in the interval $[0, t]$. Show that $\{N^*(t), t \geq 0\}$ is *not* a Poisson process.

Question no. 21
 Let $\{N_1(t), t \geq 0\}$ and $\{N_2(t), t \geq 0\}$ be two *independent* Poisson processes, with rates λ_1 and λ_2, respectively. We define $N(t) = N_1(t) - N_2(t)$.

(a) Explain why the stochastic process $\{N(t), t \geq 0\}$ is not a Poisson process.

(b) Give a formula for the probability $P[N(t_2) - N(t_1) = n]$, for $t_2 > t_1 \geq 0$ and $n \in \{0, \pm 1, \pm 2, \ldots\}$.

Question no. 22
 Suppose that $\{N(t), t \geq 0\}$ is a Poisson process with rate $\lambda > 0$ and that S is a random variable having a uniform distribution on the interval $[0, 2]$.

(a) Obtain the moment-generating function of the random variable $N(t + S)$.

Indication. If X has a Poisson distribution with parameter α, then $M_X(t) = \exp\{\alpha(e^t - 1)\}$.

(b) Calculate the mean and the variance of $N(t + S)$.

Question no. 23
(a) Is the Poisson process $\{N(t), t \geq 0\}$ an ergodic process? Justify.

(b) Is the stochastic process $\{X(t), t > 0\}$, where $X(t) := N(t)/t$, mean ergodic? Justify.

Question no. 24
 City buses arrive at a certain street corner, between 5 a.m. and 11 p.m., according to a Poisson process with rate $\lambda = 4$ per hour. Let T_1 be the

waiting time, in minutes, until the first bus (after 5 a.m.), and let M be the total number of buses between 5 a.m. and 5:15 a.m.

(a) Calculate the probability $P[T_1 \in I, M = 1]$, where $I := [a, b]$ is included in the interval $[0, 15]$.

(b) What is the variance of the waiting time between two consecutive arrivals?

(c) If a person arrives at this street corner every morning at 9:05 a.m., what is the variance of the time during which she must wait for the bus? Justify.

Question no. 25

A truck driver is waiting to join the traffic on the service road of a highway. The truck driver blocks the service road for four seconds when he joins the traffic. Suppose that vehicles arrive according to a Poisson process and that six seconds elapse, on average, between two consecutive vehicles. Answer the following questions by assuming that the truck driver makes sure that he has enough time to perform his maneuver before merging into the traffic.

(a) What is the probability that the truck driver is able to join the traffic immediately on his arrival at the intersection with the service road?

(b) What is the mathematical expectation of the gap (in seconds) between the truck and the nearest vehicle when the truck driver merges into the traffic?

(c) Calculate the average number of vehicles that the truck driver must let go by before being able to merge into the traffic.

Question no. 26

Particles are emitted by a radioactive source according to a Poisson process with rate $\lambda = \ln 5$ per hour.

(a) What is the probability that during at least one of five consecutive hours no particles are emitted?

(b) Knowing that during a given hour two particles were emitted, what is the probability that one of them was emitted during the first half-hour and the other during the second half-hour of the hour in question?

(c) In (b), if we know that the first particle was emitted over the first half-hour, what is the probability that the second particle was emitted during the first half-hour as well?

Question no. 27

A system is made up of two components. We suppose that the lifetime (in years) of each component has an exponential distribution with parameter $\lambda = 2$ and that the components operate independently. When the system goes down, the two components are then immediately replaced by new ones. We consider three cases:

1. the two components are placed in series (so that both components must function for the system to work);

2. the two components are placed in parallel (so that a single operating component is sufficient for the system to function) and the two components operate at the same time;
3. the two components are placed in parallel, but only one component operates at a time and the other is in standby.

Let $N(t)$, for $t \geq 0$, be the number of system failures in the interval $[0, t]$. Answer the following questions for each of the cases above.

(a) Is $\{N(t), t \geq 0\}$ a Poisson process? If it is, give its rate λ. If it's not, justify.

(b) What is the average time elapsed between two consecutive system failures?

Question no. 28
Let $\{N(t), t \geq 0\}$ be a Poisson process with rate $\lambda = 2$ per minute, where $N(t)$ denotes the number of customers of a newspaper salesperson in the interval $[0, t]$.

(a) Let T_k, for $k = 1, 2, \ldots$, be the arrival time of the kth customer. Calculate the probability density function of T_1, given that $T_2 = s$.

(b) Calculate the probability that at least two minutes will elapse until the salesperson's second customer arrives, from a given time instant, given that no customers have arrived in the last minute.

(c) Suppose that the probability that a given customer is a man is equal to 0.7, independently from one customer to another. Let M be the number of consecutive customers who are men before the first woman customer arrives, from some fixed time instant $t_0 > 0$. Calculate the mathematical expectation of M, given that the first customer after t_0 was a man.

Question no. 29
Suppose that $\{N(t), t \geq 0\}$ is a Poisson process with rate $\lambda = 2$.

(a) Let T_k be the arrival time of the kth event of the process $\{N(t), t \geq 0\}$, for $k = 1, 2, \ldots$. Calculate $P[T_1 + T_2 < T_3]$.

(b) Let S be a random variable having a uniform distribution on the interval $[0, 1]$ and that is independent of the process $\{N(t), t \geq 0\}$. Calculate $E[N^2(S)]$.

(c) We define $X_n = N(n^2)$, for $n = 0, 1, 2, \ldots$. Is the stochastic process $\{X_n, n = 0, 1, 2, \ldots\}$ a Markov chain? Justify by calculating the probability $P[X_{n+1} = j \mid X_n = i, X_{n-1} = i_{n-1}, \ldots, X_0 = i_0]$.

Question no. 30
We denote by $N(t)$ the number of failures of a machine in the interval $[0, t]$. We suppose that $N(0) = 0$ and that the time τ_0 until the first failure has a uniform distribution on the interval $(0, 1]$. Similarly, the time τ_{k-1} between the $(k-1)$st and the kth failure has a $U(0, 1]$ distribution, for $k = 2, 3, \ldots$. Finally, we assume that $\tau_0, \tau_1, \ldots$ are independent random variables.

(a) Calculate (i) the failure rate of the machine and (ii) $P[N(1) = 1]$.

(b) (i) Let $\tau_k^* := -\frac{1}{2} \ln \tau_k$, for $k = 0, 1, \ldots$. We define (see Ex. 5.1.2)

$$S_0 = 0 \quad \text{and} \quad S_n = \sum_{k=0}^{n-1} \tau_k^* \quad \text{for } n \geq 1$$

Finally, we set

$$N^*(t) = \max\{n \geq 0 : S_n \leq t\}$$

Show that the stochastic process $\{N^*(t), t \geq 0\}$ is a Poisson process, with rate $\lambda = 2$, by directly calculating the probability density function of τ_k^*, for all k, with the help of Proposition 1.2.2.

(ii) Use the central limit theorem to calculate (approximately) the probability $P[S_{100} \geq 40]$.

Question no. 31

In the preceding question, suppose that the random variables τ_k all have the density function

$$f(s) = se^{-s^2/2} \quad \text{for } s > 0$$

(a) (i) Now what is the failure rate of the machine and (ii) what is the value of the probability $P[N(1) = 1]$?

Indication. We have

$$\int_0^1 te^{-t^2+t}dt \simeq 0.5923$$

(b) (i) Let $\tau_k^* := \tau_k^2$, for $k = 0, 1, \ldots$. We define the random variables S_n and the stochastic process $\{N^*(t), t \geq 0\}$ as in the preceding question. Show that the process $\{N^*(t), t \geq 0\}$ is then a Poisson process with rate $\lambda = 1/2$.

(ii) Calculate approximately the probability $P[S_{25} < 40]$ with the help of the central limit theorem.

Question no. 32

A woman working in telemarketing makes telephone calls to private homes according to a Poisson process with rate $\lambda = 100$ per (working) day. We estimate that the probability that she succeeds in selling her product, on a given call, is equal to 5%, independently from one call to another. Let $N(t)$ be the number of telephone calls made in the interval $[0, t]$, where t is in (working) days, and let X be the number of sales made during *one* day.

(a) Suppose that the woman starts her working day at 9 a.m. and stops working at 7 p.m. Let τ_0 be the number of minutes between 9 a.m. and the moment of her first call of the day, and let S_0 be the duration (in minutes) of this call. We suppose that $S_0 \sim \text{Exp}(1)$ and that τ_0 and S_0 are independent random variables. What is the probability that the woman has made and finished her first call at no later than 9:06 a.m. on an arbitrary working day?

(b) Calculate $V[X \mid N(1) = 100]$.

(c) Calculate (approximately) $P[N(1) = 100 \mid X = 5]$.

Indication. Stirling's[4] formula: $n! \sim \sqrt{2\pi} n^{n+\frac{1}{2}} e^{-n}$.

(d) What is the probability that the woman will make no sales at all on exactly one day in the course of a week consisting of five working days?

Question no. 33
The breakdowns of a certain device occur according to a Poisson process with rate $\lambda = 2$ per weekday, and according to a Poisson process (independent of the first process) with rate $\lambda = 1.5$ per weekend day. Suppose that exactly four breakdowns have occurred (in all) over two consecutive days. What is the probability that both days were weekdays?

Question no. 34
Let $\{N(t), t \geq 0\}$ be a Poisson process with rate $\lambda = 2$ per minute. What is the probability that the time elapsed between at least two of the first three events of the process is smaller than or equal to one minute?

Question no. 35
Let $N(t)$ be the number of telephone calls to an emergency number in the interval $[0, t]$. We suppose that $\{N(t), t \geq 0\}$ is a Poisson process with rate $\lambda = 50$ per hour.

(a) What is, according to the model, the probability that there are more calls from 8 a.m. to 9 a.m. than from 9 a.m. to 10 a.m.?

Indication. If $X \sim \text{Poi}(50)$ and $Y \sim \text{Poi}(50)$ are independent random variables, then we have that $P[X = Y] \simeq 0.0399$.

(b) Suppose that 20% of the calls are redirected to another service. Knowing that there were 60 (independent) calls during a given hour and that the first of these calls was redirected elsewhere, what is the mathematical expectation of the number of calls that were redirected over the hour in question?

Question no. 36
From a Poisson process with rate λ, $\{N(t), t \geq 0\}$, we define the stochastic process $\{M(t), t \geq 0\}$ as follows:

$$M(t) = N(t) - t \quad \text{for } t \geq 0$$

Calculate $P[S_1 \leq 2]$, where $S_1 := \min\{t > 0: M(t) \geq 1\}$.

Question no. 37
Let
$$X(t) := \frac{N(t + \delta) - N(t)}{\delta} \quad \text{for } t \geq 0$$

where $\{N(t), t \geq 0\}$ is a Poisson process with rate λ and $\delta > 0$ is a constant.
(a) Calculate $E[X(t)]$.

---

[4] See p. 93.

(b) Calculate $\text{Cov}[X(s), X(t)]$, for $s, t \geq 0$.

(c) Is the process $\{X(t), t \geq 0\}$ wide-sense stationary? Justify.

Question no. 38

A machine is composed of three identical components placed in standby redundancy, so that the components operate (independently from each other) by turns. The lifetime (in weeks) of a component is an exponential random variable with parameter $\lambda = 1/5$. There are no spare components in stock. What is the probability that the machine will break down at some time during the next nine weeks, from the initial time, and remain down for at least a week, if we suppose that no spare components are expected to arrive in these next nine weeks?

Question no. 39

Let $\{X(t), t \geq 0\}$ be the stochastic process defined by

$$X(t) = tN(t) - [t]N([t]) \quad \text{for } t \geq 0$$

where $\{N(t), t \geq 0\}$ is a Poisson process with rate λ (> 0) and $[t]$ denotes the integer part of t.

(a) Is the process $\{X(t), t \geq 0\}$ a continuous-time Markov chain? Justify.

(b) Calculate the probability $p_{i,j}(t, t+s) := P[X(t+s) = j \mid X(t) = i]$, for $i, j \in \{0, 1, \ldots\}$ and $s, t \geq 0$.

Question no. 40

We consider the stochastic process $\{X(t), t \geq 0\}$ defined from a Poisson process with rate λ, $\{N(t), t \geq 0\}$, as follows (see Ex. 5.1.4):

$$X(t) = N(t) - tN(1) \quad \text{for } 0 \leq t \leq 1$$

(a) Let M be the number of visits of the stochastic process $\{X(t), t \geq 0\}$ to any state $x_j \geq 0$, from any state $x_i < 0$, in the interval $0 < t \leq 1$. Calculate $P[M = n \mid N(1) = n]$.

(b) We define $T = \min\{t > 0 : X(t) \geq 0\}$. Calculate the conditional probability density function $f_T(t \mid N(1) = 2)$.

Question no. 41

Consider the stochastic process $\{Y(t), t \geq 0\}$ defined in Question no. 9, where $\{N(t), t \geq 0\}$ is now a Poisson process with an arbitrary rate λ (> 0). It can be shown that we then have that $P[Y(t) = 1] = (1 + e^{-2\lambda t})/2$, for $t \geq 0$. Calculate

(a) $P[N(2) - N(1) > 1 \mid N(1) = 1]$,

(b) $P[N(t) = 0 \mid Y(t) = 1]$,

(c) $P[Y(s) = 1 \mid N(t) = 1]$, where $0 < s < t$.

Question no. 42

We set

$$X(t) = N(t + c) - N(c) \quad \text{for } t \geq 0$$

where $\{N(t), t \geq 0\}$ is a Poisson process with rate λ and c is a positive constant. Is the stochastic process $\{X(t), t \geq 0\}$ a Poisson process? Justify.

Question no. 43

Let $\{N(t), t \geq 0\}$ be a Poisson process with rate $\lambda > 0$ and let $\{X(t), t \geq 0\}$ be the stochastic process defined as follows:

$$X(t) = \lim_{\delta \downarrow 0} \frac{N(t + \delta^2) - N(t)}{\delta} \quad \text{for } t \geq 0$$

where δ is a positive constant.

(a) Calculate $\text{Cov}[X(t_1), X(t_2)]$, for $t_1, t_2 \geq 0$.

(b) Is the process $\{X(t), t \geq 0\}$ wide-sense stationary? Justify.

Question no. 44

Suppose that $\{N(t), t \geq 0\}$ is a Poisson process with rate λ. Let $X_n := N^2(n)$, for $n = 0, 1, \ldots$. Is the stochastic process $\{X_n, n = 0, 1, \ldots\}$ a discrete-time Markov chain? If it is, give its one-step transition probability matrix. If it's not, justify.

Question no. 45

Let T_k be the arrival time of the kth event of the Poisson process $\{N(t), t \geq 0\}$, with rate λ, for $k = 1, 2, \ldots$. Calculate the covariance $\text{Cov}[T_1, T_2]$.

Question no. 46

We consider a Poisson process, $\{N(t), t \geq 0\}$, with rate $\lambda = 1$. Let T_1, T_2, $\ldots$ be the arrival times of the events and let $S_1 := T_1$, $S_2 := T_2 - T_1$, $S_3 := T_3 - T_2$, etc.

(a) Calculate $P[\{S_1 < S_2 < S_3\} \cup \{S_3 < S_2 < S_1\}]$.

(b) Let $X := S_1^{1/2}$. Calculate (i) $f_X(x)$ and (ii) $P[X < S_2]$.

(c) Calculate $E[N^2(1) \mid T_6 = 5]$.

Question no. 47

Let $\{X_n, n = 0, 1, \ldots\}$ be a (discrete-time) Markov chain whose state space is the set $\mathbf{Z}$ of all integers. Suppose that the process spends an exponential time τ (in seconds) with parameter $\lambda = 1$ in each state before making a transition and that the next state visited is independent of τ. Let $N(t)$, for $t \geq 0$, be the number of transitions made in the interval $[0, t]$.

(a) What is the probability that the third transition took place before the fifth second, given that five transitions occurred during the first 10 seconds?

(b) Calculate $E[|N(5) - 1|]$.

Sections 5.2 to 5.5

Question no. 48

The failures of a certain device occur according to a nonhomogeneous Poisson process whose intensity function $\lambda(t)$ is given by

$$\lambda(t) = \begin{cases} 0.2 \text{ if } 0 \leq t \leq 10 \\ 0.3 \text{ if } t > 10 \end{cases}$$

where t is the age (in years) of the device.

(a) Calculate the probability that a five-year-old device will have exactly two failures over the next 10 years.

(b) Knowing that the device had exactly one failure in the course of the first 5 years of the 10 years considered in (a), what is the probability that this failure took place during its sixth year of use?

Question no. 49

Suppose that traffic accidents occur in a certain region according to a Poisson process with rate $\lambda = 2$ per day. Suppose also that the number M of persons involved in a given accident has a geometric distribution with parameter $p = 1/2$. That is,

$$P[M = m] = (1/2)^m \quad \text{for } m = 1, 2, \ldots$$

(a) Calculate the mean and the variance of the number of persons involved in an accident over an arbitrary week.

(b) Let T be the random variable denoting the time between the first *person* and the second *person* involved in an accident, from the initial time. Calculate the distribution function of T.

Question no. 50

Suppose that the monthly sales of a dealer of a certain luxury car constitute a conditional Poisson process such that Λ is a discrete random variable taking the values 2, 3, or 4, with probabilities 1/4, 1/2, and 1/4, respectively.

(a) If the dealer has three cars of this type in stock, what is the probability that the three cars will be sold in less than a month?

(b) Suppose that $\Lambda = 3$. Calculate $V[M \mid M \leq 1]$, where M is the number of cars sold in one month.

Question no. 51

We suppose that the traffic at a point along a certain road can be described by a Poisson process with parameter $\lambda = 2$ per minute and that 60% of the vehicles are cars, 30% are trucks, and 10% are buses. We also suppose that the number K of persons in a single vehicle is a random variable whose function p_K is

$$p_K(k) = \begin{cases} 1/2 \text{ if } k = 1 \\ 1/4 \text{ if } k = 2 \\ 1/8 \text{ if } k = 3 \\ 1/8 \text{ if } k = 4 \end{cases} \quad \text{and} \quad p_K(k) = \begin{cases} 0.9 \text{ if } k = 1 \\ 0.1 \text{ if } k = 2 \end{cases}$$

in the case of cars and trucks, respectively, and $p_K(k) = 1/50$, for $k = 1, \ldots, 50$, in the case of buses.

(a) Calculate the variance of the number of persons who pass by this point in the course of a five-minute period.

Indication. We have

$$\sum_{k=1}^{n} k^2 = \frac{n(n+1)(2n+1)}{6}$$

(b) Given that five cars passed by the point in question over a five-minute period, what is the variance of the total number of vehicles that passed by that point during these five minutes?

Indication. We assume that the number of cars is independent of the number of trucks and buses.

(c) Calculate, assuming as in (b) the independence of the vehicles, the probability that two cars will pass by this point before two vehicles that are not cars pass by there.

(d) Suppose that actually

$$\lambda = \lambda(t) = \begin{cases} t/5 & \text{if } 0 \le t \le 10 \\ 2 & \text{if } 10 < t \le 50 \\ (60 - t)/5 & \text{if } 50 < t \le 60 \end{cases}$$

Calculate the probability $P[N(60) = 100 \mid N(30) = 60]$, where $N(t)$ is the total number of vehicles in the interval $[0, t]$.

Question no. 52

Let $N(t)$ be the number of accidents at a specific intersection in the interval $[0, t]$. We suppose that $\{N(t), t \ge 0\}$ is a Poisson process with rate $\lambda_1 = 1$ per week. Moreover, the number Y_k of persons injured in the kth accident has (approximately) a Poisson distribution with parameter $\lambda_2 = 1/2$, for all k. Finally, the random variables $Y_1, Y_2, \ldots$ are independent among themselves and are also independent of the stochastic process $\{N(t), t \ge 0\}$.

(a) Calculate the probability that the total number of persons injured in the interval $[0, t]$ is greater than or equal to 2, given that $N(t) = 2$.

(b) Calculate $V[N(t)Y_1]$.

(c) Let S_k be the time instant when the kth person was injured, for $k = 1, 2, \ldots$. We set $T = S_2 - S_1$. Calculate $P[T > 0]$.

Question no. 53

Let $\{N(t), t \ge 0\}$ be a Poisson process with rate L.

(a) Suppose that L is a random variable having an exponential distribution with parameter 1 (so that $\{N(t), t \geq 0\}$ is actually a conditional Poisson process). Let T_1 be the arrival time of the first event of the stochastic process $\{N(t), t \geq 0\}$.

(i) Calculate the probability density function of T_1.

(ii) Does the random variable T_1 have the memoryless property? Justify.

(b) Let $L = 1/2$. Calculate $E[N^2(1) \mid N(1) \leq 2]$.

(c) Let $L = 1$. Calculate $P[N(3) - N(1) > 0 \mid N(2) = 1]$.

Question no. 54

Telephone calls to an emergency number arrive according to a nonhomogeneous Poisson process whose intensity function is given by

$$\lambda(t) = \begin{cases} 2 \text{ if } 0 \leq t \leq 6 & \text{(by night)} \\ 4 \text{ if } 6 < t < 24 & \text{(by day)} \end{cases}$$

where t is in hours, and $\lambda(t) = \lambda(t - 24)$, for $t \geq 24$. Furthermore, the duration (in minutes) of a call received at night has a uniform distribution on the interval $(0, 2]$, whereas the duration of a call received during the day has a uniform distribution on the interval $(0, 3]$. Finally, the durations of calls are independent random variables.

(a) Calculate the probability that an arbitrary telephone call received at night will be longer than a given call received during the day.

(b) Calculate the variance of the number of calls received in the course of a given week.

(c) Let D be the total duration of the calls received during a given day. Calculate the variance of D.

Question no. 55

Let $\{N(t), t \geq 0\}$ be a Poisson process with rate $\lambda = 2$. Suppose that all the events that occur in the intervals $(2k, 2k + 1]$, where $k \in \{0, 1, 2, \dots\}$, are counted, whereas the probability of counting an event occurring in an interval of the form $(2k + 1, 2k + 2]$ is equal to $1/2$. Let $N_1(t)$ be the number of events counted in the interval $[0, t]$.

(a) Calculate $P[N_1(2.5) \geq 2]$.

(b) Calculate $V[N_1(2) - 2N_1(1)]$.

(c) Let S_1 be the arrival time of the first counted event. Calculate

$$P[S_1 \leq s \mid N_1(2) = 1] \quad \forall \, s \in (0, 2]$$

Question no. 56

In the preceding question, suppose that $\lambda = 1$ and that the probability of counting an event occurring in an interval of the form $(2k + 1, 2k + 2]$ is equal to $(1/2)^k$, for $k \in \{0, 1, 2, \dots\}$.

(a) Calculate $P[N_1(5) \geq 5 \mid N_1(2) = 2]$.

(b) Calculate $\text{Cov}[N_1(5), N_1(2)]$.

(c) Let M be the total number of counted events in the intervals of the form $(2k+1, 2k+2]$. Calculate $E[M]$.

(d) Let S be the arrival time of the first event in an interval of the form $(2k, 2k+1]$. Calculate $f_S(s \mid N(3) - N(2) + N(1) = 1)$.

Question no. 57

Let $\{N_1(t), t \geq 0\}$ be a (homogeneous) Poisson process with rate $\lambda = 1$, and let $\{N_2(t), t \geq 0\}$ be a nonhomogeneous Poisson process whose intensity function is

$$\lambda(t) = \begin{cases} 1 \text{ if } 0 \leq t \leq 1 \\ 2 \text{ if } t > 1 \end{cases}$$

We suppose that the two stochastic processes are independent. Moreover, let $\{Y(t), t \geq 0\}$ be a compound Poisson process defined by

$$Y(t) = \sum_{i=1}^{N_1(t)} X_i \qquad (\text{and } Y(0) = 0 \text{ if } N_1(t) = 0)$$

where X_i has a Poisson distribution with parameter 1, for all i. Calculate

(a) $P[T_{1,2} < T_{2,2}, N_1(1) = N_2(1) = 0]$, where $T_{m,n}$ denotes the arrival time of the nth event of the process $\{N_m(t), t \geq 0\}$, for $m = 1, 2$ and $n = 1, 2, \ldots$,

(b) $V[N_2(1)(N_2(2) - N_2(1))]$,

(c) (i) $P[Y(1) < N_1(1) \leq 1]$; (ii) $V[Y(5) \mid N_1(5) \leq 2]$.

Question no. 58

Suppose that the intensity function $\lambda(t)$ of the nonhomogeneous Poisson process $\{N(t), t \geq 0\}$ is given by

$$\lambda(t) = \frac{t}{t^2 + 1} \lambda \quad \text{for } t \geq 0$$

where $\lambda > 0$ and t is in minutes.

(a) Let T_1 be the arrival time of the first event of the stochastic process $\{N(t), t \geq 0\}$. Calculate $P[T_1 \leq s \mid N(1) = 1]$, for $s \in (0, 1]$.

(b) Suppose that $\lambda = 2$ and that $N(5) \geq 2$. Calculate the probability that, at time $t_0 = 10$, at least five minutes have elapsed since the penultimate event occurred.

Question no. 59

In the preceding question, suppose that

$$\lambda(t) = 1 + \frac{1}{t+1} \quad \text{for } t \geq 0$$

where t is in minutes.

(a) Calculate $f_{T_1}(s \mid N(1) = 1)$, for $s \in (0, 1]$.

(b) Suppose that $N(5) \geq 3$. Calculate the probability that, at time $t_0 = 10$, at least five minutes have elapsed since the antepenultimate (that is, the one before the penultimate) event occurred.

Question no. 60

The stochastic process $\{Y(t), t \geq 0\}$ is a compound Poisson process defined by

$$Y(t) = \sum_{k=1}^{N(t)} X_k \qquad (\text{and } Y(t) = 0 \text{ if } N(t) = 0)$$

where X_k has a geometric distribution with parameter $1/4$, for $k = 1, 2, \ldots$, and $\{N(t), t \geq 0\}$ is a Poisson process with rate $\lambda = 3$.

(a) Calculate $E[Y(t) \mid Y(t) > 0]$, for $t > 0$.

(b) Calculate approximately $P[Y(10) > 100]$ with the help of the central limit theorem.

Question no. 61

Suppose that, in the preceding question, X_k is a discrete random variable such that $P[X_k = 1] = P[X_k = 2] = 1/2$, for $k = 1, 2, \ldots$, and that $\{N(t), t \geq 0\}$ is a Poisson process with rate $\lambda = 2$.

(a) Calculate $E[Y^2(t) \mid Y(t) > 0]$, for $t > 0$.

(b) Use the central limit theorem to calculate (approximately) the probability $P[Y(20) < 50]$.

Question no. 62

The (independent) visitors of a certain Web site may be divided into two groups: those who arrived on this site voluntarily (type I) and those who arrived there by chance or by error (type II). Let $N(t)$ be the total number of visitors in the interval $[0, t]$. We suppose that $\{N(t), t \geq 0\}$ is a Poisson process with rate $\lambda = 10$ per hour, and that 80% of the visitors are of type I (and 20% of type II).

(a) Calculate the mean and the variance of the number of visitors of type I, from a given time instant, before a second type II visitor accesses this site.

(b) Calculate the variance of the total time spent on this site by the visitors arrived in the interval $[0, 1]$ if the time (in minutes) X_I (respectively, X_{II}) that a type I (resp., type II) visitor spends on the site in question is an exponential random variable with parameter $1/5$ (resp., 2). Moreover, we assume that X_I and X_{II} are independent random variables.

(c) Suppose that, actually, $\{N(t), t \geq 0\}$ is a nonhomogeneous Poisson process whose intensity function is

$$\lambda(t) = \begin{cases} 5 & \text{if } 0 \leq t \leq 7 \\ 20 & \text{if } 7 < t < 24 \end{cases}$$

and $\lambda(t+24n) = \lambda(t)$, for $n = 1, 2, \ldots$. Given that exactly one visitor accessed this site between 6 a.m. and 8 a.m., what is the distribution function of the random variable S denoting the arrival time of this visitor?

Question no. 63

At night, vehicles circulate on a certain highway with separate roadways according to a Poisson process with parameter $\lambda = 2$ per minute (in each direction). Due to an accident, traffic must be stopped in one direction. Suppose that 60% of the vehicles are cars, 30% are trucks, and 10% are semitrailers. Suppose also that the length of a car is equal to 5 m, that of a truck is equal to 10 m, and that of a semitrailer is equal to 20 m.

(a) From what moment is there a 10% probability that the length of the queue of stopped vehicles is greater than or equal to one kilometer?

(b) Give an exact formula for the distribution of the length of the queue of stopped vehicles after t minutes.

Indication. Neglect the distance between the stopped vehicles.

Question no. 64

During the rainy season, we estimate that showers, which significantly increase the flow of a certain river, occur according to a Poisson process with rate $\lambda = 4$ per day. Every shower, independently from the others, increases the river flow during T days, where T is a random variable having a uniform distribution on the interval $[3, 9]$.

(a) Calculate the mean and the variance of the number of showers that significantly increase the flow of the river
    (i) six days after the beginning of the rainy season,
    (ii) t_0 days after the beginning of the rainy season, where $t_0 \geq 9$.

(b) Suppose that every (significant) shower increases the river flow by a quantity X (in m^3/s) having an exponential distribution with parameter $1/10$, independently from the other showers and from the number of significant showers. Suppose also that there is a risk of flooding when the increase in the river flow reaches the critical threshold of 310 m^3/s. Calculate approximately the probability of flooding 10 days after the beginning of the rainy season.

Question no. 65

Independent visitors to a certain Web site (having infinite capacity) arrive according to a Poisson process with rate $\lambda = 30$ per minute. The time that a given visitor spends on the site in question is an exponential random variable with mean equal to five minutes. Let $Y(t)$ be the number of visitors at time $t \geq 0$. The stochastic process $\{Y(t), t \geq 0\}$ is a filtered Poisson process.

(a) What is the appropriate response function?

(b) Calculate $E[Y(t)]$ and $V[Y(t)]$.

Question no. 66

Let $\{N_1(t), t \geq 0\}$ and $\{N_2(t), t \geq 0\}$ be independent Poisson processes, with rates λ_1 and λ_2, respectively.

(a) We denote by $T_{2,1}$ the arrival time of the first event of the stochastic process $\{N_2(t), t \geq 0\}$. Calculate $E[N_1(T_{2,1}) \mid N_1(T_{2,1}) < 2]$.

(b) We define $M(t) = N_1(t) + N_2(t)$ and we set

$$X_k = \begin{cases} 1 \text{ if the } k\text{th event of the process } \{M(t), t \geq 0\} \\ \quad \text{is an event of } \{N_1(t), t \geq 0\} \\[2mm] 0 \text{ if the } k\text{th event of the process } \{M(t), t \geq 0\} \\ \quad \text{is an event of } \{N_2(t), t \geq 0\} \end{cases}$$

Calculate $V[Y(t)]$, where

$$Y(t) := \sum_{k=1}^{M(t)} X_k \qquad (\text{and } Y(t) = 0 \text{ if } M(t) = 0)$$

(c) Suppose that $\{N_2(t), t \geq 0\}$ is rather a nonhomogeneous Poisson process whose intensity function is

$$\lambda_2(t) = \begin{cases} 1 & \text{if } 0 \leq t < 1 \\ 1 + t^{-1} & \text{if } t \geq 1 \end{cases}$$

Calculate $P[T_{2,1} \leq 2]$, where $T_{2,1}$ is defined in (a).

Section 5.6

Question no. 67

A system is composed of two components that operate at alternate times. When component i, for $i = 1, 2$, starts to operate, it is active during X_i days, where X_i is an exponential random variable with parameter λ_i and is independent of what happened previously. The state of the components is checked only at the beginning of each day. If we notice that component i is down, then we set the other component going, and component i will be repaired (in less than one day).

(a) Let N_i be the number of consecutive days during which component i is responsible for the functioning of the system, for $i = 1, 2$. What is the probability distribution of N_i?

(b) Suppose that the two components are identical. That is, $\lambda_1 = \lambda_2 := \lambda$. At what rate do the components relieve each other (over a long period)?

(c) If $\lambda_1 = 1/10$ and $\lambda_2 = 1/12$, what proportion of time, when we consider a long period, is component 1 responsible for the functioning of the system?

Question no. 68

A woman makes long-distance calls with the help of her cell phone according to a Poisson process with rate λ. Suppose that for each long-distance call billed, the next call is free and that we fix the origin at the moment when a call has just been billed. Let $N(t)$ be the number of calls billed in the interval $(0, t]$.

(a) Find the probability density function of the random variables $\tau_0, \tau_1, \ldots,$ where τ_0 is the time until the first billed call, and τ_i is the time between the $(i-1)$st and the ith billed call, for $i \geq 1$.

(b) Calculate the probability $P[N(t) = n]$, for $n = 0, 1, \ldots$.

(c) What is the average time elapsed at time t since a call has been billed?

Question no. 69

A machine is made up of two independent components placed in series. The lifetime of each component is uniformly distributed over the interval $[0, 1]$. As soon as the machine breaks down, the component that caused the failure is replaced by a new one. Let $N(t)$ be the number of replacements in the interval $[0, t]$.

(a) Is the stochastic process $\{N(t), t \geq 0\}$ a continuous-time Markov chain? Justify.

(b) Is $\{N(t), t \geq 0\}$ a renewal process? Justify.

(c) Let S_1 be the time of the first replacement. Calculate the probability density function of S_1.

Question no. 70 (See Question no. 98, p. 171)

The lifetime of a certain machine is a random variable having an exponential distribution with parameter λ. When the machine breaks down, there is a probability equal to p (respectively, $1 - p$) that the failure is of type I (resp., II). In the case of a type I failure, the machine is out of use for an exponential time, with mean equal to $1/\mu$ time unit(s). To repair a type II failure, two independent operations must be performed. Each operation takes an exponential time with mean equal to $1/\mu$.

(a) Use the results on regenerative processes to calculate the probability that the machine will be in working state at a (large enough) given time instant.

(b) What is the average age of the machine at time t? That is, what is the average time elapsed at time t since the most recent failure has been repaired? Assume that $\lambda = \mu$.

Question no. 71

We consider a discrete-time Markov chain whose state space is the set $\{0, 1, 2\}$ and whose one-step transition probability matrix is

$$\mathbf{P} = \begin{bmatrix} 0 & 1/2 & 1/2 \\ 1/2 & 0 & 1/2 \\ 1 & 0 & 0 \end{bmatrix}$$

We say that a renewal took place when the initial state 0 is revisited.

(a) What is the average number of transitions needed for a renewal to take place?

(b) Let $N(t)$, for $t > 0$, be the number of renewals in the interval $[0, t]$, where t is in seconds. If we suppose that every transition of the Markov chain takes one second, calculate

    (i) the distribution of the random variable $N(6.5)$,

    (ii) the probability $P[N(90) < 40]$ (approximately).

Question no. 72

We consider a birth and death process, $\{X(t), t \geq 0\}$, whose state space is the set $\{0, 1\}$ and for which

$$\lambda_0 = \lambda \quad \text{and} \quad \mu_1 = \mu$$

Moreover, we suppose that $X(0) = 0$. We say that a renewal occurred when the initial state 0 is revisited. Let $N(t)$, for $t \geq 0$, be the number of renewals in the interval $[0, t]$.

(a) Let τ_{n-1} be the time between the $(n-1)$st and the nth renewal, for $n \geq 1$ (τ_0 being the time until the first renewal). Find the probability density function of τ_{n-1} if $\lambda = \mu$.

(b) Calculate approximately the probability $P[N(50) \leq 15]$ if $\lambda = 1/3$ and $\mu = 1$.

(c) Calculate the average time elapsed at a fixed time instant t_0 since a renewal occurred if $\lambda = 2$ and $\mu = 1$.

Question no. 73

Let $\{X(t), t \geq 0\}$ be a birth and death process for which

$$\lambda_n = \lambda \quad \forall\, n \geq 0 \quad \text{and} \quad \mu_1 = \mu, \ \mu_n = 0 \quad \forall\, n \geq 2$$

We suppose that $X(0) = 0$ and we say that a renewal took place when the process revisits the initial state 0. We denote by $N(t)$, for $t \geq 0$, the number of renewals in the interval $[0, t]$.

(a) Let T_1 be the time elapsed until the first renewal, and let $M(t)$ be the number of deaths in the interval $[0, t]$. Calculate

    (i) $E[T_1]$,

    (ii) $P[\lim_{t \to \infty} M(t) = k]$, for $k \in \{0, 1, 2, \dots\}$.

(b) Is the stochastic process $\{X(t), t \geq 0\}$ a regenerative process? Justify.

(c) Suppose now that $\lambda_0 = \lambda$ and $\lambda_n = 0$ if $n \geq 1$. Calculate the proportion of time that the process $\{X(t), t \geq 0\}$ spends in state 0, over a long period.

Question no. 74

Let $\{N(t), t \geq 0\}$ be a renewal process for which the time τ between the successive events is a continuous random variable whose probability density function is given by

$$f_\tau(s) = se^{-s} \quad \text{for } s \geq 0$$

That is, τ has a gamma distribution with parameters $\alpha = 2$ and $\lambda = 1$.

(a) Show that the renewal function, $m_N(t)$, is given by

$$m_N(t) = \frac{1}{4}(e^{-2t} + 2t - 1) \quad \text{for } t \geq 0$$

Indication. We have

$$\sum_{n=1}^{\infty} \frac{x^{2n-1}}{(2n-1)!} = \frac{1}{2}(e^x - e^{-x})$$

(b) Calculate approximately the probability $P[T_{N(100)+1} < 101.5]$.

(c) Let $X(t) := (-1)^{N(t)}$, for $t \geq 0$. The stochastic process $\{X(t), t \geq 0\}$ *is a regenerative process. Calculate the limiting probability that $X(t) = 1$.*

Question no. 75
    Consider the renewal process $\{N(t), t \geq 0\}$ for which the time τ between the consecutive events is a continuous random variable having the following probability density function:

$$f_\tau(s) = 2s \quad \text{for } 0 \leq s \leq 1$$

(a) Calculate the renewal function, $m_N(t)$, for $0 \leq t \leq 1$.

Indication. The general solution of the second-order ordinary differential equation $y''(x) = ky(x)$ is

$$y(x) = c_1 e^{k^{1/2}x} + c_2 e^{-k^{1/2}x}$$

where $k \neq 0$, and c_1 and c_2 are constants.

(b) According to Markov's inequality, what is the maximum value of the probability $P[T_{N(1/2)+1} \geq 1]$?

(c) Suppose that we receive a reward of \$1 per time unit when the age, $A(t)$, of the renewal process is greater than or equal to $1/2$, and of \$0 when $A(t) < 1/2$. Calculate the average reward per time unit over a long period.

Question no. 76
    The time between the successive renewals, for a certain renewal process $\{N(t), t \geq 0\}$, is a continuous random variable whose probability density function is the following:

$$f(s) = \begin{cases} 1/2 & \text{if } 0 < s < 1/2 \\ 3/2 & \text{if } 1/2 \leq s < 1 \end{cases}$$

(a) We can show that $m_N(t) = e^{t/2} - 1$, for $t \in (0, 1/2)$. Use this result to calculate $m_N(t)$, for $t \in [1/2, 1)$.

Indication. The general solution of the ordinary differential equation

$$y'(x) + P(x)y(x) = Q(x)$$

is

$$y(x) = e^{-\int P(x)dx} \left(\int Q(x)e^{\int P(x)dx} \, dx + \text{constant} \right)$$

(b) Use the central limit theorem to calculate approximately $P[T_{25} > 15]$, where T_{25} is the time of the 25th renewal of the process $\{N(t), t \geq 0\}$.

(c) Let $\{X(t), t \geq 0\}$ be a regenerative process whose state space is the set $\{1, 2\}$. We suppose that the time that the process spends in state 1 is a random variable Y_1 such that

$$f_{Y_1}(y) = \begin{cases} 1/2 & \text{if } 0 < y < 1/2 \\ 3/2 & \text{if } 1/2 \leq y < 1 \end{cases}$$

while the probability density function of the time Y_2 that the process spends in state 2 is

$$f_{Y_2}(y) = \begin{cases} 3/2 & \text{if } 0 < y < 1/2 \\ 1/2 & \text{if } 1/2 \leq y < 1 \end{cases}$$

Calculate $\lim_{t \to \infty} P[X(t) = 1]$.

Question no. 77

We consider a system composed of two subsystems placed in series. The first subsystem comprises a single component (component no. 1), whereas the second subsystem comprises two components placed in parallel. Let S_k be the lifetime of component no. k, for $k = 1, 2, 3$. We assume that the continuous random variables S_k are independent.

(a) Let $N(t)$ be the number of system failures in the interval $[0, t]$. In what case(s) will the stochastic process $\{N(t), t \geq 0\}$ be a renewal process if the random variables S_k do not all have an exponential distribution? Justify.

(b) In the particular case when the variable S_k has an exponential distribution with parameter $\lambda = 1$, for all k, the process $\{N(t), t \geq 0\}$ *is* a renewal process. Calculate the probability density function of the time τ between two consecutive renewals.

Question no. 78

Suppose that $\{N(t), t \geq 0\}$ is a renewal process for which the time τ between two consecutive renewals is a continuous random variable such that

$$f_\tau(s) = \begin{cases} \dfrac{e^s}{e - 1} & \text{if } 0 < s < 1 \\ 0 & \text{elsewhere} \end{cases}$$

(a) Calculate the renewal function, $m_N(t)$, for $0 < t < 1$.

Remark. See the indication for Question no. 76.

(b) If we receive a reward of $1 at the moment of the nth renewal if the duration of the cycle has been greater than $1/2$ (and $0 otherwise), what is the average reward per time unit over a long period?

Question no. 79

We consider a system made up of two subsystems placed in parallel. The first subsystem is composed of two components (components nos. 1 and 2) placed in parallel, while the second subsystem comprises a single component (component no. 3). Let S_k denote the lifetime of component no. k, for $k = 1, 2, 3$. The continuous random variables S_k are assumed to be independent.

(a) Suppose that components nos. 1 and 2 operate at the same time, from the initial time, whereas component no. 3 is in standby and starts operating when the first subsystem fails. When the system breaks down, the three components are replaced by new ones. Let $N(t)$, for $t \geq 0$, be the number of system failures in the interval $[0, t]$. Then $\{N(t), t \geq 0\}$ *is* a renewal process. Let τ be the time between two consecutive renewals. Calculate the mean and the variance of τ if $S_k \sim U(0, 1)$, for $k = 1, 2, 3$.

(b) Suppose that we consider only the first subsystem and that the two components are actually placed in series. When this subsystem fails, the two components are replaced by new ones. As in (a), the process $\{N(t), t \geq 0\}$ is a renewal process. Calculate the renewal function $m_N(t)$, for $0 < t < 1$, if $S_k \sim U(0, 1)$, for $k = 1, 2$.

Indication. The general solution of the differential equation

$$y''(x) - 2y'(x) + 2y(x) + 2 = 0$$

is

$$y(x) = -1 + c_1 e^x \cos x + c_2 e^x \sin x$$

where c_1 and c_2 are constants.

(c) Suppose that in (b) we replace only the failed component when the subsystem breaks down and that $S_k \sim Exp(2)$, for $k = 1, 2$.
    (i) Calculate the mean of $T_{N(t)+1} - t$.
    (ii) Deduce from it the value of $m_N(t)$, for $t > 0$.

Question no. 80

Is a nonhomogeneous Poisson process with intensity function $\lambda(t) = t$, for all $t \geq 0$, a renewal process? If it is, give the distribution of the random variables τ_k. If it's not, justify.

Question no. 81

Use the renewal equation to find the distribution of the random variables τ_k, taking their values in the interval $[0, \pi/2]$, of a renewal process for which $m_N(t) = t^2/2$, for $0 \leq t \leq \pi/2$.

Question no. 82

Let $\{X(t), t \geq 0\}$ be a birth and death process with state space $\{0, 1, 2\}$ and having the following birth and death rates: $\lambda_0 = \lambda_1 = \lambda > 0$, μ_1 unknown, and $\mu_2 = \mu > 0$. For what value(s) of μ_1 is the process regenerative? Justify.

Question no. 83

Suppose that $\{N(t), t \geq 0\}$ is a Poisson process with rate $\lambda > 0$. For a fixed time instant $t > 0$, we consider the random variables $A(t)$ and $D(t)$ (see p. 280).

(a) Calculate the distribution of both $A(t)$ and $D(t)$.

(b) What is the distribution of $A(t) + D(t)$?

(c) Suppose now that we consider the Poisson process over the entire real line, that is, $\{N(t), t \in \mathbb{R}\}$. What then is the distribution of $A(t) + D(t)$?

(d) In (c), we can interpret the sum $A(t) + D(t)$ as being the length of the interval, between two events, which contains the fixed time instant t. Explain why the distribution of this sum is not an exponential distribution with parameter λ.

Question no. 84

Let $\{B(t), t \geq 0\}$ be a standard Brownian motion, that is, a Wiener process with drift coefficient $\mu = 0$ and diffusion coefficient $\sigma^2 = 1$. Suppose that when $B(t) = a$ (> 0) or $b = -a$, the process spends an exponential time with parameter $\lambda = 2$ in this state (a or $b = -a$). Then, it starts again from state 0.

(a) What fraction of time does the process spend in state a or $b = -a$, over a long period?

Indication. Let $m_1(x)$ be the average time that the process, starting from $x \in (b, a)$, takes to reach a or b. We can show that the function $m_1(x)$ satisfies the ordinary differential equation

$$\frac{1}{2}m_1''(x) = -1 \qquad (\text{with } m_1(b) = m_1(a) = 0)$$

(b) Answer the question in (a) if $b = -\infty$ rather than $-a$.

Question no. 85

Suppose that the time between two consecutive events for the renewal process $\{N(t), t \geq 0\}$ is equal to 1 with probability $1/2$ and equal to 2 with probability $1/2$.

(a) Give a general formula for the probability $P[N(2n) = k]$, where n and k are positive integers.

(b) Calculate $m_N(3)$.

(c) Let $I(t) := 1$ if $N(t) = [t]$ and $I(t) = 0$ otherwise, where $[\ ]$ denotes the integer part. Calculate the variance of $I(t)$.

6

Queueing Theory

6.1 Introduction

In this chapter, we will consider continuous-time and discrete-state stochastic processes, $\{X(t), t \geq 0\}$, where $X(t)$ represents the number of persons in a *queueing system* at time t. We suppose that the *customers* who arrive in the system come to receive some service or to perform a certain task (for example, to withdraw money from an automated teller machine). There can be one or many *servers* or *service stations*. The process $\{X(t), t \geq 0\}$ is a model for a *queue* or a *queueing phenomenon*. If we want to be precise, the queue should designate the customers who are waiting to be served, that is, who are *queueing*, while the queueing system includes all the customers in the system. Since *queue* is the standard expression for this type of process, we will use these two expressions interchangeably. Moreover, it is clear that the queueing models do not apply only to the case when we are interested in the number of *persons* who are waiting in line. The *customers* in the system may be, for example, airplanes that are landing or are waiting for the landing authorization, or machines that have been sent to a repair shop, etc.

Kendall[1] proposed, in a research paper published in 1953, a notation to classify the various queueing models. The most general notation is of the form $A/S/s/c/p/D$, where

A denotes the distribution of the time between two successive arrivals,

S denotes the distribution of the service time of customers,

s is the number of servers in the system,

c is the capacity of the system,

p is the size of the population from which the customers come,

D designates the *service policy*, called the *discipline*, of the queue.

---

[1] David George Kendall, retired professor of the University of Cambridge, in England.

We suppose that the times τ_n between the arrivals of successive customers are independent and identically distributed random variables. Similarly, the service times S_n of the customers are random variables assumed to be i.i.d. and independent of the τ_n's. Actually, we could consider the case when these variables, particularly the S_n's, are not independent among themselves.

The most commonly used distributions for the random variables τ_n and S_n, and the corresponding notations for A or S, are the following:

M exponential with parameter λ or μ;

E_k Erlang (or gamma) with parameters k and λ or μ;

D *degenerate* or *deterministic* (if τ_n or S_n is a constant);

G general (this case includes, in particular, the uniform distribution).

Remarks. i) We write M when τ_n (respectively, S_n) has an exponential distribution, because the arrivals of customers in the system (resp., the departures from the system *in equilibrium*) then constitute a Poisson process, which is a *Markovian* process.

ii) We can use the notation GI for *general independent*, rather than G, to be more precise.

The number s of servers is a positive integer, or sometimes infinity. (For example, if the *customers* are persons arriving in a park and staying there some time before leaving for home or elsewhere, in which case, the customers do not have to wait to be *served*.)

By default, the capacity of the system is infinite. Similarly, the size of the population from which the customers come is assumed to be infinite. If c (or p) is not equal to infinity, its value must be specified. On the other hand, when $c = p = \infty$, we may omit these quantities in the notation.

Finally, the queue discipline is, by default, that of *first-come, first-served*, which we denote by *FCFS* or by *FIFO*, for *first-in, first-out*. We may also omit this default discipline in the notation. In all other cases, the service policy used must be indicated. We can have *LIFO*, that is, *last-in, first-out*. The customers may also be served at random (*RANDOM*). Sometimes one or more special customers are receiving priority service, etc.

In this book, except in the penultimate subsection of the current chapter and some exercises, we will limit ourselves to the case when the random variables τ_n and S_n have exponential distributions, with parameters λ and μ, respectively. That is, we will only study models of the form M/M. Moreover, in the text, the service policy will be the default one (*FIFO*). However, in the exercises, we will often modify this service policy.

For all the queueing systems that we will consider, we may assume that the limit

$$\pi_n := \lim_{t \to \infty} P[X(t) = n] \tag{6.1}$$

exists, for all $n \geq 0$. Thus, π_n designates the *limiting probability* that there are exactly n customers in the system. Moreover, π_n is also the *proportion of time* when the number of customers in the system is equal to n, over a long period.

The quantities of interest when we study a particular queueing system are above all the *average number* of customers in the system, when it is *in equilibrium* (or *in stationary regime*), and the *average time* that an arbitrary customer spends in the system. We introduce the following notations:

$\bar{N}$ is the average number of customers in the system (in equilibrium);

$\bar{N}_Q$ is the average number of customers who are waiting in line;

$\bar{N}_S$ is the average number of customers being served;

$\bar{T}$ is the average time that an arbitrary customer spends in the system;

$\bar{Q}$ is the average waiting time of an arbitrary customer;

$\bar{S}$ is the average service time of an arbitrary customer.

Often, we must content ourselves with expressing these quantities in terms of the limiting probabilities π_n. Moreover, notice that $\bar{N} = \bar{N}_Q + \bar{N}_S$ and $\bar{T} = \bar{Q} + \bar{S}$.

Let $N(t)$, for $t \geq 0$, be the number of customers who arrive in the system in the interval $[0, t]$. Given that the random variables τ_n are independent and identically distributed, the process $\{N(t), t \geq 0\}$ is a renewal process. We denote by λ_a the *average arrival rate* of customers in the system. That is (see Prop. 5.6.7),

$$\lambda_a := \lim_{t \to \infty} \frac{N(t)}{t} = \frac{1}{E[\tau_n]} \tag{6.2}$$

Remarks. i) If $\tau_n \sim \text{Exp}(\lambda)$, then we obtain $\lambda_a = \lambda$.

ii) We can also define λ_e, namely, the *average entering rate* of customers into the system. If all the arriving customers enter the system, then $\lambda_e = \lambda_a$. However, if the capacity of the system is finite, or if some customers refuse to enter the system if it is too full when they arrive, etc., then λ_e will be smaller than λ_a. Suppose, for example, that the customers arrive according to a Poisson process with rate λ. If the system capacity is equal to c ($<\infty$) customers, then we may write that the rate λ_e is given by $\lambda (1 - \pi_c)$. Indeed, in this case, $(1 - \pi_c)$ is the (limiting) probability that an arriving customer will enter the system.

We can establish a relation between the quantities $\bar{N}$ and $\bar{T}$ by using a *cost equation*. Suppose that the customers entering the system pay a certain amount of money. Let λ_g be the *average earning rate* of the system. Then, after a long enough time t_0, the average amount of money the system earns is approximately equal to $\lambda_g \cdot t_0$. On the other hand, this quantity is also approximately equal to $\bar{M} \cdot \lambda_e \cdot t_0$, where $\bar{M}$ is the average amount of money a customer who enters the system pays, and λ_e is defined above. The equality between the two expressions for the average amount of money the system earns, over a long period, can be justified rigorously. We then obtain the following proposition.

Proposition 6.1.1. *We have*

$$\lambda_g = \lambda_e \cdot \bar{M} \tag{6.3}$$

Corollary 6.1.1. *If the customers pay $1 per time unit that they spend in the system (waiting to be served or being served), then Eq. (6.3) becomes*

$$\bar{N} = \lambda_e \cdot \bar{T} \tag{6.4}$$

Remarks. i) The formula above is known as *Little's[2] formula.*

ii) Little's formula may be rewritten as follows: if t is large enough (for the process to be in equilibrium), then we have

$$E[X(t)] = \lambda_e E[T] \tag{6.5}$$

where T is the total time that an arbitrary customer who enters the system will spend in this system. Actually, Eq. (6.5) is valid under very general conditions, in particular, for all the systems studied in this book, but in some cases it is not correct. Moreover, we can prove the following result:

$$\lim_{t_0 \to \infty} \frac{1}{t_0} \int_0^{t_0} X(t) \, dt = E[X(t)] = \lambda_e E[T] \tag{6.6}$$

That is, the stochastic process $\{X(t), t \geq 0\}$ is, in the cases of interest to us, *mean ergodic* (see p. 56).

iii) If every customer pays $1 per time unit while being served (but not while she is waiting in queue) instead, then we obtain

$$\bar{N}_S = \lambda_e \cdot \bar{S} \tag{6.7}$$

It follows that we also have

$$\bar{N}_Q = \lambda_e \cdot \bar{Q} \tag{6.8}$$

When the times between the arrivals of successive customers and the service times of customers are independent exponential random variables, the

---

[2] John D.C. Little, professor at the Sloan School of Management of the Massachusetts Institute of Technology, in the United States.

process $\{X(t), t \geq 0\}$ is a *continuous-time Markov chain*. Moreover, if we assume (which will be the case, in general) that the customers arrive one at a time and are served one at a time, $\{X(t), t \geq 0\}$ is then a *birth and death process*. We may therefore appeal to the results that were proved in Chapter 3 concerning this type of process, particularly Theorem 3.3.4, which gives us the limiting probabilities of the process.

6.2 Queues with a single server

6.2.1 The model $M/M/1$

We first consider a queueing system with a *single* server, in which the customers arrive according to a *Poisson process* with rate λ, and the *service times* are independent *exponential* random variables, with mean equal to $1/\mu$. We suppose that the system capacity is infinite, as well as the population from which the customers come. Finally, the queue discipline is that of first-come, first-served. We can therefore denote this model simply by $M/M/1$.

The stochastic process $\{X(t), t \geq 0\}$ is an irreducible birth and death process. We will calculate the limiting probabilities π_n as we did in Chapter 3. The *balance equations* of the system (see p. 140) are the following:

state j departure rate from j = arrival rate to j

$$0 \qquad \lambda \pi_0 = \mu \pi_1$$
$$n \, (\geq 1) \qquad (\lambda + \mu)\pi_n = \lambda \pi_{n-1} + \mu \pi_{n+1}$$

We have (see p. 141)

$$\Pi_0 := 1 \quad \text{and} \quad \Pi_n = \underbrace{\frac{\lambda\lambda\cdots\lambda}{\mu\mu\cdots\mu}}_{n\times} = \left(\frac{\lambda}{\mu}\right)^n \quad \text{for } n = 1, 2, \ldots \quad (6.9)$$

If $\lambda < \mu$, the process $\{X(t), t \geq 0\}$ is positive recurrent, and Theorem 3.3.4 then enables us to write that

$$\pi_n = \frac{(\lambda/\mu)^n}{\sum_{k=0}^{\infty}(\lambda/\mu)^k} = \frac{(\lambda/\mu)^n}{[1 - (\lambda/\mu)]^{-1}} \quad \text{for } n = 0, 1, \ldots \quad (6.10)$$

That is,

$$\pi_n = \left(\frac{\lambda}{\mu}\right)^n \left(1 - \frac{\lambda}{\mu}\right) \quad \forall \, n \geq 0 \quad (6.11)$$

We can now calculate the quantities of interest. We already know that $\bar{S} = 1/\mu$. Moreover, because here $\lambda_e = \lambda$, we may write that

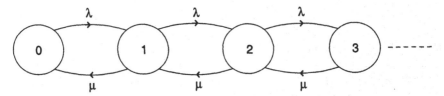

Fig. 6.1. State-transition diagram for the model $M/M/1$.

$$\bar{N}_S = \lambda_e \cdot \bar{S} = \frac{\lambda}{\mu} \tag{6.12}$$

Remarks. i) Since there is only one server, we could have directly calculated $\bar{N}_S$ as follows:

$$\bar{N}_S = 1 - \pi_0 = 1 - \left(1 - \frac{\lambda}{\mu}\right) = \frac{\lambda}{\mu} \tag{6.13}$$

because the *random variable* N_S denoting the number of persons who are being served, when the system is in equilibrium, here has a Bernoulli distribution with parameter $p_0 := 1 - \pi_0$.

ii) The quantity $\rho := \lambda/\mu$ is sometimes called the *traffic intensity* of the system, or the *utilization rate* of the system. We see that the limiting probabilities exist if and only if $\rho < 1$, which is logical, because this parameter gives us the average number of arrivals in the system during a time period corresponding to the mean service time of an arbitrary customer. If $\rho \geq 1$, the length of the queue increases indefinitely.

iii) From the balance equations of the system, we can draw the corresponding *state transition diagram* (and vice versa). Each possible state is represented by a circle, and the possible transitions between the states by arrows. Moreover, we indicate above or under each arrow the rate at which the corresponding transition takes place (see Fig. 6.1).

Next, we have

$$\bar{N} := \sum_{n=0}^{\infty} n\,\pi_n = \left(\frac{\lambda}{\mu}\right) \sum_{n=1}^{\infty} n \left(\frac{\lambda}{\mu}\right)^{n-1} \left(1 - \frac{\lambda}{\mu}\right)$$

$$= \left(\frac{\lambda}{\mu}\right) E[Z], \qquad \text{where } Z \sim \text{Geom}\left(1 - \frac{\lambda}{\mu}\right)$$

$$= \left(\frac{\lambda}{\mu}\right) \frac{\mu}{\mu - \lambda} = \frac{\lambda}{\mu - \lambda} \tag{6.14}$$

We can then write that

$$\bar{N}_Q = \bar{N} - \bar{N}_S = \frac{\lambda}{\mu - \lambda} - \frac{\lambda}{\mu} = \frac{\lambda^2}{\mu(\mu - \lambda)} \tag{6.15}$$

We deduce from this formula that

$$\bar{Q} = \frac{\bar{N}_Q}{\lambda_e} = \frac{\lambda}{\mu(\mu - \lambda)} \tag{6.16}$$

which implies that

$$\bar{T} = \bar{Q} + \bar{S} = \frac{\lambda}{\mu(\mu - \lambda)} + \frac{1}{\mu} = \frac{1}{\mu - \lambda} \tag{6.17}$$

Generally, it is difficult to find the exact distribution of the random variable T that designates the total time that an arriving customer will spend in the system (so that $\bar{T} = E[T]$). For the $M/M/1$ queue, we can explicitly determine this distribution. To do so, we need a result, which we will now prove.

Proposition 6.2.1. *Let α_n be the probability that an arbitrary customer finds n customers in the system (in equilibrium) upon arrival. If the customers arrive according to a Poisson process, then we have*

$$\pi_n = \alpha_n \quad \forall\, n \geq 0 \tag{6.18}$$

Proof. It suffices to use the fact that the Poisson process has *independent increments*. Suppose that the customer in question arrives at time t. Let $F_\epsilon =$ the customer arrives in the interval $[t, t + \epsilon)$. We have

$$
\begin{aligned}
\alpha_n &= \lim_{t \to \infty} \lim_{\epsilon \downarrow 0} P[X(t^-) = n \mid F_\epsilon] = \lim_{t \to \infty} \lim_{\epsilon \downarrow 0} \frac{P[\{X(t^-) = n\} \cap F_\epsilon]}{P[F_\epsilon]} \\
&= \lim_{t \to \infty} \lim_{\epsilon \downarrow 0} \frac{P[F_\epsilon \mid X(t^-) = n] P[X(t^-) = n]}{P[F_\epsilon]} \\
&\stackrel{\text{ind.}}{=} \lim_{t \to \infty} \lim_{\epsilon \downarrow 0} \frac{P[F_\epsilon] P[X(t^-) = n]}{P[F_\epsilon]} = \lim_{t \to \infty} \lim_{\epsilon \downarrow 0} P[X(t^-) = n] \\
&= \lim_{t \to \infty} P[X(t^-) = n] = \pi_n \quad \square
\end{aligned} \tag{6.19}
$$

Remark. It can also be shown that, for any system in which the customers arrive one at a time and are served one at a time, we have

$$\alpha_n = \delta_n \quad \forall\, n \geq 0 \tag{6.20}$$

where δ_n is the long-term proportion of customers who leave behind them n customers in the system when they depart. This follows from the fact that the *transition rate* from n to $n + 1$ is equal to the transition rate from $n + 1$ to n, over a long period.

Proposition 6.2.2. *For an M/M/1 queue, we have $T \sim Exp(\mu - \lambda)$.*

Proof. Let R be the number of customers in the system when a new customer arrives. We have

$$P[T \le t] = \sum_{r=0}^{\infty} P[T \le t \mid R = r]P[R = r] \qquad (6.21)$$

Moreover, by the *memoryless property* of the exponential distribution, we may write that

$$T \mid \{R = r\} \sim G(r+1, \mu) \qquad (6.22)$$

Finally, since the customers arrive according to a Poisson process, we have (by the preceding proposition)

$$P[R = r] \equiv \alpha_r = \pi_r = \left(\frac{\lambda}{\mu}\right)^r \left(1 - \frac{\lambda}{\mu}\right) \qquad (6.23)$$

It follows that

$$P[T \le t] = \sum_{r=0}^{\infty} \left[\int_0^t \mu e^{-\mu s} \frac{(\mu s)^r}{r!} \, ds\right] \left(\frac{\lambda}{\mu}\right)^r \left(1 - \frac{\lambda}{\mu}\right)$$

$$= \sum_{r=0}^{\infty} \int_0^t (\mu - \lambda)e^{-\mu s} \frac{(\lambda s)^r}{r!} \, ds \qquad (6.24)$$

Interchanging the summation and the integral, we obtain

$$P[T \le t] = \int_0^t (\mu - \lambda)e^{-\mu s} \underbrace{\sum_{r=0}^{\infty} \frac{(\lambda s)^r}{r!}}_{e^{\lambda s}} \, ds = \int_0^t (\mu - \lambda)e^{-(\mu-\lambda)s} \, ds \quad (6.25)$$

$$\implies \quad f_T(t) = (\mu - \lambda)e^{-(\mu-\lambda)t} \quad \text{for } t \ge 0 \quad \square \qquad (6.26)$$

We can also calculate the distribution of the waiting time, Q, of an arbitrary customer arriving in the system. This random variable is of *mixed* type. Indeed, if the customer arrives while the system is empty, we have that $Q = 0$. On the other hand, if there are $R = r \ge 1$ customer(s) in the system upon his arrival, then Q has a $G(r, \mu)$ distribution. Since $P[R = 0] = \pi_0 = 1 - \lambda/\mu$, by proceeding as above, we find that

$$P[Q \le t] = \sum_{r=0}^{\infty} P[Q \le t \mid R = r]P[R = r] \qquad (6.27)$$

$$= 1 - \left(\frac{\lambda}{\mu}\right) + \left(\frac{\lambda}{\mu}\right)\left(1 - e^{-(\mu-\lambda)t}\right) = 1 - \left(\frac{\lambda}{\mu}\right)e^{-(\mu-\lambda)t}$$

Thus, we have $P[Q = 0] = 1 - \lambda/\mu$ and

$$P[0 < Q \le t] = \left(\frac{\lambda}{\mu}\right)\left(1 - e^{-(\mu-\lambda)t}\right) \quad \text{if } t > 0 \tag{6.28}$$

Moreover, we calculate

$$P[Q \le t \mid Q > 0] = \frac{P[0 < Q \le t]}{P[Q > 0]} = 1 - e^{-(\mu-\lambda)t} \quad \text{if } t > 0 \tag{6.29}$$

That is, $Q \mid \{Q > 0\} \sim \text{Exp}(\mu - \lambda)$. We may therefore write that

$$Q \mid \{Q > 0\} \overset{\text{d}}{=} T \tag{6.30}$$

which follows directly from the fact that

$$P[R = r \mid R > 0] = \frac{\pi_r}{1 - \pi_0} = \frac{\left(\frac{\lambda}{\mu}\right)^r \left(1 - \frac{\lambda}{\mu}\right)}{\lambda/\mu} = \left(\frac{\lambda}{\mu}\right)^{r-1}\left(1 - \frac{\lambda}{\mu}\right)$$
$$= \pi_{r-1} = P[R = r - 1] \quad \forall\, r \ge 1 \tag{6.31}$$

Remark. Since the random variables Q and S are independent (by assumption), we can check that T has an exponential distribution with parameter $\mu - \lambda$ by convoluting the probability density functions of Q and S. Because Q is a mixed-type random variable, it is actually easier to make use of the formula (with $\rho = \lambda/\mu$)

$$F_T(t) = \int_0^\infty F_Q(t - s)\, f_S(s)\, ds = \int_0^t \left(1 - \rho\, e^{-(\mu-\lambda)(t-s)}\right)\mu e^{-\mu s}\, ds$$

$$= 1 - e^{-\mu t} - \rho\mu e^{-(\mu-\lambda)t} \int_0^t e^{-\lambda s}\, ds$$

$$= 1 - e^{-\mu t} - e^{-(\mu-\lambda)t}\left(1 - e^{-\lambda t}\right) = 1 - e^{-(\mu-\lambda)t} \quad \forall\, t \ge 0 \tag{6.32}$$

The variance of the random variable N designating the number of customers present in the system in equilibrium (and whose mean value is $\bar{N}$) is easily obtained by noticing [see Eq. (6.11)] that

$$N + 1 \sim \text{Geom}(1 - \rho) \tag{6.33}$$

It follows that

$$V[N] = V[N + 1] = \frac{\rho}{(1 - \rho)^2} = \frac{\lambda\mu}{(\mu - \lambda)^2} \tag{6.34}$$

Note that the mean $\bar{N}$ can be expressed as follows:

$$\bar{N} \equiv E[N] = \frac{1}{1 - \rho} - 1 = \frac{\rho}{1 - \rho} \tag{6.35}$$

In the case of the random variable N_S, we mentioned above that it has a Bernoulli distribution with parameter $1 - \pi_0 = \rho$. Then we directly have

$$V[N_S] = \rho(1 - \rho) = \frac{\lambda}{\mu}\left(1 - \frac{\lambda}{\mu}\right) \tag{6.36}$$

To obtain the variance of N_Q, we will use the following relation between N_Q and N:

$$N_Q = \begin{cases} 0 & \text{if } N = 0 \text{ or } 1 \\ N - 1 & \text{if } N \geq 2 \end{cases} \tag{6.37}$$

It follows that

$$E[N_Q^2] = \sum_{k=1}^{\infty} k^2 \pi_{k+1} = \rho^2 \sum_{k=1}^{\infty} k^2 \pi_{k-1} = \rho^2 E[Z^2] \tag{6.38}$$

where $Z \sim \text{Geom}(1 - \rho)$, from which we deduce that

$$E[N_Q^2] = \rho^2 \left(\frac{\rho}{(1 - \rho)^2} + \frac{1}{(1 - \rho)^2}\right) \tag{6.39}$$

Since the mean value of N_Q [see Eq. (6.15)] is given by

$$E[N_Q] = \frac{\lambda^2}{\mu(\mu - \lambda)} = \frac{\rho^2}{1 - \rho} \tag{6.40}$$

we obtain that

$$V[N_Q] = \rho^2 \left(\frac{1 + \rho}{(1 - \rho)^2}\right) - \frac{\rho^4}{(1 - \rho)^2} = \frac{\rho^2(1 + \rho - \rho^2)}{(1 - \rho)^2} \tag{6.41}$$

Remark. Given that $N = N_Q + N_S$, we can use the formula

$$V[N] = V[N_Q] + V[N_S] + 2\,\text{Cov}[N_Q, N_S] \tag{6.42}$$

to calculate the covariance (and then the correlation coefficient) of the random variables N_Q and N_S.

Now, we have shown that $T \sim \text{Exp}(\mu - \lambda)$. We then have

$$V[T] = \frac{1}{(\mu - \lambda)^2} \tag{6.43}$$

Similarly, $S \sim \text{Exp}(\mu)$ (by assumption), so that $V[S] = 1/\mu^2$. Finally, by independence of the random variables Q and S, we obtain that

$$V[Q] = V[T] - V[S] = \frac{1}{(\mu - \lambda)^2} - \frac{1}{\mu^2} = \frac{2\mu\lambda - \lambda^2}{\mu^2(\mu - \lambda)^2} \tag{6.44}$$

Remark. We can calculate $V[Q]$ without making use of the independence of Q and S. We deduce indeed from the formula (6.30) that

$$E[Q^2] = E[Q^2 \mid Q > 0]P[Q > 0] = E[T^2](1 - \pi_0) = E[T^2]\rho = \frac{2\rho}{(\mu - \lambda)^2}$$
$$(6.45)$$

from which we obtain that

$$V[Q] \stackrel{(6.16)}{=} \frac{2\rho}{(\mu - \lambda)^2} - \frac{\lambda^2}{\mu^2(\mu - \lambda)^2} = \frac{2\lambda\mu - \lambda^2}{\mu^2(\mu - \lambda)^2} \qquad (6.46)$$

Example 6.2.1. Let K be the number of arrivals in the system during the service period of an arbitrary customer. We will calculate the distribution of K. First, we have

$$P[K = 0] = P[S < \tau] = \frac{\mu}{\mu + \lambda}$$

where the service time $S \sim \text{Exp}(\mu)$ and the time τ needed for a customer to arrive, having an exponential $\text{Exp}(\lambda)$ distribution, are independent random variables. Then, by the memoryless property of the exponential distribution, we may write that

$$P[K = k] = \left(1 - \frac{\mu}{\mu + \lambda}\right)^k \left(\frac{\mu}{\mu + \lambda}\right) \qquad \text{for } k = 1, 2, \ldots$$

That is,

$$K + 1 \sim \text{Geom}\left(\frac{\mu}{\mu + \lambda}\right)$$

from which we deduce that

$$E[K] = \left(\frac{\mu + \lambda}{\mu}\right) - 1 = \frac{\lambda}{\mu}$$

Thus, the average number of arrivals during the service period of a given customer is equal to the average number of arrivals in the course of a period corresponding to the mean service time of an arbitrary customer (see p. 320). This result is also easily proved as follows:

$$E[K] = E[E[K \mid S]] = E[\lambda S] = \lambda E[S] = \frac{\lambda}{\mu}$$

Example 6.2.2. The conditional distribution of the random variable N, namely, the number of customers in the system in stationary regime, knowing that $N \leq m$, is given by

$$p_N(n \mid N \leq m) = \frac{\pi_n}{\sum_{k=0}^m \pi_k} = \frac{(\lambda/\mu)^n}{\sum_{k=0}^m (\lambda/\mu)^k} \qquad \text{for } n = 0, 1, \ldots, m$$

Note that the condition $N \leq m$ does not mean that the system capacity is equal to m. It rather means that, at a *given time instant*, there were at most m customers in the system (in equilibrium).

Example 6.2.3. The queueing system $M/M/1$ is modified as follows: after having been served, an arbitrary customer returns to the end of the queue with probability $p \in (0,1)$.

Remark. A given customer may return any number of times to the end of the queue.

(a) (i) Write the balance equations of the system.
      (ii) Calculate, if they exist, the limiting probabilities π_i.

(b) Calculate, in terms of the π_i's (when they exist),
      (i) the average time that an arbitrary customer spends in the system,
      (ii) the variance of the number of customers in the system, given that it is not empty.

Solution. (a) (i) The balance equations of the system are the following:

$$\text{state } j \quad \underline{\text{departure rate from } j = \text{arrival rate to } j}$$

$$0 \qquad\qquad \lambda\pi_0 = \mu(1-p)\pi_1$$
$$n\ (\geq 1) \qquad [\lambda + \mu(1-p)]\pi_n = \lambda\pi_{n-1} + \mu(1-p)\pi_{n+1}$$

(ii) The process considered is an irreducible birth and death process. We first calculate

$$S_1 := \sum_{k=1}^{\infty} \frac{\lambda_0\lambda_1 \cdots \lambda_{k-1}}{\mu_1\mu_2 \cdots \mu_k} = \sum_{k=1}^{\infty} \frac{\lambda^k}{\mu^k(1-p)^k}$$

The infinite sum converges if and only if $\lambda < \mu(1-p)$. In this case, we find that

$$S_1 = \frac{\lambda}{\mu(1-p) - \lambda}$$

Then, the π_i's exist and are given by

$$\pi_i = \frac{\lambda^i}{\mu^i(1-p)^i}\left(1 + \frac{\lambda}{\mu(1-p)-\lambda}\right)^{-1} = \left(\frac{\lambda}{\mu(1-p)}\right)^i\left(1 - \frac{\lambda}{\mu(1-p)}\right)$$

for $i = 0, 1, \ldots$.

Remark. The result is obtained at once by noticing that the process considered is equivalent to an $M/M/1$ queueing system for which the service rate is $(1-p)\mu$.

(b) (i) We seek $\bar{T}$. By the preceding remark, we may write (see the formula (6.17), p. 321)

$$\bar{T} = \frac{1}{\mu(1-p) - \lambda}$$

(ii) Let N be the number of customers in stationary regime. We have (see the formula (6.33), p. 323)

$$N + 1 \sim \text{Geom}\left(1 - \frac{\lambda}{\mu(1-p)}\right)$$

Let $M := N \mid \{N > 0\}$. We seek $V[M]$. By the memoryless property of the geometric distribution, we can write that $M \sim \text{Geom}\left(1 - \frac{\lambda}{\mu(1-p)}\right)$. It follows that

$$V[M] = \frac{\lambda}{\mu(1-p)} \bigg/ \left(1 - \frac{\lambda}{\mu(1-p)}\right)^2 = \frac{\lambda\mu(1-p)}{[\mu(1-p) - \lambda]^2}$$

6.2.2 The model $M/M/1/c$

Although the $M/M/1$ queue is very useful to model various phenomena, it is more realistic to suppose that the system capacity is an integer $c < \infty$. For $j = 0, 1, \ldots, c-1$, the balance equations of the system remain the same as when $c = \infty$. However, when the system is in state c, it can only leave it because of the departure of the customer being served. In addition, this state can only be entered from $c-1$, with the arrival of a new customer. We thus have

state j	departure rate from j = arrival rate to j
0	$\lambda\pi_0 = \mu\pi_1$
$1 \leq k \leq c-1$	$(\lambda + \mu)\pi_k = \lambda\pi_{k-1} + \mu\pi_{k+1}$
c	$\mu\pi_c = \lambda\pi_{c-1}$

The process $\{X(t), t \geq 0\}$ remains a birth and death process. Moreover, given that the number of states is *finite*, the limiting probabilities exist regardless of the values the (positive) parameters λ and μ take.

As in the case when the system capacity is infinite, we find (see p. 319) that

$$\Pi_k = \left(\frac{\lambda}{\mu}\right)^k \quad \text{for } k = 0, 1, \ldots, c \tag{6.47}$$

It follows, if $\rho := \lambda/\mu \neq 1$, that

$$\sum_{k=0}^{c} \Pi_k = \sum_{k=0}^{c} \left(\frac{\lambda}{\mu}\right)^k = \sum_{k=0}^{c} \rho^k = \frac{1 - \rho^{c+1}}{1 - \rho} \tag{6.48}$$

When $\lambda = \mu$, we have that $\rho = 1$, $\Pi_k = 1$, and $\sum_{k=0}^{c} \Pi_k = c+1$, from which we calculate

$$\pi_j = \frac{\Pi_j}{\sum_{k=0}^{c} \Pi_k} = \begin{cases} \dfrac{\rho^j(1-\rho)}{1 - \rho^{c+1}} & \text{if } \rho \neq 1 \\[2ex] \dfrac{1}{c+1} & \text{if } \rho = 1 \end{cases} \tag{6.49}$$

for $j = 0, 1, \ldots, c$.

Remarks. i) We see that if $\lambda = \mu$, then the $c + 1$ possible states of the system in equilibrium are *equally likely.* Moreover, when c tends to infinity, the probability π_j decreases to 0, for every finite j. This confirms the fact that, in the $M/M/1/\infty$ model, the queue length increases indefinitely if $\lambda = \mu$, so that there is no stationary regime.

ii) If $\lambda > \mu$, the limiting probabilities exist. However, the larger the ratio $\rho = \lambda/\mu$ is, the more π_c increases to 1 (and π_j decreases to 0, for $j = 0, 1, \dots, c-1$), which is logical.

iii) Even if, in practice, the capacity c cannot be infinite, the $M/M/1$ model is a good approximation of reality if the probability π_c that the system is full is very small.

With the help of the formula (6.49), we can calculate the value of $\bar{N}$.

Proposition 6.2.3. *In the case of the $M/M/1/c$ queue, the average number of customers in the system in equilibrium is given by*

$$\bar{N} = \begin{cases} \dfrac{\rho}{1-\rho} - \dfrac{(c+1)\rho^{c+1}}{1-\rho^{c+1}} & \text{if } \rho \neq 1 \\ \\ c/2 & \text{if } \rho = 1 \end{cases} \tag{6.50}$$

Proof. First, when $\rho = 1$, we have

$$\bar{N} = \sum_{k=0}^{c} k \frac{1}{c+1} = \frac{1}{c+1} \sum_{k=0}^{c} k = \frac{1}{c+1} \frac{c(c+1)}{2} = \frac{c}{2} \tag{6.51}$$

When $\rho \neq 1$, we must evaluate the finite sum

$$\bar{N} = \sum_{k=0}^{c} k \frac{\rho^k(1-\rho)}{1-\rho^{c+1}} = \frac{1}{1-\rho^{c+1}} \sum_{k=0}^{c} k\rho^k(1-\rho) \tag{6.52}$$

Let $X := Z - 1$, where Z has a geometric distribution with parameter $1 - \rho$. We find that the probability mass function of X is given by

$$p_X(k) = \rho^k(1-\rho) \quad \text{for } k = 0, 1, \dots \tag{6.53}$$

It follows that

$$\sum_{k=0}^{\infty} \rho^k(1-\rho) = 1 \quad \text{and} \quad \sum_{k=0}^{\infty} k\rho^k(1-\rho) = \frac{1}{1-\rho} - 1 = \frac{\rho}{1-\rho} \tag{6.54}$$

Making use of these formulas, we may write that

$$\sum_{k=0}^{c} k\rho^k(1-\rho) = \frac{\rho}{1-\rho} - \sum_{k=c+1}^{\infty} k\rho^k(1-\rho) \tag{6.55}$$

and

$$\sum_{k=c+1}^{\infty} k\rho^k(1-\rho) = \sum_{k=c+1}^{\infty} [k-(c+1)+(c+1)]\rho^k(1-\rho)$$

$$= \sum_{k=c+1}^{\infty} [k-(c+1)]\rho^k(1-\rho) + \sum_{k=c+1}^{\infty} (c+1)\rho^k(1-\rho)$$

$$= \rho^{c+1}\sum_{m=0}^{\infty} m\rho^m(1-\rho) + (c+1)\rho^{c+1}\sum_{m=0}^{\infty} \rho^m(1-\rho)$$

$$= \rho^{c+1}\frac{\rho}{1-\rho} + (c+1)\rho^{c+1} \tag{6.56}$$

so that

$$\bar{N} = \frac{1}{1-\rho^{c+1}}\left\{\frac{\rho}{1-\rho} - \rho^{c+1}\frac{\rho}{1-\rho} - (c+1)\rho^{c+1}\right\}$$

$$= \frac{1}{1-\rho^{c+1}}\left\{(1-\rho^{c+1})\left(\frac{\rho}{1-\rho}\right) - (c+1)\rho^{c+1}\right\} \quad \square \tag{6.57}$$

Remarks. i) We easily find that

$$\lim_{c\to\infty} \bar{N} = \begin{cases} \dfrac{\rho}{1-\rho} & \text{if } \rho < 1 \\ \infty & \text{if } \rho \geq 1 \end{cases} \tag{6.58}$$

which corresponds to the results obtained in the preceding subsection.

ii) When the capacity c of the system is very limited (for example, when $c = 2, 3,$ or 4), once we have calculated the π_j's, we can directly obtain $\bar{N}$ from the definition of the mathematical expectation of a discrete random variable: $\bar{N} := \sum_{k=0}^{c} k\,\pi_k$. We can then also calculate the variance of the random variable N as follows:

$$V[N] := \sum_{k=0}^{c}(k-E[N])^2\pi_k = \sum_{k=0}^{c}k^2\pi_k - (E[N])^2 \tag{6.59}$$

iii) When $\rho = 1$, we find, using the formula

$$\sum_{k=0}^{c} k^2 = \frac{c(c+1)(2c+1)}{6} \tag{6.60}$$

that

$$E[N^2] := \sum_{k=0}^{c} k^2\pi_k = \sum_{k=0}^{c} k^2\frac{1}{c+1} = \frac{c(2c+1)}{6} \tag{6.61}$$

so that

$$V[N] = \frac{c(2c+1)}{6} - \left(\frac{c}{2}\right)^2 = \frac{c(c+2)}{12} \tag{6.62}$$

Now, as in the case of the $M/M/1/\infty$ queue, we have

$$\bar{N}_S = \sum_{k=1}^{c} 1 \cdot \pi_k = 1 - \pi_0 \tag{6.63}$$

and

$$\bar{N}_Q = \bar{N} - 1 + \pi_0 \tag{6.64}$$

Next, the average *entering* rate of customers into the system (in equilibrium) is given by

$$\lambda_e = \lambda(1 - \pi_c) \tag{6.65}$$

because the customers always arrive according to a Poisson process with rate λ but can enter the system only if it is not full (or *saturated*), that is, if it is in one the following states: $0, 1, \dots, c-1$. Using both this fact and Little's formula (see p. 318), we may write that the average time that a customer *entering* the system spends in this system is equal to

$$\bar{T} = \frac{\bar{N}}{\lambda(1 - \pi_c)} \tag{6.66}$$

For the customers entering the system, we still have that $\bar{S} = 1/\mu$ (by assumption). Then

$$\bar{Q} = \frac{\bar{N}}{\lambda(1 - \pi_c)} - \frac{1}{\mu} \tag{6.67}$$

Remark. If we consider an *arbitrary* customer arriving in the system, the average time that she will spend in it is then $\bar{T} = \bar{N}/\lambda$, as noted previously. Indeed, in this case the random variable T is of mixed type, and we may write that

$$E[T] = E[T \mid T = 0]P[T = 0] + E[T \mid T > 0]P[T > 0]$$
$$= 0 \times \pi_c + \frac{\bar{N}}{\lambda(1 - \pi_c)}(1 - \pi_c) = \frac{\bar{N}}{\lambda} \tag{6.68}$$

We also have

$$\bar{S} = \frac{1}{\mu}(1 - \pi_c) \implies \bar{Q} = \frac{\bar{N}}{\lambda} - \frac{1}{\mu}(1 - \pi_c) \tag{6.69}$$

Example 6.2.4. We consider a queueing system with a single server and finite capacity $c = 3$, in which customers arrive according to a Poisson process with rate λ and the service times are independent exponential random variables, with parameter μ. When the system is full, the customer who arrived last will be served before the one standing in line in front of him.

The balance equations of the system are the following:

state j departure rate from j = arrival rate to j

0	$\lambda \pi_0 = \mu \pi_1$
1	$(\lambda + \mu)\pi_1 = \lambda \pi_0 + \mu \pi_2$
2	$(\lambda + \mu)\pi_2 = \lambda \pi_1 + \mu \pi_3$
3	$\mu \pi_3 = \lambda \pi_2$

Notice that these equations are the same as the ones obtained in the case of the $M/M/1/3$ system, although the queue discipline is not the default one (that is, first-come, first-served). When $\mu = \lambda$, we know that the solution of these equations, under the condition $\sum_{j=0}^{3} \pi_j = 1$, is $\pi_j = 1/4$, for $j = 0, 1, 2, 3$.

Suppose that a customer arrives and finds exactly one person in the system. We will calculate the mathematical expectation of the total time T_1 that this new customer will spend in the system if $\mu = \lambda$. Let K be the number of customers who will arrive after the customer in question but will be served before her (if the case may be). By the memoryless property of the exponential distribution, we may write that

$$K + 1 \sim \text{Geom}(1/2)$$

It follows that

$$E[T_1] = \sum_{k=0}^{\infty} E[T_1 \mid K = k]P[K = k] \stackrel{\mu = \lambda}{=} \sum_{k=0}^{\infty} \frac{2+k}{\lambda} \left(\frac{1}{2}\right)^{k+1}$$

$$= \frac{2}{\lambda} \sum_{k=0}^{\infty} \left(\frac{1}{2}\right)^{k+1} + \frac{1}{\lambda} \sum_{k=0}^{\infty} k \left(\frac{1}{2}\right)^{k+1}$$

$$= \frac{2}{\lambda} P[K < \infty] + \frac{1}{\lambda} E[K] = \frac{2}{\lambda} + \frac{1}{\lambda} \left(\frac{1}{1/2} - 1\right) = \frac{3}{\lambda}$$

Finally, let T_2 be the total time that an *entering* customer will spend in the system, and let L be the number of customers already present in the system upon his arrival. We may write that

$$E[T_2] = \sum_{l=0}^{2} E[T_2 \mid L = l]P[L = l]$$

Given that (in stationary regime)

$$P[L = l] = P[N = l \mid N \leq 2] = \frac{P[N = l]}{P[N \leq 2]} = \frac{1/4}{3/4} = \frac{1}{3} \quad \text{for } l = 0, 1, 2$$

we then deduce from what precedes that

$$E[T_2] \overset{\mu=\lambda}{=} \frac{1}{3}\left(\frac{1}{\lambda} + \frac{3}{\lambda} + \frac{2}{\lambda}\right) = \frac{2}{\lambda}$$

Notice that when $L = 2$, it is as if the customer we are interested in had arrived at the moment when there was exactly one customer in an $M/M/1/3$ system with service policy *FIFO* (first-in, first-out).

6.3 Queues with many servers

6.3.1 The model $M/M/s$

An important generalization of the $M/M/1$ model is obtained by supposing that there are s servers in the system and that they all serve at an exponential rate μ. The other basic assumptions that were made in the description of the $M/M/1$ model remain valid. Thus, the customers arrive in the system according to a Poisson process with rate λ. The capacity of the system is infinite, and the service policy is that by default, namely, first-come, first-served.

We suppose that the arriving customers form a single queue and that the customer at the front of the queue advances to the first server who becomes available. A system with this waiting discipline is clearly more efficient than one in which there is a queue in front of each server, since, in this case, there could be one or more idle servers while some customers are waiting in line before other servers.

Remark. A *real* waiting line in which the customers stand one behind the other need not be formed. It suffices that the customers arriving in the system *take a number*, or that the tasks to be accomplished by the servers be numbered according to their arrival order in the system.

Since the customers arrive one at a time and are served one at a time, the process $\{X(t), t \geq 0\}$, where $X(t)$ represents the number of customers in the system at time t, is a birth and death process. Note that two arbitrary customers cannot leave the system exactly at the same time instant, because the service times are *continuous* random variables. The balance equations of the system are the following (see Fig. 6.2 for the $M/M/2$ model):

state j	departure rate from j = arrival rate to j
0	$\lambda\pi_0 = \mu\pi_1$
$0 < k < s$	$(\lambda + k\mu)\pi_k = (k+1)\mu\pi_{k+1} + \lambda\pi_{k-1}$
$k \geq s$	$(\lambda + s\mu)\pi_k = s\mu\pi_{k+1} + \lambda\pi_{k-1}$

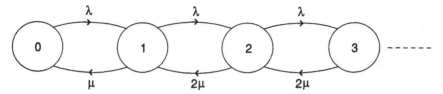

Fig. 6.2. State-transition diagram for the model $M/M/2$.

We can solve this system of equations under the condition $\sum_{k=0}^{\infty} \pi_k = 1$. As in the case of the $M/M/1$ model, it is, however, simpler to use Theorem 3.3.4. We first calculate the quantities Π_k, for $k = 1, 2, \dots$. The cases when $k \leq s$ and when $k > s$ must be considered separately. We have

$$\mu_k = \begin{cases} k\mu \ \text{if } 1 \leq k \leq s \\ s\mu \ \text{if } k > s \end{cases} \tag{6.70}$$

Then

$$\Pi_k = \frac{\lambda \times \lambda \times \cdots \times \lambda}{\mu \times 2\mu \times \cdots \times k\mu} = \frac{1}{k!}\left(\frac{\lambda}{\mu}\right)^k \quad \text{for } k = (0), 1, 2, \dots, s \tag{6.71}$$

and, for $k = s+1, s+2, \dots,$

$$\Pi_k = \frac{\lambda \times \lambda \times \cdots \times \lambda \times \lambda \times \cdots \times \lambda}{\mu \times 2\mu \times \cdots \times s\mu \times \underbrace{s\mu \times \cdots \times s\mu}_{(k-s) \text{ times}}} = \frac{1}{s!s^{k-s}}\left(\frac{\lambda}{\mu}\right)^k \tag{6.72}$$

Given that the stochastic process $\{X(t), t \geq 0\}$ is irreducible and that

$$S_1 := \sum_{k=1}^{\infty} \Pi_k < \infty \quad \Longleftrightarrow \quad \rho := \frac{\lambda}{\mu} < s \tag{6.73}$$

we can indeed appeal to Theorem 3.3.4. We first have [see Eq. (3.301)]

$$\pi_0 = \left[1 + \sum_{k=1}^{s} \frac{\rho^k}{k!} + \frac{s^s}{s!}\sum_{k=s+1}^{\infty}\left(\frac{\rho}{s}\right)^k\right]^{-1} = \left[\sum_{k=0}^{s-1}\frac{\rho^k}{k!} + \frac{s^s}{s!}\sum_{k=s}^{\infty}\left(\frac{\rho}{s}\right)^k\right]^{-1}$$

$$\overset{\rho \leq s}{=} \left[\sum_{k=0}^{s-1}\frac{\rho^k}{k!} + \frac{\rho^s}{s!}\frac{s}{(s-\rho)}\right]^{-1} \tag{6.74}$$

Once π_0 has been calculated, we set

$$\pi_k = \begin{cases} \dfrac{\rho^k}{k!}\pi_0 \quad \text{if } k = 0, 1, \dots, s \\[2ex] \dfrac{\rho^k}{s!s^{k-s}}\pi_0 \ \text{if } k = s+1, s+2, \dots \end{cases} \tag{6.75}$$

We will now calculate, in terms of π_0, the average number of waiting customers in the system in stationary regime. We have

$$
\bar{N}_Q = \sum_{k=s+1}^{\infty} (k-s)\pi_k = \sum_{k=s+1}^{\infty} (k-s)\frac{\rho^k}{s!s^{k-s}}\pi_0 = \frac{s^s}{s!}\pi_0 \sum_{k=s+1}^{\infty} (k-s)\left(\frac{\rho}{s}\right)^k
$$

$$(6.76)$$

Next,

$$
\sum_{k=s+1}^{\infty} (k-s)\left(\frac{\rho}{s}\right)^k = \left(\frac{\rho}{s}\right)^{s+1} \sum_{j=1}^{\infty} j \left(\frac{\rho}{s}\right)^{j-1} = \left(\frac{\rho}{s}\right)^{s+1}\left(1-\frac{\rho}{s}\right)^{-1}\frac{1}{1-(\rho/s)}
$$

$$(6.77)$$

from which we obtain

$$
\bar{N}_Q = \frac{\rho^{s+1}}{s!s}\frac{s^2}{(s-\rho)^2}\pi_0 = \frac{\rho^{s+1}}{s!s}\frac{1}{(1-\xi)^2}\pi_0
$$

$$(6.78)$$

where $\xi := \rho/s$. We can also write that

$$
\bar{N}_Q = \frac{1}{(1-\xi)^2}\pi_{s+1} = \frac{\xi}{(1-\xi)^2}\pi_s
$$

$$(6.79)$$

From $\bar{N}_Q$ and Little's formula, we deduce all the other quantities of interest. Since the average entering rate of customers into the system is $\lambda_e = \lambda$, we may write that

$$
\bar{Q} = \frac{\bar{N}_Q}{\lambda} \quad \Longrightarrow \quad \bar{T} = \frac{\rho^{s+1}}{\lambda s!}\frac{s}{(s-\rho)^2}\pi_0 + \frac{1}{\mu}
$$

$$(6.80)$$

because $\bar{S} = 1/\mu$ for every server (by assumption).

Finally, we have

$$
\bar{N}_S = \lambda\bar{S} = \rho \quad \text{and} \quad \bar{N} = \lambda\bar{T} = \bar{N}_Q + \rho
$$

$$(6.81)$$

Remarks. i) We can also calculate, in particular, the probability π_b that all the servers are busy. We have

$$
\pi_b = \sum_{k \geq s}^{\infty} \pi_k = \sum_{k=s}^{\infty} \frac{\rho^k}{s!s^{k-s}}\pi_0 = \frac{s^s}{s!}\pi_0 \sum_{k=s}^{\infty} \left(\frac{\rho}{s}\right)^k
$$

$$
\overset{\rho \leq s}{=} \frac{s^s}{s!}\pi_0 \left(\frac{\rho}{s}\right)^s \frac{1}{1-(\rho/s)} = \frac{s\rho^s}{s!(s-\rho)}\pi_0
$$

$$(6.82)$$

ii) If the number s of servers tends to infinity, then we find that

$$
\pi_0 \longrightarrow e^{-\lambda/\mu} \quad \text{and} \quad \pi_k \longrightarrow \frac{(\lambda/\mu)^k}{k!}e^{-\lambda/\mu} \quad \text{for } k = 1, 2, \ldots
$$

$$(6.83)$$

That is, in the case of the $M/M/\infty$ model, we have

$$\pi_k = P[Y = k], \quad \text{where } Y \sim \text{Poi}(\lambda/\mu) \tag{6.84}$$

Since $\bar{N}_Q = \bar{Q} = 0$ (because there is no waiting time), we then directly obtain

$$\bar{N} = E[Y] = \frac{\lambda}{\mu} = \bar{N}_S \quad \text{and} \quad \bar{T} = \bar{S} = \frac{1}{\mu} \tag{6.85}$$

As we did for the $M/M/1$ model, we can explicitly find the distribution of the random variable Q, namely, the time that an arbitrary customer spends waiting in line. This variable is of mixed type. We have

$$P[Q = 0] = 1 - \pi_b \tag{6.86}$$

and, for $t > 0$,

$$P[0 < Q \le t] = \sum_{r=s}^{\infty} P[0 < Q \le t \mid R = r]P[R = r] \tag{6.87}$$

where R is the number of customers in the system upon the arrival of the customer of interest. When at least s customers are in the system, the time needed for some customer to depart is a random variable having an exponential distribution with parameter $s\mu$. It follows that

$$Q \mid \{R = r\} \sim G(r - s + 1, s\mu) \quad \text{for } r = s, s + 1, \ldots \tag{6.88}$$

because the customer must wait until $r - s + 1$ persons ahead of him have left the system before starting to be served. Making use of the fact that

$$P[R = r] = \pi_r \tag{6.89}$$

(because the arrivals constitute a Poisson process) and proceeding as in the case when $s = 1$, we find that

$$P[0 < Q \le t] = \pi_b \left(1 - e^{(\lambda - s\mu)t}\right) \quad \text{for } t > 0 \tag{6.90}$$

Finally, adding the probabilities $P[Q = 0]$ and $P[0 < Q \le t]$, we obtain the following proposition.

Proposition 6.3.1. *The distribution function of the random variable Q in an $M/M/s$ queueing system (for which $\lambda < s\mu$) is given by*

$$F_Q(t) \equiv P[Q \le t] = 1 - \pi_b e^{(\lambda - s\mu)t} \quad \text{for } t \ge 0 \tag{6.91}$$

Remarks. i) With $s = 1$, we calculate

$$\pi_b = \frac{\rho}{1 - \rho}\left(1 + \frac{\rho}{1 - \rho}\right)^{-1} = \rho \qquad (6.92)$$

so that

$$P[Q \le t] \overset{s=1}{=} 1 - \rho\, e^{(\lambda - \mu)t} \quad \text{for } t \ge 0 \qquad (6.93)$$

which corresponds to the formula (6.27).

ii) To obtain the distribution of the total time, T, that an arbitrary customer spends in an $M/M/s$ system in equilibrium, we would have to compute the convolution product of the probability density functions of the random variables Q and $S \sim \text{Exp}(\mu)$.

Example 6.3.1. The number K of idle servers in an $M/M/s$ system, in stationary regime, may be expressed as follows:

$$K = s - N_S$$

where N_S is the number of customers being served. We then deduce from Eq. (6.81) that

$$E[K] = s - \bar{N}_S = s - \rho$$

6.3.2 The model $M/M/s/c$ and loss systems

As we mentioned in Subsection 6.2.2, in reality the capacity c of a waiting system is generally finite. In the case of the $M/M/s/c$ model, we can use the results on birth and death processes to calculate the limiting probabilities of the process.

Example 6.3.2. Consider the queueing system $M/M/2/3$ for which $\lambda = 2$ and $\mu = 4$. The balance equations of the system are

state j departure rate from j = arrival rate to j

0	$2\pi_0 \overset{(0)}{=} 4\pi_1$
1	$(2 + 4)\pi_1 \overset{(1)}{=} 2\pi_0 + (2 \times 4)\pi_2$
2	$(2 + 2 \times 4)\pi_2 \overset{(2)}{=} 2\pi_1 + (2 \times 4)\pi_3$
3	$(2 \times 4)\pi_3 \overset{(3)}{=} 2\pi_2$

We can directly solve this system of linear equations. Equation (0) yields $\pi_1 = \frac{1}{2}\pi_0$. Substituting into (1), we obtain $\pi_2 = \frac{1}{4}\pi_1 = \frac{1}{8}\pi_0$. Next, we deduce from Eq. (3) that $\pi_3 = \frac{1}{4}\pi_2 = \frac{1}{32}\pi_0$. We can then write that

$$\pi_0 + \frac{1}{2}\pi_0 + \frac{1}{8}\pi_0 + \frac{1}{32}\pi_0 = 1 \quad \Longrightarrow \quad \pi_0 = \frac{32}{53}, \ \pi_1 = \frac{16}{53}, \ \pi_2 = \frac{4}{53}, \ \pi_3 = \frac{1}{53}$$

Note that this solution also satisfies Eq. (2).

The average number of customers in the system in equilibrium is given by

$$\bar{N} = \sum_{k=0}^{3} k\, \pi_k = \frac{16}{53} + 2 \times \frac{4}{53} + 3 \times \frac{1}{53} = \frac{27}{53}$$

from which the average time that an *entering* customer spends in the system is

$$\bar{T} = \frac{\bar{N}}{\lambda_e} = \frac{27/53}{2(1 - \pi_3)} = \frac{27/53}{2(52/53)} = \frac{27}{104} \simeq 0.2596$$

We also find, in particular, that the limiting probability that the system is not empty is given by $1 - \pi_0 = \frac{21}{53}$.

Particular case: The model $M/M/s/s$

When the capacity of the system is equal to the number of servers in the system, no waiting line is formed. The customers who arrive and find all the servers busy do not enter the system and are thus *lost*. Such a system is called a *no-wait system* or a *loss system*.

Let $X(t)$, for $t \geq 0$, be the number of customers in the queueing system $M/M/s/s$ at time t. We find that the balance equations of the system are given by

state j	departure rate from j = arrival rate to j
0	$\lambda \pi_0 = \mu \pi_1$
$0 < k < s$	$(\lambda + k\mu)\pi_k = (k+1)\mu\, \pi_{k+1} + \lambda \pi_{k-1}$
s	$s\mu\, \pi_s = \lambda \pi_{s-1}$

The birth and death process $\{X(t), t \geq 0\}$ is irreducible and, since

$$\Pi_k = \frac{\lambda \times \lambda \times \cdots \times \lambda}{\mu \times 2\mu \times \cdots \times k\mu} = \frac{1}{k!}\left(\frac{\lambda}{\mu}\right)^k \quad \text{for } k = 1, 2, \ldots, s \tag{6.94}$$

the sum S_1 is given by

$$S_1 := \sum_{k=1}^{s} \Pi_k = \sum_{k=1}^{s} \frac{1}{k!}\left(\frac{\lambda}{\mu}\right)^k \tag{6.95}$$

This sum is finite for all (positive) values of λ and μ. Theorem 3.3.4 then enables us to write that

$$\pi_k = \frac{\rho^k/k!}{\sum_{j=0}^{s} \rho^j/j!} \quad \text{for } k = 0, 1, \ldots, s \tag{6.96}$$

where $\rho := \lambda/\mu$. Thus, the π_k's correspond to the probabilities $P[Y = k]$ of a random variable Y having a *truncated* Poisson distribution. Indeed, if $Y \sim$ Poi(ρ), then we have

$$P[Y = k \mid Y \leq s] = \frac{e^{-\rho}\rho^k/k!}{\sum_{j=0}^{s} e^{-\rho}\rho^j/j!} = \frac{\rho^k/k!}{\sum_{j=0}^{s} \rho^j/j!} \quad \text{for } k = 0, 1, \dots, s \tag{6.97}$$

Note that in the particular case when $\lambda = \mu$, the limiting probabilities π_k do not depend on λ (or μ):

$$\pi_k \stackrel{\lambda=\mu}{=} \frac{1/k!}{\sum_{j=0}^{s} 1/j!} \quad \text{for } k = 0, 1, \dots, s \tag{6.98}$$

Moreover, in the general case, the probability π_b that all servers are busy is simply

$$\pi_b = \pi_s = \frac{\rho^s/s!}{\sum_{j=0}^{s} \rho^j/j!} \quad \text{for all } \rho > 0 \tag{6.99}$$

This formula is known as *Erlang's formula*.

Finally, the average entering rate of customers into the system is $\lambda_e = \lambda(1 - \pi_s)$. It follows that

$$\bar{N} = \lambda_e \bar{T} = \lambda(1 - \pi_s)\frac{1}{\mu} \tag{6.100}$$

Since there is no waiting period in this system, as in the $M/M/\infty$ model, we have that $\bar{N} = \bar{N}_S$, $\bar{T} = \bar{S}$, and $\bar{N}_Q = \bar{Q} = 0$.

Remarks. i) If the number of servers tends to infinity, we obtain

$$\lim_{s \to \infty} \sum_{j=0}^{s} \rho^j/j! = e^\rho = e^{\lambda/\mu} \tag{6.101}$$

and we retrieve the formula (6.83):

$$\lim_{s \to \infty} \pi_k = \frac{(\lambda/\mu)^k}{k!}e^{-\lambda/\mu} \quad \text{for } k = 0, 1, \dots \tag{6.102}$$

ii) Note that the limiting probabilities may be expressed as follows:

$$\pi_k = \frac{(\lambda E[S])^k /k!}{\sum_{j=0}^{s} (\lambda E[S])^j /j!} \quad \text{for } k = 0, 1, \dots, s \tag{6.103}$$

Now, it can be shown that the formula (6.103) is also valid for the more general model $M/G/s/s$ (known as *Erlang's loss system*), in which the service time is an arbitrary nonnegative random variable. This result is very interesting, because it enables us to treat problems for which the service time does not have an exponential distribution.

Example 6.3.3. We consider the queueing system $M/G/2/2$ in which the arrivals constitute a Poisson process with rate $\lambda = 2$ per hour.

(a) What is the probability π_0 that there are no customers in the system at a (large enough) time t if we suppose that the service time S (in hours) is an exponential random variable with mean equal to $1/4$? What is the average number of customers in the system in equilibrium?

(b) Calculate the value of π_0 if S is a continuous random variable whose probability density function is

$$f_S(s) = 64se^{-8s} \quad \text{for } s \geq 0$$

Solution. (a) When $S \sim \text{Exp}(\mu = 4)$, we have the $M/M/2/2$ model with $\rho = 2/4 = 1/2$. We seek

$$\pi_0 = \frac{1}{\sum_{j=0}^{2}(1/2)^j/j!} = \frac{1}{1 + \frac{1}{2} + \frac{1}{8}} = \frac{8}{13}$$

We also find that

$$\pi_1 = \frac{1}{2}\pi_0 = \frac{4}{13} \quad \text{and} \quad \pi_2 = \frac{1}{8}\pi_0 = \frac{1}{13}$$

It follows that

$$\bar{N} = 0 \times \frac{8}{13} + 1 \times \frac{4}{13} + 2 \times \frac{1}{13} = \frac{6}{13}$$

(b) We can check that the density function above is that of a random variable having a gamma distribution with parameters $\alpha = 2$ and $\lambda^* = 8$. It follows that $E[S] = \alpha/\lambda^* = 1/4$. Since the mean of S is equal to that of a random variable having an $\text{Exp}(4)$ distribution, we can conclude that the value of π_0 is the same as that in (a). Actually, we can assert that all the limiting probabilities π_k are the same as the π_k's in (a). Consequently, we have that $\bar{N} = 6/13$ as well. If we do not recognize the distribution of the service time S, then we must calculate the mean $E[S]$ by integrating by parts, or by proceeding as follows:

$$E[S] := \int_0^\infty s \cdot 64se^{-8s}ds = \int_0^\infty 64s^2 e^{-8s}ds$$

$$\overset{t=8s}{=} \frac{1}{8}\int_0^\infty t^2 e^{-t}dt = \frac{1}{8}\Gamma(3) = \frac{2!}{8} = \frac{1}{4}$$

Remark. The queueing systems $M/M/s$ $(= M/M/s/\infty)$ and $M/M/s/s$ are the two extreme cases that can be considered. The model $M/M/s/c$, with $s < c < \infty$, is the one that can most often represent reality well, because there is generally some space where potential customers can wait until being served, but this space is not infinite. If the capacity of a queueing system is finite, and if this system is part of a *network* of queues (see the next subsection), then we say that the possibility of *blocking* of the network exists.

Moreover, we found that for the system $M/M/s$ to attain a stationary regime, the condition $\lambda < s\mu$ must be satisfied. That is, the arrival rate of customers in the system must be smaller than the rate at which the customers are served when the s servers are busy. Otherwise, the length of the queue increases indefinitely. However, in practice, some arriving customers will not enter the system if they deem that the queue length is too long. Some may also decide to leave the system before having been served if they consider that they have already spent too much time waiting in line.

To make the model $M/M/s$ more realistic, we may therefore suppose that the customers are *impatient*. A first possibility is to arrange things so that the probability that an arriving customer decides to stay in the system and wait for her turn depends on the queue length upon her arrival. For example, in Exercise no. 4, p. 346, we suppose that the probability r_n that an arriving customer who finds n persons in the system decides to stay is given by $1/(n+1)$. This assumption leads to a particularly simple solution. We could also suppose that $r_n = \kappa^n$, where $\kappa \in (0, 1]$, etc.

When the potential customers decide by themselves not to enter the system, we speak of a priori *impatience*. We call a posteriori *impatience* the case when the customers, once entered into the system, decide to leave before having been served, or even before their service is completed. This situation may be expressed as follows: an arbitrary customer, who entered the system, decides to leave it if $Q > t_0$ or if $T > t_1$, where t_0 and t_1 are constants fixed in advance. We can also imagine that the time that an arbitrary customer is willing to spend in the system (or waiting in line) is a random variable having a given distribution (an exponential distribution, for example).

Example 6.3.4. (a) Suppose that, in the loss system $M/M/K/K$, for which $\lambda = \mu$, the quantity K is a random variable such that

$$K = \begin{cases} 1 \text{ with probability } 1/2 \\ 2 \text{ with probability } 1/2 \end{cases}$$

(i) Calculate the average number of customers in the system in stationary regime.

(ii) For which value of K is the average profit (per time unit) larger if each customer pays $\$x$ per time unit and each server costs $\$y$ per time unit?

(b) Redo question (i) of part (a) if K is instead a random variable such that $P[K = k] = 1/3$, for $k = 1$, 2, and 3, and if the capacity of the system is $c = 3$, so that potential customers can wait to be served (when $K = 1$ or 2).

Solution. (a) (i) If $K = 1$, we have (see the formula (6.96) or (6.98)):

$$\pi_0 = \frac{1}{1 + \frac{\lambda}{\mu}} \stackrel{\lambda = \mu}{=} \frac{1}{2} \quad \text{and} \quad \pi_1 = \frac{\lambda/\mu}{1 + \frac{\lambda}{\mu}} = \frac{1}{2}$$

so that $E[N \mid K = 1] = 1/2$. When $K = 2$, we find that

$$\pi_0 = \frac{1}{1 + \frac{\lambda}{\mu} + \left(\frac{\lambda}{\mu}\right)^2 \frac{1}{2}} \stackrel{\lambda=\mu}{=} \frac{2}{5}, \quad \pi_1 = \pi_0, \quad \text{and} \quad \pi_2 = \frac{1}{2}\pi_0$$

It follows that $E[N \mid K = 2] = 4/5$. Finally, we have

$$\bar{N} \equiv E[N] = \frac{1}{2}\{E[N \mid K = 1] + E[N \mid K = 2]\} = \frac{13}{20}$$

(ii) If $K = 1$, the average profit per time unit, Pr, is given by $\$\left(\frac{1}{2}x - y\right)$, while $Pr = \$\left(\frac{4}{5}x - 2\,y\right)$ when $K = 2$. Therefore, the value $K = 1$ is the one for which the average profit is larger if and only if

$$\frac{1}{2}x - y > \frac{4}{5}x - 2y \quad \Longleftrightarrow \quad y > \frac{3}{10}x$$

(b) First, if $K = 1$ (and $c = 3$), we have (see the formula (6.49)) $\pi_i = 1/4$, for $i = 0, 1, 2, 3$. Then $E[N \mid K = 1] = 3/2$. Next, if $K = 2$, we write the balance equations of the system (see Ex. 6.3.2):

state j | departure rate from j | $=$ | arrival rate to j

$$
\begin{array}{cll}
0 & \lambda\pi_0 & \stackrel{(1)}{=} \mu\pi_1 \\
1 & (\lambda + \mu)\pi_1 & \stackrel{(2)}{=} \lambda\pi_0 + 2\mu\pi_2 \\
2 & (\lambda + 2\mu)\pi_2 & \stackrel{(3)}{=} \lambda\pi_1 + 2\mu\pi_3 \\
3 & 2\mu\pi_3 & \stackrel{(4)}{=} \lambda\pi_2
\end{array}
$$

When $\lambda = \mu$, we find that (1) implies that $\pi_0 = \pi_1$. Furthermore, (2) then implies that $\pi_2 = \pi_1/2$. It follows, from (4), that $\pi_3 = \pi_1/4$. Making use of the condition $\pi_0 + \pi_1 + \pi_2 + \pi_3 = 1$, we obtain that

$$\pi_0 = \pi_1 = \frac{4}{11}, \quad \pi_2 = \frac{2}{11}, \quad \text{and} \quad \pi_3 = \frac{1}{11} \quad \Longrightarrow \quad E[N \mid K = 2] = 1$$

Finally, with $K = 3$, we calculate

$$\pi_0 = \frac{1}{1 + \frac{\lambda}{\mu} + \left(\frac{\lambda}{\mu}\right)^2 \frac{1}{2} + \left(\frac{\lambda}{\mu}\right)^3 \frac{1}{3!}} \stackrel{\lambda=\mu}{=} \frac{6}{16}, \quad \pi_1 = \pi_0, \quad \pi_2 = \frac{1}{2}\pi_0, \quad \text{and} \quad \pi_3 = \frac{1}{6}\pi_0$$

so that

$$E[N \mid K = 3] = 0 \times \frac{6}{16} + 1 \times \frac{6}{16} + 2 \times \frac{3}{16} + 3 \times \frac{1}{16} = \frac{15}{16}$$

The average number of customers in the system (in stationary regime) is therefore

$$E[N] = \frac{1}{3}\left(\frac{3}{2} + 1 + \frac{15}{16}\right) = \frac{55}{48} \simeq 1.146$$

6.3.3 Networks of queues

We consider a network made up of k queueing systems. In the ith system, there are s_i servers (each of them serving only one customer at a time), for $i = 1, 2, \ldots, k$. We suppose that the capacity of each system is infinite and that customers, coming from outside the network, arrive in system i according to a Poisson process with rate θ_i. The k Poisson processes are independent.

After having left the queueing system i, an arbitrary customer goes to system $j \in \{1, 2, \ldots, k\}$ with probability $p_{i,j}$, so that the probability that the customer leaves the network (after having been served in system i) is given by

$$p_{i,0} := 1 - \sum_{j=1}^{k} p_{i,j} \geq 0 \qquad (6.104)$$

We assume that the probability that a given customer remains indefinitely in the network is equal to *zero*. Finally, the service times are independent exponential random variables with rates μ_i, for $i = 1, \ldots, k$, and are also independent of the times between the successive arrivals.

Remarks. i) A network of this type is said to be *open*, because the customers can enter and leave the system. We could also consider the case when the network is *closed*, that is, the number of customers is constant, and they move indefinitely inside the network.

ii) Notice that the probability $p_{i,i}$ may be strictly positive, for any i. That is, it is possible that a customer, after having departed system i, returns to this system immediately.

Let $\mathbf{X}(t) := (X_1(t), \ldots, X_k(t))$, where $X_i(t)$ designates the number of customers in system i at time t, for $i = 1, \ldots, k$. We want to obtain the distribution of $\mathbf{X}(t)$ in stationary regime. Let λ_j be the *total* rate at which customers arrive in system j. Since the arrival rate into a queueing system must be equal to the departure rate from this system, the λ_j's are the solution of the system of equations

$$\lambda_j = \theta_j + \sum_{i=1}^{k} \lambda_i \, p_{i,j} \quad \text{for } j = 1, 2, \ldots, k \qquad (6.105)$$

Once this system has been solved, the next theorem, known as *Jackson's*[3] *theorem*, gives us the solution to our problem.

Theorem 6.3.1. *Let* $N_i := \lim_{t \to \infty} X_i(t)$. *If* $\lambda_i < s_i \mu_i$, *for* $i = 1, \ldots, k$, *then*

$$\lim_{t \to \infty} P[\mathbf{X}(t) = \mathbf{n}] = P\left[\bigcap_{i=1}^{k} \{N_i = n_i\}\right] \qquad (6.106)$$

---

[3] James R. Jackson, emeritus professor at UCLA (University of California, Los Angeles), in the United States.

where $\mathbf{n} := (n_1, \ldots, n_k)$, *and* N_i *is the number of customers in an* $M/M/s_i$ *queueing system in stationary regime.*

Remarks. i) Under the assumption that all customers eventually leave the network, it can be shown that the system (6.105) has a *unique* solution.

ii) The statement of the theorem is surprising, because it implies that the random variables N_i are *independent*. If an arbitrary customer cannot return to a system he already departed, then the arrivals in each system constitute Poisson processes, because the *departure process* of an $M/M/s$ queue is (in stationary regime) a Poisson process with rate λ (if $\lambda < s\mu$). Moreover, these Poisson processes are independent. In this case, the result of the theorem is easily proved. However, when a customer may be in the same system more than once, it can be shown that the arrival processes are no longer Poisson processes. Indeed, then the increments of these processes are no longer independent. Now, according to the theorem, the random variables are nevertheless independent.

iii) When there is a single server per system, so that $s_i \equiv 1$, the network described above is called a *Jackson network*. We then deduce from the formula (6.11) that

$$\lim_{t \to \infty} P[(X_1(t), \ldots, X_k(t)) = (n_1, \ldots, n_k)] = \prod_{i=1}^{k} \left(\frac{\lambda_i}{\mu_i} \right)^{n_i} \left(1 - \frac{\lambda_i}{\mu_i} \right) \quad (6.107)$$

(if $\lambda_i < \mu_i$, for all i). The average number of customers in the network in equilibrium is then given by

$$\bar{N} = \sum_{i=1}^{k} \bar{N}_i = \sum_{i=1}^{k} \frac{\lambda_i}{\mu_i - \lambda_i} \quad (6.108)$$

Finally, given that the average entering rate of customers (coming from the outside) into the network is

$$\lambda_e = \sum_{i=1}^{k} \theta_i \quad (6.109)$$

the average time that an arbitrary customer spends in the network is

$$\bar{T} = \frac{\bar{N}}{\lambda_e} = \frac{\bar{N}}{\sum_{i=1}^{k} \theta_i} \quad (6.110)$$

Example 6.3.5. The simplest example of a Jackson network is that of a *sequential system* in which there are two servers and the arriving customers must necessarily go the first server, and next directly to the second one. Then they leave the system. We suppose that the first queueing system is an $M/M/1$ model, with service rate μ_1. Thus, the arrivals in the network constitute a Poisson process with rate λ. Similarly, we suppose that the capacity of the second queueing system is infinite and that the server performs the desired service (to only one customer at a time) at an exponential rate μ_2. Finally, all the service times and the times between the successive arrivals are independent random variables.

Since the customers cannot find themselves more than once in front of the same server, we can assert, if $\lambda < \mu_1$, that the departure process of the first system (in stationary regime) is also a Poisson process with rate λ, so that the second queueing system is an $M/M/1$ model as well. We deduce from Jackson's theorem that

$$\pi_{n_1,n_2} := \lim_{t \to \infty} P[(X_1(t), X_2(t)) = (n_1, n_2)]$$

$$= \left(\frac{\lambda}{\mu_1}\right)^{n_1} \left(1 - \frac{\lambda}{\mu_1}\right) \left(\frac{\lambda}{\mu_2}\right)^{n_2} \left(1 - \frac{\lambda}{\mu_2}\right) \qquad (6.111)$$

for n_1 and $n_2 \in \{0, 1, \dots\}$.

We can show that the formula above is valid by checking that the joint limiting probabilities π_{n_1,n_2} satisfy the balance equations of the network. Suppose, to simplify further, that $\mu_1 = \mu_2 := \mu$. These balance equations are then

state (i,j)	departure rate from (i,j) = arrival rate to (i,j)
$(0,0)$	$\lambda\pi_{0,0} = \mu\pi_{0,1}$
$(n_1,0), n_1 > 0$	$(\lambda + \mu)\pi_{n_1,0} = \mu\pi_{n_1,1} + \lambda\pi_{n_1-1,0}$
$(0,n_2), n_2 > 0$	$(\lambda + \mu)\pi_{0,n_2} = \mu(\pi_{0,n_2+1} + \pi_{1,n_2-1})$
$(n_1,n_2), n_1 n_2 > 0$	$(\lambda + 2\mu)\pi_{n_1,n_2} = \mu(\pi_{n_1,n_2+1} + \pi_{n_1+1,n_2-1})$
	$+ \lambda\pi_{n_1-1,n_2}$

When $\mu_1 = \mu_2 = \mu$, we can rewrite the formula (6.111) as follows:

$$\pi_{n_1,n_2} = \left(\frac{\lambda}{\mu}\right)^{n_1+n_2} \left(1 - \frac{\lambda}{\mu}\right)^2 \qquad (6.112)$$

We have, in particular,

$$(\lambda + 2\mu)\pi_{n_1,n_2} = \mu(\pi_{n_1,n_2+1} + \pi_{n_1+1,n_2-1}) + \lambda\pi_{n_1-1,n_2}$$

$$\Longleftrightarrow$$

$$(\lambda + 2\mu)\left(\frac{\lambda}{\mu}\right)^{n_1+n_2} = \mu\left[\left(\frac{\lambda}{\mu}\right)^{n_1+n_2+1} + \left(\frac{\lambda}{\mu}\right)^{n_1+n_2}\right] + \lambda\left(\frac{\lambda}{\mu}\right)^{n_1-1+n_2}$$

$$\Longleftrightarrow$$

$$\lambda + 2\mu = \mu\left[\left(\frac{\lambda}{\mu}\right) + 1\right] + \lambda\left(\frac{\lambda}{\mu}\right)^{-1} = \lambda + \mu + \mu \qquad (6.113)$$

By the *uniqueness* of the solution, under the condition $\sum_{(n_1,n_2)} \pi_{n_1,n_2} = 1$, we can then conclude that the probabilities π_{n_1,n_2} are indeed given by the formula (6.111).

Remarks. i) When $\mu_1 = \mu_2 = \mu$, the limiting probabilities π_{n_1,n_2} depend only on the sum $n_1 + n_2$. However, the probability π_{n_1,n_2} is *not* equal to the probability that there will be exactly $n_1 + n_2$ customers in the network in stationary regime. Indeed, we calculate, for example,

$$\lim_{t\to\infty} P[X_1(t) + X_2(t) = 1] \stackrel{\text{sym.}}{=} 2 \lim_{t\to\infty} P[X_1(t) = 1, X_2(t) = 0]$$

$$= 2 \left(\frac{\lambda}{\mu}\right) \left(1 - \frac{\lambda}{\mu}\right)^2 \tag{6.114}$$

Moreover, we have

$$\sum_{n_1+n_2=0}^{\infty} \left(\frac{\lambda}{\mu}\right)^{n_1+n_2} \left(1 - \frac{\lambda}{\mu}\right)^2 = \left(1 - \frac{\lambda}{\mu}\right)^2 \sum_{n_1+n_2=0}^{\infty} \left(\frac{\lambda}{\mu}\right)^{n_1+n_2}$$

$$= \left(1 - \frac{\lambda}{\mu}\right)^2 \frac{1}{1 - \frac{\lambda}{\mu}} = 1 - \frac{\lambda}{\mu} < 1 \tag{6.115}$$

while

$$\sum_{n_1+n_2=0}^{\infty} \lim_{t\to\infty} P[X_1(t) + X_2(t) = n_1 + n_2] = 1 \tag{6.116}$$

ii) We can also write that

$$\pi_{n_1,n_2} \stackrel{\mu_1=\mu_2}{=} \lim_{t\to\infty} P[X_1(t) = n_1 + n_2, X_2(t) = 0]$$

$$= \lim_{t\to\infty} P[X_1(t) = 0, X_2(t) = n_1 + n_2] \tag{6.117}$$

More generally, we have

$$\pi_{n_1,n_2} \stackrel{\mu_1=\mu_2}{=} \lim_{t\to\infty} P[X_1(t) = i, X_2(t) = j] \tag{6.118}$$

for all nonnegative integers i and j such that $i + j = n_1 + n_2$. Thus, when $\mu_1 = \mu_2$, all the possible distributions of the $n_1 + n_2$ customers between the two servers are equally likely.

iii) Finally, the network described in this example is different from the $M/M/2$ model, even if the two service rates are equal. Indeed, when there are exactly two customers in this network, both these customers may stand in front of server 1 (or server 2), whereas, in the case of the $M/M/2$ model, there must be one customer in front of each server.

6.4 Exercises

Remark. In the following exercises, we assume that the service times are random variables that are independent among themselves and are independent of the times between successive arrivals.

Section 6.2

Question no. 1
Calculate the average number of arrivals in the system during the service period of an arbitrary customer for the queueing model $M/M/1/3$.

Question no. 2
Drivers stop to fill up their cars at a service station according to a Poisson process with rate $\lambda = 15$ per hour. The service station has only one gasoline pump, and there is room for only two waiting cars. We suppose that the average service time is equal to two minutes.

(a) Calculate $\bar{N}$ and $\bar{N}_Q$. Why is $\bar{N} \neq \bar{N}_Q + 1$?

(b) If we suppose that an arriving driver who finds the three spaces occupied will go to another service station, what proportion of potential customers is lost?

Question no. 3
Airplanes arrive at an airport having a single runway according to a Poisson process with rate $\lambda = 18$ per hour. The time during which the runway is used by a landing airplane has an exponential distribution with mean equal to two minutes (from the moment it receives the landing authorization).

(a) Knowing that there is at most one airplane in the system (at a given time instant), what is the probability that an arriving airplane will have to wait before being allowed to land?

(b) Given that an airplane has been waiting for the authorization to land for the last 5 minutes, what is the probability that it will have landed and cleared the runway in the next 10 minutes?

Question no. 4
We suppose that the probability that an arriving customer in an $M/M/1$ queueing system decides to stay and wait until being served is given by $1/(n + 1)$, where n is the number of customers in the system at the time when the customer in question arrives, for $n = 0, 1, 2, \ldots$.

(a) Calculate $\bar{N}$ and $\bar{N}_Q$.

(b) What is the percentage of customers who decide not to enter the system?

Question no. 5
We wish to compare two maintenance policies for the airplanes of a certain airline company. In the case of policy A (respectively, B), the airplanes arrive to the maintenance shop according to a Poisson process with rate $\lambda_A = 1$

(resp., $\lambda_B = 1/4$) per day. Moreover, when policy A (resp., B) is used, the service time (in days) is an exponential random variable with parameter $\mu_A = 2$ (resp., the sum of four independent exponential random variables, each of them with parameter $\mu_B = 2$). In both cases, maintenance work is performed on only one airplane at a time.

(a) What is the better policy? Justify your answer by calculating the average number of airplanes in the maintenance shop (in stationary regime) in each case.

Indication. The average number of customers in a queueing system $M/G/1$ (after a long enough time) is given by

$$\bar{N} = \lambda E[S] + \frac{\lambda^2 E[S^2]}{2(1 - \lambda E[S])}$$

where S is the service time and λ is the average arrival rate of customers.

(b) Let N be the number of airplanes in the maintenance shop in stationary regime. Calculate the distribution of N if policy A is used, given that there are two or three airplanes in the shop (at a particular time instant).

(c) If policy A is used, what is the average time that an airplane, which has already been in the maintenance shop for two days, will spend in the shop overall?

Question no. 6

We consider a queueing system in which there are two types of customers, both types arriving according to a Poisson process with rate λ. The customers of type I always enter the system. However, the type II customers only enter the system if there is no more than *one* customer in the system when they arrive. There is a single server and the service time has an exponential distribution with parameter μ.

(a) Write the balance equations of the system.

(b) Calculate the limiting probability that an arbitrary type II customer enters the system if $\lambda = 1$ and $\mu = 2$.

Indication. The system considered is a birth and death process.

(c) Calculate the average time that a given arriving customer of type II will spend in the system if $\lambda = 1$ and $\mu = 2$.

Question no. 7

We consider a waiting system with a single server and finite capacity $c = 3$, in which the customers arrive according to a Poisson process with rate λ and the service times are independent exponential random variables with parameter $\mu = 2\lambda$. When the system is full, a fair coin is tossed to determine whether the second or third customer will be the next one to be served.

(a) Write the balance equations for this system, and calculate the limiting probabilities π_j.

(b) Calculate the average time that an arriving customer who finds (exactly) one customer in the system will spend in it.

(c) Calculate the average time that a customer who enters the system will spend in it.

Question no. 8

Customers arrive at a service facility according to a Poisson process with rate $\lambda = 10$ per hour. The (only) server is able to serve up to three customers at a time. The service time (in hours) has an exponential distribution with parameter $\mu = 5$, regardless of the number of customers (namely 1, 2, or 3) being served at the same time. However, an arbitrary customer is not served immediately if the server is busy upon her arrival. Moreover, we suppose that the service times are independent random variables and that there can be at most three customers waiting at any time. We define the states

> $0'$: nobody is being served
> 0: the server is busy; nobody is waiting
> n: there is (are) n customer(s) waiting, for $n = 1, 2, 3$

(a) Write the balance equations of the system.

(b) Calculate the limiting probabilities π_j, for all states j.

(c) What is the probability that an arriving customer will be served alone?

Question no. 9 (Modification of the preceding question)

Customers arrive at a service facility according to a Poisson process with rate λ. The (only) server is able to serve up to two customers at a time. However, an arbitrary customer is not served immediately if the server is busy upon his arrival. The service time has an exponential distribution with parameter μ_i when the service is provided to i customer(s) at a time, for $i = 1, 2$. Moreover, we suppose that the service times are independent random variables and that there can be at most two customers waiting at any time. We define the states

> 0 : nobody is being served
> n_i: there is (are) n customer(s) waiting and i customer(s) being served

for $n = 0, 1, 2$ and $i = 1, 2$.

(a) Write the balance equations of the system. Do not solve them.

(b) Calculate, in terms of the limiting probabilities, the probability that
(i) the server is serving two customers at a time, given that he is busy,
(ii) an arriving customer who enters the system will not be served alone.

Question no. 10

Customers arrive into a queueing system according to a Poisson process with rate λ. There is a single server, who cannot serve more than one customer at a time. However, the larger the number of customers in the system is, the faster the server works. More precisely, we suppose that the service time has an exponential distribution with parameter $\mu(k) = \frac{k}{k+1}\mu$ when there are k customers in the system, for $k = 1, 2, \ldots$. Moreover, we suppose that the service times are independent random variables and that $\lambda = 1$ and $\mu = 2$.

(a) Calculate the limiting probabilities π_n, for $n = 0, 1, \ldots$.

(b) Let $X(t)$ be the number of customers in the system at time t, for $t \geq 0$. Calculate $E[X(t) \mid X(t) \leq 2]$ when the system is in equilibrium.

(c) Let T_0 be the time spent in the system by a customer who arrived, at time t_0, while the system was empty. Calculate the expected value of T_0, given that the following customer arrived at time $t_0 + 1$ and the customer in question had already left the system.

Question no. 11

Drivers arrive according to a Poisson process with rate λ to fill up their cars at a service station where there are two employees who serve at the exponential rates μ_1 and μ_2, respectively. However, only one employee works at a time serving gasoline. Moreover, there is space for only one waiting car. We suppose that

- when the system is empty and a customer arrives, employee no. 1 fills up the car, ·
- when employee no. 1 (respectively, no. 2) finishes filling up a car and another car is waiting, there is a probability equal to p_1 (resp., p_2) that this employee will service the customer waiting to be served, independently from one time to another.

Finally, we suppose that the service times are independent random variables.

(a) Let $X(t)$ be the state of the system at time t. Define a state space in such a way that the stochastic process $\{X(t), t \geq 0\}$ is a continuous-time Markov chain.

(b) Write the balance equations of the process.

(c) Calculate, in terms of the limiting probabilities, the probability that

(i) an arbitrary customer entering the system will be served by employee no. 2,

(ii) two customers arriving consecutively will be served by different employees, given that the first of these customers arrived while there was exactly one car, being filled up by employee no. 1, in the system.

Question no. 12

We consider the queueing system $M/M/1$. However, the customers do not have to wait, because the server is able to serve all the customers at one time, at an exponential rate μ, regardless of the number of customers being served.

Calculate

(a) the limiting probability, π_n, that there are n customers being served, for $n = 0, 1, \ldots$,

(b) the variance of the number of customers being served when the system is in equilibrium.

Question no. 13

Suppose that customers arrive at a service facility with a single server according to a Poisson process with rate λ. The server waits until there are four customers in the system before beginning to serve them, all at once. The service times are independent random variables, all having a uniform distribution on the interval $(0, 1)$. Moreover, the system capacity is equal to four customers. What fraction of time, π_i, are there i customer(s) in the system, over a long period?

Indication. Use the results on renewal processes.

Question no. 14

Suppose that the times between the arrivals of consecutive customers in a certain queueing system are independent random variables uniformly distributed over the interval $(0, 1)$. The service time is exponentially distributed, with parameter μ. Finally, the (only) server is able to serve all the customers at once, so that there is no waiting. Calculate the limiting probability that the server is busy.

Indication. Use the results on renewal processes.

Question no. 15

We modify the $M/M/1/4$ queueing system as follows: the server always waits until there are at least two customers in the system before serving them, *two at a time*, at an exponential rate μ.

(a) Write the balance equations of the system.

(b) Calculate the limiting probabilities, π_n, for $n = 0, 1, 2, 3, 4$, in the case when $\lambda = \mu$, where λ is the average arrival rate of the customers.

(c) With the help of the limiting probabilities calculated in (b), find

(i) the probability that the system is not empty at the moment when the server has just finished serving two customers,

(ii) the variance of the number $X(t)$ of customers in the system (in equilibrium) at time t, given that $X(t) \leq 2$.

Section 6.3

Question no. 16

Customers arrive according to a Poisson process with rate λ at a bank where two clerks work. Clerk 1 (respectively, 2) serves at an exponential rate μ_1 (resp., μ_2). We suppose that the customers form a single queue and that, when the system is empty, an arriving customer will go to clerk 1 (resp., 2) with probability p_1 (resp., $1 - p_1$). On the other hand, when a customer must wait, she will eventually be served by the first available clerk. We also suppose that an arbitrary customer can enter the bank only if there are no more than 10 customers waiting in line. We say that the system is in state $n = 0, 2, \ldots$ if there are n customers in the bank, and in state 1_1 (resp., 1_2) if there is exactly one customer in the bank and if this customer is being served by clerk 1 (resp., 2).

(a) Write the balance equations of the system.

(b) In terms of the limiting probabilities, what is the probability that an entering customer will be served by clerk 1?

Question no. 17

In a certain garage, there are three mechanics per work shift of eight hours. The garage is open 24 hours a day. Customers arrive according to a Poisson process with rate $\lambda = 2.5$ per hour. The time a mechanic takes to perform an arbitrary task is an exponential random variable with mean equal to 30 minutes.

(a) What proportion of time are all the mechanics busy?

(b) How much time, on average, must a customer wait for his car to be ready?

Question no. 18

In a small train station, there are two counters where the travelers can buy their tickets, but the customers form a single waiting line. During the slack hours, only one counter is manned continuously by a clerk. When there is at least one customer waiting to be served, the second clerk opens his counter. When this second clerk finishes serving a customer and there is nobody waiting, he goes back to attending to other tasks. We suppose that the clerks both serve in a random time having an exponential distribution with parameter μ and that, during the slack hours, the customers arrive according to a Poisson process with rate λ. We also suppose that the slack period lasts long enough for the process to reach a stationary regime. The state $X(t)$ of the system is defined as being the total number of customers present in the system at time t.

(a) Calculate the limiting probabilities of the process $\{X(t), t \geq 0\}$ if $\lambda < 2\mu$.

(b) Write the balance equation for the state 1_2 corresponding to the case when only the second clerk is busy (serving a customer).

(c) What fraction of time is the second counter open?

Question no. 19

Customers arrive at a hairdresser's salon according to a Poisson process with rate $\lambda = 8$ per hour. There are two chairs, and the two hairdressers' service times are exponential random variables with means equal to 15 minutes. Moreover, currently, there is no room where potential customers could wait for their turn to have their hair cut.

(a) The owner considers the possibility of enlarging the salon, so that she could install an additional chair and hire a third hairdresser. This would increase her operation costs by $20 per hour. If each customer pays $10, would the enlargement be profitable? Justify.

Indication. Calculate the average rate at which customers enter the salon.

(b) Another possibility the owner considered consists of enlarging the salon to install a chair where *one* potential customer could wait to be served. In this case, the increase in operation costs would be equal to $5 per hour. Would this possibility be profitable? Justify.

Question no. 20

We consider the queueing system $M/M/2/3$ (see Example 6.3.2).

(a) Write the balance equations of the system, and calculate the limiting probabilities π_j if $\lambda = 2\mu$.

(b) Let T^* be the total time that an entering customer will spend in the system. Calculate the expected value of T^* if $\mu = 1$.

(c) Suppose that the customers form two waiting lines, by standing at random in front of either server. Calculate, with $\mu = 1$, the average time that an arbitrary customer who enters the system, and finds two customers already present, will spend in this system if we assume that the number of customers in each queue is then a random variable having a binomial distribution with parameters $n = 2$ and $p = 1/2$.

Question no. 21

We consider a queueing system with two servers. The customers arrive according to a Poisson process with rate $\lambda = 1$, and the system capacity is equal to four customers. The service times are independent random variables having an exponential distribution. Each server is able to serve two customers at a time. If a server attends to only one customer, he does so at rate $\mu = 2$, whereas the service rate is equal to 1 when two customers are served at the same time.

Indication. If two customers are served together, then they will leave the system at the same time. Moreover, if there are two customers in the system, then one of the servers may be free.

(a) Write the balance equations of the system.

(b) Let T^* be the total time that a given customer entering the system will spend in it.

   (i) Calculate, in terms of the limiting probabilities, and supposing that no customers arrive during the service period of the customer in question, the distribution function of T^*.

   (ii) Under the same assumption as in (i), does the random variable T^* have the memoryless property? Justify.

Question no. 22

Customers arrive according to a Poisson process with rate $\lambda = 5$ per hour in a system with two servers. The probability that an arbitrary customer goes to server no. 1 (respectively, no. 2) is equal to 3/4 (resp., 1/4). The service times (in hours) are independent exponential random variables with parameters $\mu_1 = 6$ and $\mu_2 = 4$, respectively. A customer who goes to server no. 2 immediately leaves the system after having been served. On the other hand, after having been served by server no. 1, a customer (independently from one time to another)

$$\begin{cases} \text{leaves the system} & \text{with probability } 1/2 \\ \text{goes to server no. 2} & \text{with probability } 2/5 \\ \text{returns in front of server no. 1} & \text{with probability } 1/10 \end{cases}$$

Moreover, there is no limit on the number of customers who can be in the system at any time.

Let (n, m) be the state of the system when there are n customers in front of server no. 1 and m customers in front of server no. 2.

(a) Calculate $\pi_{(n,m)}$, for all $n, m \geq 0$.

(b) Calculate the average number of customers in the system at a large enough time instant, given that the system is not empty at the time in question.

(c) What is the average time that an arbitrary customer who arrives in the system and goes to server no. 1 will spend being served by this server before leaving the system if we suppose that the customer in question never goes to server no. 2?

Question no. 23

Customers arrive according to a Poisson process with rate λ outside a bank where there are two automated teller machines (ATM). The two ATMs are not identical. We estimate that 30% of the customers use only ATM no. 1, while 20% of the customers use only ATM no. 2. The other customers (50%) make use of either ATM indifferently. The service times at each ATM are independent exponential random variables with parameter μ. Finally, there is space for a single waiting customer. We define the states

0: the system is empty
(n, m): there are n customers for ATM no. 1 and m customers
for ATM no. 2, for $1 \leq n + m \leq 2$
3: the system is full

(a) Write the balance equations of the system. Do not solve them.

(b) Calculate, in terms of the limiting probabilities,

(i) the variance of the number of customers who are waiting to use an ATM,

(ii) the average time that an arbitrary customer, who enters the system and wishes to use ATM no. 2, will spend in the system.

Question no. 24

We consider a queueing system in which ordinary customers arrive according to a Poisson process with rate λ and are served in a random time having an exponential distribution, with parameter μ, by either of two servers. Furthermore, there is a special customer who, when she arrives in the system, is immediately served by server no. 1, at an exponential rate μ_s. If an ordinary customer is being served by server no. 1 when the special customer arrives, then this customer is returned to the head of the queue. We suppose that the service times are independent random variables and that the special customer spends an exponential time (independent of the service times), with parameter λ_s, outside the system between two consecutive visits.

(a) Suppose that if an arbitrary customer is returned to the queue, then he will resume being served as soon as either server becomes available. Define an appropriate state space, and write the balance equations of the system.

(b) Suppose that the system capacity is $c = 2$, but that if a customer is displaced by the special customer, then she will wait, a few steps behind, until server no. 1 becomes available to resume being served by this server (whether server no. 2 is free or not). Define a state space such that the stochastic process $\{X(t), t \geq 0\}$, where $X(t)$ represents the state of the system at time t, is a continuous-time Markov chain.

(c) Suppose that the system capacity is $c = 2$ and that, if a customer is displaced by the special customer, then he will go to server no. 2 only if this server is free upon the arrival of the special customer in the system. Otherwise, he will wait, a few steps behind, before server no. 1 becomes available. Let K be the number of times that a given customer, who has started receiving service from server no. 1, will be displaced by the special customer. Calculate $P[K = 1]$ in terms of the limiting probabilities of the system (with an appropriate state space).

Question no. 25

Let N be the number of customers in an $M/G/2/2$ (loss) system after a time long enough for the system to be in stationary regime.

(a) Calculate $V[N \mid N > 0]$ if the service time, S, has a uniform distribution on the interval $(0, 1)$ and if the average arrival rate of customers in the system is $\lambda = 4$.

(b) Calculate $V[N \mid X = 1/4]$ if S has an exponential distribution with parameter $1/X$, where $X \sim U(0, 1)$ and $\lambda = 2$.

Question no. 26

Suppose that we modify the $M/M/2$ queueing system as follows: when a server is free, he assists (if needed) the other server, so that the service time, S, has an exponential distribution with parameter 2μ. If a new customer arrives while a customer is being served by the two servers at the same time, then one the servers starts serving the new customer.

(a) Calculate the limiting probabilities if we suppose that $\lambda < 2\mu$.

(b) Suppose that the system capacity is $c = 2$ and that $\lambda = \mu$. Calculate the average number of customers in the system in stationary regime, given that the system is not full.

Question no. 27

Let $X(t)$ be the number of customers at time $t \geq 0$ in a queueing system with s servers and finite capacity c, for which the time τ between two consecutive arrivals is a random variable such that $f_\tau(t) = 2te^{-t^2}$, for $t \geq 0$. We assume that the times between the arrivals of customers are independent and identically distributed random variables. Similarly, the service times are independent random variables having, for each server, the same probability density function as τ. We define the stochastic process $\{Y(t), t \geq 0\}$ by

$$Y(t) = X(g(t)) \quad \text{for } t \geq 0$$

where $g(t)$ is a one-to-one function of t. Find a function g such that $\{Y(t), t \geq 0\}$ is an $M/M/s$ queueing system with $\lambda = \mu = 1$ (and finite capacity c). Justify.

Question no. 28

We consider the loss system $M/G/2/2$. Suppose that the service times have independent exponential distributions with parameter Θ, where Θ is a random variable such that

$$f_\Theta(\theta) = \frac{3}{8}(2\theta^2 + 2\theta + 1) \quad \text{for } 0 < \theta < 1$$

Suppose also that the average arrival rate of customers is $\lambda = 1$. Calculate, assuming they exist, the limiting probabilities π_i, for $i = 0, 1, 2$.

Question no. 29

In the $M/M/2$ queueing system, we define the random variable S as being the first time that both servers are busy. Let
F = exactly two customers arriving in the interval $(0, t]$.
Calculate $P[S \leq t \mid F]$.

Question no. 30

A hairdresser and her assistant operate a salon. There are two types of customers: those of type I prefer to have their hair cut by the hairdresser but are willing to be served by her assistant, while those of type II want to be served by the assistant only. The type I (respectively, type II) customers arrive at the salon according to a Poisson process with rate λ_1 (resp., λ_2). Moreover, the two Poisson processes are independent. Finally, the hairdresser (resp., the assistant) serves in a random time having an exponential distribution with parameter μ_1 (resp., μ_2), and the service times are independent random variables. Answer the following questions, supposing that there is no room where potential customers can wait until being served:

(a) define a state space that enables you to answer part (c);

(b) write the balance equations;

(c) give, in terms of the limiting probabilities,

(i) the average number of customers in the system;
(ii) the average time that an arbitrary entering customer will spend in the system.

Question no. 31

Redo the preceding question, supposing that the system capacity is infinite and that

(a) only the potential type I customers are willing to wait until being served (those of type II go away if the assistant is busy),

(b) only the potential type II customers are willing to wait until being served (those of type I go away if the hairdresser and her assistant are busy),

(c) all the potential customers are willing to wait until being served.

Question no. 32

Suppose that in the queueing system $M/M/2/c$, with $c = 3$, server no. 1 serves only one person at a time, at rate μ_1, while server no. 2 can serve one or two persons at a time, from any time instant, at rate μ_2. Moreover, when server no. 1 is free, an arriving customer will go to this server.

(a) Let $X(t)$ be the number of persons in the system at time t. Define a state space such that the stochastic process $\{X(t), t \geq 0\}$ is a continuous-time Markov chain.

Remark. Server no. 1 may be free while server no. 2 serves two customers at a time. That is, the two customers finish their service period with server no. 2.

(b) Write the balance equations of the system. Do not solve them.

(c) In terms of the limiting probabilities, what fraction of time does server no. 2 serve two customers at a time, given that she is busy?

Question no. 33

Redo the preceding question, supposing that the system capacity is instead $c = 4$ and that server no. 2 can serve one or two persons at a time (at rate μ_2) but only from the *same* time instant.

Appendix A: Statistical Tables

Table A.1. Distribution Function of the Binomial Distribution

n	x	p					
		0.05	0.10	0.20	0.25	0.40	0.50
2	0	0.9025	0.8100	0.6400	0.5625	0.3600	0.2500
	1	0.9975	0.9900	0.9600	0.9375	0.8400	0.7500
3	0	0.8574	0.7290	0.5120	0.4219	0.2160	0.1250
	1	0.9927	0.9720	0.8960	0.8438	0.6480	0.5000
	2	0.9999	0.9990	0.9920	0.9844	0.9360	0.8750
4	0	0.8145	0.6561	0.4096	0.3164	0.1296	0.0625
	1	0.9860	0.9477	0.8192	0.7383	0.4752	0.3125
	2	0.9995	0.9963	0.9728	0.9493	0.8208	0.6875
	3	1.0000	0.9999	0.9984	0.9961	0.9744	0.9375
5	0	0.7738	0.5905	0.3277	0.2373	0.0778	0.0313
	1	0.9774	0.9185	0.7373	0.6328	0.3370	0.1875
	2	0.9988	0.9914	0.9421	0.8965	0.6826	0.5000
	3	1.0000	0.9995	0.9933	0.9844	0.9130	0.8125
	4	1.0000	1.0000	0.9997	0.9990	0.9898	0.9688
10	0	0.5987	0.3487	0.1074	0.0563	0.0060	0.0010
	1	0.9139	0.7361	0.3758	0.2440	0.0464	0.0107
	2	0.9885	0.9298	0.6778	0.5256	0.1673	0.0547
	3	0.9990	0.9872	0.8791	0.7759	0.3823	0.1719
	4	0.9999	0.9984	0.9672	0.9219	0.6331	0.3770
	5	1.0000	0.9999	0.9936	0.9803	0.8338	0.6230
	6		1.0000	0.9991	0.9965	0.9452	0.8281
	7			0.9999	0.9996	0.9877	0.9453
	8			1.0000	1.0000	0.9983	0.9893
	9					0.9999	0.9990
15	0	0.4633	0.2059	0.0352	0.0134	0.0005	0.0000
	1	0.8290	0.5490	0.1671	0.0802	0.0052	0.0005
	2	0.9638	0.8159	0.3980	0.2361	0.0271	0.0037
	3	0.9945	0.9444	0.6482	0.4613	0.0905	0.0176

Table A.1. Continued

n	x	0.05	0.10	0.20	0.25	0.40	0.50
15	4	0.9994	0.9873	0.8358	0.6865	0.2173	0.0592
	5	0.9999	0.9977	0.9389	0.8516	0.4032	0.1509
	6	1.0000	0.9997	0.9819	0.9434	0.6098	0.3036
	7		1.0000	0.9958	0.9827	0.7869	0.5000
	8			0.9992	0.9958	0.9050	0.6964
	9			0.9999	0.9992	0.9662	0.8491
	10			1.0000	0.9999	0.9907	0.9408
	11				1.0000	0.9981	0.9824
	12					0.9997	0.9963
	13					1.0000	0.9995
	14						1.0000
20	0	0.3585	0.1216	0.0115	0.0032	0.0000	
	1	0.7358	0.3917	0.0692	0.0243	0.0005	0.0000
	2	0.9245	0.6769	0.2061	0.0913	0.0036	0.0002
	3	0.9841	0.8670	0.4114	0.2252	0.0160	0.0013
	4	0.9974	0.9568	0.6296	0.4148	0.0510	0.0059
	5	0.9997	0.9887	0.8042	0.6172	0.1256	0.0207
	6	1.0000	0.9976	0.9133	0.7858	0.2500	0.0577
	7		0.9996	0.9679	0.8982	0.4159	0.1316
	8		0.9999	0.9900	0.9591	0.5956	0.2517
	9		1.0000	0.9974	0.9861	0.7553	0.4119
	10			0.9994	0.9961	0.8725	0.5881
	11			0.9999	0.9991	0.9435	0.7483
	12			1.0000	0.9998	0.9790	0.8684
	13				1.0000	0.9935	0.9423
	14					0.9984	0.9793
	15					0.9997	0.9941
	16					1.0000	0.9987
	17						0.9998
	18						1.0000

Table A.2. Distribution Function of the Poisson Distribution

x	λ							
	0.5	1	1.5	2	5	10	15	20
0	0.6065	0.3679	0.2231	0.1353	0.0067	0.0000		
1	0.9098	0.7358	0.5578	0.4060	0.0404	0.0005		
2	0.9856	0.9197	0.8088	0.6767	0.1247	0.0028	0.0000	
3	0.9982	0.9810	0.9344	0.8571	0.2650	0.0103	0.0002	
4	0.9998	0.9963	0.9814	0.9473	0.4405	0.0293	0.0009	0.0000
5	1.0000	0.9994	0.9955	0.9834	0.6160	0.0671	0.0028	0.0001
6		0.9999	0.9991	0.9955	0.7622	0.1301	0.0076	0.0003
7		1.0000	0.9998	0.9989	0.8666	0.2202	0.0180	0.0008
8			1.0000	0.9998	0.9319	0.3328	0.0374	0.0021
9				1.0000	0.9682	0.4579	0.0699	0.0050
10					0.9863	0.5830	0.1185	0.0108
11					0.9945	0.6968	0.1848	0.0214
12					0.9980	0.7916	0.2676	0.0390
13					0.9993	0.8645	0.3632	0.0661
14					0.9998	0.9165	0.4657	0.1049
15					0.9999	0.9513	0.5681	0.1565
16					1.0000	0.9730	0.6641	0.2211
17						0.9857	0.7489	0.2970
18						0.9928	0.8195	0.3814
19						0.9965	0.8752	0.4703
20						0.9984	0.9170	0.5591
21						0.9993	0.9469	0.6437
22						0.9997	0.9673	0.7206
23						0.9999	0.9805	0.7875
24						1.0000	0.9888	0.8432
25							0.9938	0.8878
26							0.9967	0.9221
27							0.9983	0.9475
28							0.9991	0.9657
29							0.9996	0.9782
30							0.9998	0.9865
31							0.9999	0.9919
32							1.0000	0.9953

Table A.3. Distribution Function of the $N(0,1)$ Distribution

z	+0.00	+0.01	+0.02	+0.03	+0.04	+0.05	+0.06	+0.07	+0.08	+0.09
0.0	0.5000	0.5040	0.5080	0.5120	0.5160	0.5199	0.5239	0.5279	0.5319	0.5359
0.1	0.5398	0.5438	0.5478	0.5517	0.5557	0.5596	0.5636	0.5675	0.5714	0.5753
0.2	0.5793	0.5832	0.5871	0.5910	0.5948	0.5987	0.6026	0.6064	0.6103	0.6141
0.3	0.6179	0.6217	0.6255	0.6293	0.6331	0.6368	0.6406	0.6443	0.6480	0.6517
0.4	0.6554	0.6591	0.6628	0.6664	0.6700	0.6736	0.6772	0.6808	0.6844	0.6879
0.5	0.6915	0.6950	0.6985	0.7019	0.7054	0.7088	0.7123	0.7157	0.7190	0.7224
0.6	0.7257	0.7291	0.7324	0.7357	0.7389	0.7422	0.7454	0.7486	0.7517	0.7549
0.7	0.7580	0.7611	0.7642	0.7673	0.7704	0.7734	0.7764	0.7794	0.7823	0.7852
0.8	0.7881	0.7910	0.7939	0.7967	0.7995	0.8023	0.8051	0.8078	0.8106	0.8133
0.9	0.8159	0.8186	0.8212	0.8238	0.8264	0.8289	0.8315	0.8340	0.8365	0.8389
1.0	0.8413	0.8438	0.8461	0.8485	0.8508	0.8531	0.8554	0.8577	0.8599	0.8621
1.1	0.8643	0.8665	0.8686	0.8708	0.8729	0.8749	0.8770	0.8790	0.8810	0.8830
1.2	0.8849	0.8869	0.8888	0.8907	0.8925	0.8944	0.8962	0.8980	0.8997	0.9015
1.3	0.9032	0.9049	0.9066	0.9082	0.9099	0.9115	0.9131	0.9147	0.9162	0.9177
1.4	0.9192	0.9207	0.9222	0.9236	0.9251	0.9265	0.9279	0.9292	0.9306	0.9319
1.5	0.9332	0.9345	0.9357	0.9370	0.9382	0.9394	0.9406	0.9418	0.9429	0.9441
1.6	0.9452	0.9463	0.9474	0.9484	0.9495	0.9505	0.9515	0.9525	0.9535	0.9545
1.7	0.9554	0.9564	0.9573	0.9582	0.9591	0.9599	0.9608	0.9616	0.9625	0.9633
1.8	0.9641	0.9649	0.9656	0.9664	0.9671	0.9678	0.9686	0.9693	0.9699	0.9706
1.9	0.9713	0.9719	0.9726	0.9732	0.9738	0.9744	0.9750	0.9756	0.9761	0.9767
2.0	0.9772	0.9778	0.9783	0.9788	0.9793	0.9798	0.9803	0.9808	0.9812	0.9817
2.1	0.9821	0.9826	0.9830	0.9834	0.9838	0.9842	0.9846	0.9850	0.9854	0.9857
2.2	0.9861	0.9864	0.9868	0.9871	0.9875	0.9878	0.9881	0.9884	0.9887	0.9890
2.3	0.9893	0.9896	0.9898	0.9901	0.9904	0.9906	0.9909	0.9911	0.9913	0.9916
2.4	0.9918	0.9920	0.9922	0.9925	0.9927	0.9929	0.9931	0.9932	0.9934	0.9936
2.5	0.9938	0.9940	0.9941	0.9943	0.9945	0.9946	0.9948	0.9949	0.9951	0.9952
2.6	0.9953	0.9955	0.9956	0.9957	0.9959	0.9960	0.9961	0.9962	0.9963	0.9964
2.7	0.9965	0.9966	0.9967	0.9968	0.9969	0.9970	0.9971	0.9972	0.9973	0.9974
2.8	0.9974	0.9975	0.9976	0.9977	0.9977	0.9978	0.9979	0.9979	0.9980	0.9981
2.9	0.9981	0.9982	0.9982	0.9983	0.9984	0.9984	0.9985	0.9985	0.9986	0.9986

Table A.3. Continued

z	+0.00	+0.01	+0.02	+0.03	+0.04	+0.05	+0.06	+0.07	+0.08	+0.09
3.0	0.9987	0.9987	0.9987	0.9988	0.9988	0.9989	0.9989	0.9989	0.9990	0.9990
3.1	0.9990	0.9991	0.9991	0.9991	0.9992	0.9992	0.9992	0.9992	0.9993	0.9993
3.2	0.9993	0.9993	0.9994	0.9994	0.9994	0.9994	0.9994	0.9995	0.9995	0.9995
3.3	0.9995	0.9995	0.9995	0.9996	0.9996	0.9996	0.9996	0.9996	0.9996	0.9997
3.4	0.9997	0.9997	0.9997	0.9997	0.9997	0.9997	0.9997	0.9997	0.9997	0.9998
3.5	0.9998	0.9998	0.9998	0.9998	0.9998	0.9998	0.9998	0.9998	0.9998	0.9998
3.6	0.9998	0.9998	0.9999	0.9999	0.9999	0.9999	0.9999	0.9999	0.9999	0.9999
3.7	0.9999	0.9999	0.9999	0.9999	0.9999	0.9999	0.9999	0.9999	0.9999	0.9999
3.8	0.9999	0.9999	0.9999	0.9999	0.9999	0.9999	0.9999	0.9999	0.9999	0.9999
3.9	1.0000	1.0000	1.0000	1.0000	1.0000	1.0000	1.0000	1.0000	1.0000	1.0000

Appendix B: Answers to Even-Numbered Exercises

Chapter 1

2. (a) $\simeq 0.0062$; (b) $\simeq 0.80$.
4. $1 - \pi/8$ $(\simeq 0.6073)$.
6. $379/2187$ $(\simeq 0.1733)$.
8. $244/495$ $(\simeq 0.4929)$.
10. $1/3$.
12.

$$\frac{2}{c} - \frac{c}{e^c - (1+c)}$$

16. (a) $Y \sim U[0, 1]$; (b) $g(x) = 2F_X(x) + 1$.
18. (a) $2/3$; (b) $1/3$; (c) $8/3$.
20. (a) $\simeq 0.3591$; (b) $\simeq 0.3769$; (c) $\simeq 0.1750$.
22. (a) $\simeq 0.4354$; (b) $\simeq \$781.80$; (c) $\simeq 81.5\%$.
24. (b) (i) 0.1837; (ii) 2; (c) $\sqrt{\pi/2}$.
26. (a)

$$f_{Y_1, Y_2}(y_1, y_2) = \frac{1}{2\pi\sigma^2} \exp\left\{ -\frac{1}{2\sigma^2} \left(5y_1^2 + 2y_2^2 - 6y_1 y_2 - 2y_1 \mu + 2\mu^2 \right) \right\}$$

for all $(y_1, y_2) \in \mathbb{R}^2$; (b) $3\sigma^2$.
32. (b) $V[X]/2$.
34. $f_{Z|X}(z \mid x) = f_Y(z - x)$.
36. (a)

$$f_Z(z) = \frac{\left(\sum_{k=1}^n a_k \right)/\pi}{z^2 + \left(\sum_{k=1}^n a_k \right)^2} \quad \text{for } z \in \mathbb{R}$$

(b) $f_Z(z)$ does *not* tend to a Gaussian density. The central limit theorem does not apply, because $V[X_k] = \infty \ \forall \ k$ (and, actually, $E[X_k]$ does not exist).
38. (a) $Y/2$; (b) $1/2$.
40. (a) $1/8$; (b) $7/144$; (c) $\simeq 0.9667$.

42. (a)

$$f_X(x) = \frac{1}{x^2} - e^{-x}\left(\frac{1}{x} + \frac{1}{x^2}\right) \quad \text{if } x > 0$$

(b) e^{-1}; (c) $1/Y^2$; (d) ∞.

44. (a)

$$F_{Y|X_1}(y \mid x_1) = \begin{cases} 0 \text{ if } y < x_1 \\ y \text{ if } x_1 \le y < 1 \\ 1 \text{ if } y \ge 1 \end{cases}$$

(b) $\frac{1}{2}(1 + x_1^2)$; (c) $1/45$; (d) $1/30$.

46. (a) 3/4; (b) 13/24; (c) 0; (d) $X_1^2 + \frac{3}{2}X_1 + \frac{1}{2}$.

48. $1 - e^{-1}$ (≈ 0.6321).

50.

$$f_{X_1 X_2 \cdots X_{30}}(x) \simeq \frac{1}{\sqrt{5\pi}\, x} \exp\left\{-\frac{(\ln x - 15)^2}{5}\right\} \quad \text{for } x > 0$$

54. (a) $X_1 + X_2 + \frac{1}{2}$; (b) $\frac{1}{2}(X_1 + X_2 + X_3)$; (c) $1/18$.

Chapter 2

2. (a) $p(1 - p)$; (b) 0 if $n \ne m$ and 1 if $n = m$.

4. $t/(2x^2)$.

6. The increments are not independent, but they are stationary.

8. (a)

$$f(x; t) = f_Y\left(-\frac{\ln x}{t}\right)\frac{1}{tx} \quad \text{for } x \in (0, 1]$$

(b) $E[X(t)] = 1/(1 + t)$ and $R_X(t_1, t_2) = E[X(t_1 + t_2)]$.

10. No, since $E[X(t)]$ ($= (1 - e^{-t})/t$) depends on t.

12. No, because $E[Y(t)]$ ($= t$) depends on t.

14. $E[X(t)X(t + s)] = E[Y^6 t(t + s)] = t(t + s)/7 \ne R_X(s)$. Consequently, the process is not WSS and therefore not SSS.

16. $S_Y(\omega) = 2(1 - \cos\omega)S_X(\omega)$.

18. $h(0)q(t)$.

20. $C_X(k_1, k_2) = C_X(k_2 - k_1) = 0$ if $k_2 - k_1 \ne 0$ and $C_X(k_2 - k_1) = p(1 - p)$ if $k_2 - k_1 = 0$. Since $m_X(k) \equiv p$, the process is WSS. We have that $C_X(0) < \infty$ and $\lim_{|k| \to \infty} C_X(k) = 0$. It follows that the process is mean ergodic.

22. Yes, because $C_X(0) = R_X(0) = 1 < \infty$ and $\lim_{|s| \to \infty} C_X(s) = 0$.

24. $c/2T$.

26. (a) $e^{-\omega^2}$; (b) $(Y, Z) \sim N(\mu_Y = 0, \mu_Z = 0; \sigma_Y^2 = 2, \sigma_Z^2 = 2; \rho = 0)$.

28. (a) and (b) No, because $E[Y(t)]$ ($= e^{-t}$) is not a constant.

30. $m(y_0; t_0) = y_0/2$ and $v(y_0; t_0) = y_0^2$.

32. (a) $N(0, 2)$; (b) $2e^{-8s}$; (c) $4\left(1 + e^{-8s}\right)$.

34. (a) $N(0, 1)$; (b) $U(1, 2)$.

Chapter 3

2. (a) $1/4$; (b) (i) 0; (ii) no, because $E[Y(t)]\ (= t/2)$ is not a constant; (iii) $1/2$.
4. $11/12$.
6. $2/3$.
8. 0.1465.
10.
$$\begin{bmatrix} 1/4 & 1/2 & 1/4 \\ 1/2 & 1/2 & 0 \\ 1/4 & 1/2 & 1/4 \end{bmatrix}$$

12. $2/3$.
14. (a) Aperiodic, because it is irreducible and $p_{0,0} > 0$; (b)

$$\pi_i = \left(\frac{p}{q}\right)^i \left(\frac{1-(p/q)}{1-(p/q)^5}\right) \quad \text{for } i = 0,1,2,3,4$$

22. (b) $q - p$ if $p < 1/2$; 0 if $p = 1/2$; $p - q$ if $p > 1/2$; (c) 0 if $p \leq 1/2$;
$(q/p)^{k-1}[1 - (q/p)]$, for $k = 1, 2, \ldots$, if $p > 1/2$.
24. (a) A single class, which is recurrent and aperiodic; (b) $\pi_i = 1/5$, for all i;
(c) no.
28. (b)

$$\frac{1}{(k+1)!} \left(\frac{1}{e-1}\right)$$

30. (a) $0 < \alpha \leq 1$; (b)

$$\pi_0 = \frac{4\alpha}{4\alpha+3}, \quad \pi_1 = \frac{2}{4\alpha+3}, \quad \pi_2 = \frac{1}{4\alpha+3}$$

32. (a) 1 (because the chain is irreducible and $p_{0,0} > 0$); (b) $\pi_0 = 3/7, \pi_1 = 2/7, \pi_2 = \pi_3 = 1/7$; (c) (i) $\simeq 0.1716$; (ii) 1.
34. (a) $\mathbf{P}^{(n)} = \mathbf{P}^{(2)}$, where

$$\mathbf{P}^{(2)} = \begin{bmatrix} 1-p & 0 & p \\ 0 & 1 & 0 \\ 1-p & 0 & p \end{bmatrix}$$

if n is even, and $\mathbf{P}^{(n)} = \mathbf{P}$ if n is odd; (b) $d = 2$; (c) (i) $\pi_0 = (1-p)/2$; (ii) no, because this limit does not exist (since $d = 2$).
36. (a) $\pi_0 = 2/3$ and $\pi_j = (1/4)^j$, for $j = 1, 2, \ldots$; (b) 1.
38. (a) $\pi_0 = \pi_3 = 1/6, \pi_1 = \pi_2 = 1/3$; (b) $d = 2$; (c) $2/3$.
40. (a) (i)

$$\begin{bmatrix} 1 & 0 & 0 & 0 & 0 \\ 1/4 & 1/2 & 1/4 & 0 & 0 \\ 0 & 1/4 & 1/2 & 1/4 & 0 \\ 0 & 0 & 1/4 & 1/2 & 1/4 \\ 0 & 0 & 0 & 0 & 1 \end{bmatrix}$$

(ii) yes; $\pi_0^* = \pi_4^* = 1/2$ and $\pi_j^* = 0$, for $j = 1, 2, 3$; (b) $1/4$.

42. (a) $0 \leq \alpha \leq 1$, $0 < \beta \leq 1$; (b) $\pi_0 = \pi_2 = 2/5$ and $\pi_1 = 1/5$; (c) (i) $2/3$; (ii) $1/4 + \alpha/2$.

44. (a) $\alpha \neq 1$, $\beta \neq 0$; (b) $\pi_0 = 1/9$, $\pi_1 = 5/9$, and $\pi_2 = 1/3$; (c) $\alpha = 1/2$ and $\beta = 1$; (d) 3.

46.

$$\lim_{n \to \infty} p_{1,j}^{(n)} = \begin{cases} 1 \text{ if } j = 1 \\ 0 \text{ if } j = 0 \text{ or } 2 \end{cases}$$

and

$$\lim_{n \to \infty} p_{0,j}^{(n)} = \lim_{n \to \infty} p_{2,j}^{(n)} = \begin{cases} 2/3 \text{ if } j = 0 \\ 0 \text{ if } j = 1 \\ 1/3 \text{ if } j = 2 \end{cases}$$

48. (a) $1/2$; (b) we have

$$2(\pi_0^*)^4 + (\pi_0^*)^2 - 4\pi_0^* + 1 = 0$$

We can check that the value $(1/2)^2 = 1/4$ is not a solution of the equation, so that $\pi_0^* \neq \pi_0^2$.

50.

$$\lim_{n \to \infty} p_{1,j}^{(n)} = \begin{cases} 1 \text{ if } j = 1 \\ 0 \text{ otherwise} \end{cases}$$

and

$$\lim_{n \to \infty} p_{0,j}^{(n)} = \lim_{n \to \infty} p_{2,j}^{(n)} = \lim_{n \to \infty} p_{3,j}^{(n)} = \begin{cases} 2/5 \text{ if } j = 0 \\ 0 \text{ if } j = 1 \\ 1/5 \text{ if } j = 2 \\ 2/5 \text{ if } j = 3 \end{cases}$$

52. (a) $-1 + \sqrt{2}$ ($\simeq 0.4142$); (b) for example, $p_0 = p_2 = p_3 = 1/3$ and $p_1 = 0$.

54. (a) The state space is $\{0, 1, 2\}$, and the matrix $\mathbf{P}$ is given by

$$\begin{bmatrix} (1 - p_1)(1 - p_2) & p_1 + p_2 - 2p_1 p_2 & p_1 p_2 \\ 1 - p & p & 0 \\ (1 - p_1)(1 - p_2) & p_1 + p_2 - 2p_1 p_2 & p_1 p_2 \end{bmatrix}$$

where $p_i := e^{-\lambda_i}$, for $i = 1, 2$, and

$$p := p_1 \frac{\lambda_2}{\lambda_1 + \lambda_2} + p_2 \frac{\lambda_1}{\lambda_1 + \lambda_2}$$

(b) Yes, by the memoryless property of the exponential distribution. We calculate

$$\mathbf{P} = \begin{bmatrix} (1 - p_1)(1 - p_2) & p_1(1 - p_2) & (1 - p_1)p_2 & p_1 p_2 \\ 1 - p_1 & p_1 & 0 & 0 \\ 1 - p_2 & 0 & p_2 & 0 \\ (1 - p_1)(1 - p_2) & p_1(1 - p_2) & (1 - p_1)p_2 & p_1 p_2 \end{bmatrix}$$

56. (a) There are $3^2 = 9$ possible states: $0 = (0,0)$, $1 = (0,1)$, $2 = (0,2)$, $3 = (1,0)$, $4 = (1,1)$, $5 = (1,2)$, $6 = (2,0)$, $7 = (2,1)$, $8 = (2,2)$. We find that

$$\mathbf{P} = \begin{bmatrix} 2p & p & 0 & p & 0 & 0 & 0 & 0 & 0 \\ p & p & p & 0 & p & 0 & 0 & 0 & 0 \\ 0 & p & 2p & 0 & 0 & p & 0 & 0 & 0 \\ p & 0 & 0 & p & p & 0 & p & 0 & 0 \\ 0 & p & 0 & p & 0 & p & 0 & p & 0 \\ 0 & 0 & p & 0 & p & p & 0 & 0 & p \\ 0 & 0 & 0 & p & 0 & 0 & 2p & p & 0 \\ 0 & 0 & 0 & 0 & p & 0 & p & p & p \\ 0 & 0 & 0 & 0 & 0 & p & 0 & p & 2p \end{bmatrix}$$

where $p = 1/4$; (b) $\pi_i = 1/9$, for all i.

58. n.

60. (a) $P[N = 1] = (1/2)^{n-1}$ and $P[N = i] = (1/2)^{n-i+1}$, for $i = 2, 3, \ldots, n$;
(b) $(n-1)/(2\mu)$; (c) $G(\alpha = n - 1, \lambda = 2\mu)$.

62. (a)

$$\mathbf{P} = \begin{bmatrix} 1/N & 1/N & 1/N & \ldots & 1/N & 1/N & 1/N \\ 0 & 2/N & 1/N & \ldots & 1/N & 1/N & 1/N \\ \cdots & \cdots & \cdots & \cdots & \cdots & \cdots & \cdots \\ 0 & 0 & 0 & \ldots & 0 & (N-1)/N & 1/N \\ 0 & 0 & 0 & \ldots & 0 & 0 & 1 \end{bmatrix}$$

(b) (i) We have

$$m_i = 1 + \frac{i}{N} m_i + \frac{1}{N}(m_{i+1} + \ldots + m_N) \quad \text{for } i = 1, \ldots, N-1$$

We find, with $m_N = 0$, that $m_i \equiv N$; (ii) $m_i = E[\text{Geom}(p := 1/N)] = N$, for $i = 1, \ldots, N-1$.

64. (a) $\simeq 0.00002656$; (b) $\simeq 0.8794$; (c) $\simeq -\$20$.

66. (a)

$$\mathbf{P} = \begin{array}{c} \\ -4 \\ -2 \\ -1 \\ 0 \\ 1 \\ \\ \end{array} \begin{bmatrix} \ddots & \ddots & & & & \vdots \\ & 1/2 & 0 & 0 & 0 & 1/2 \\ & & 1/2 & 0 & 0 & 1/2 \\ & & & 1/2 & 0 & 1/2 \\ & & & & 1/2 & 0 & 1/2 \\ & & & & & 1/2 & 0 & 1/2 \\ \vdots & & & & & \ddots & \ddots & \ddots \end{bmatrix}$$

(b) $\pi_{-4} = \pi_{-2} = 1/16$, $\pi_{-1} = 1/8$, and $\pi_0 = \pi_1 = \pi_2 = 1/4$; (c) $3/8$.

68. $\pi_j = p_j \; \forall \, j$.

70. (a) $9/4$.

72. (c) $n/2^{n-1}$.

76. (a) $30/37$; (b) $16/37$; (c) $4/19$.

78. (a) $\simeq 0.1170$; (b) $\simeq 0.0756$.

82. (a) $1 + \lambda t$; (b) $\pi_0 = 1$ and $\pi_k = 0$, for $k = 1, 2, \ldots$; (c) $(\lambda t)^2/(1 + \lambda t)^2$.

84. (b) $e^{-4\lambda}(e^{2\lambda} - 1)$; (c) the π_j's do not exist, because $\{X(u), u \geq 0\}$ is a pure birth process.

86. (a) $\alpha = 0$; (b) $\pi_0 = 10/27$, $\pi_1 = 8/27$, and $\pi_2 = 9/27$.

88. (a) No, because the time τ that the process spends in either state does not have an exponential distribution, since $P[\tau > t] = \frac{1}{3}e^{-t} + \frac{2}{3}e^{-2t}$; (b) $\simeq 0.2677$.

90. $12(1 - e^{-2t})/(3 - 2e^{-2})$.

92. (a) $\lambda_0 = 2\lambda$ and $\lambda_n = \mu_n = \lambda$, for $n = 1, 2, \ldots$; (b) the π_j's do not exist.

94. (a) $\lambda_0 = 2\lambda$, $\lambda_1 = \lambda/2$, $\mu_1 = 3\lambda/2$, and $\mu_2 = 2\lambda$; (b) $1/2$; (c) $\pi_0 = 3/8$, $\pi_1 = 1/2$, and $\pi_2 = 1/8$.

96.

$$\frac{\mu^{c-k}}{(\lambda + \mu)^c} \left(\lambda e^{-(\mu+\lambda)t} + \mu\right)^k \left(1 - e^{-(\mu+\lambda)t}\right)^{c-k}$$

98. (a) We define the states

> 0: the machine is functioning
> 1: failure of type 1
> 2: failure of type 2; first repairing operation
> 3: failure of type 2; second repairing operation

(b) $\mu/[\mu + (2 - \alpha)\lambda]$.

Chapter 4

2. (a) 0; (b) $t + \min\{t^2, t + \tau\} + \min\{t, (t+\tau)^2\} + t^2$; (c) (i) yes; (ii) no, because $\text{Cov}[X(t), X(t + \tau)]$ depends on t; (iii) no, since $\text{Cov}[X(t), X(t + \tau)] \neq \sigma^2 t$; (d)

$$\rho_{B(t),B(t^2)} = \begin{cases} t^{1/2} & \text{if } 0 < t \leq 1 \\ t^{-1/2} & \text{if } t > 1 \end{cases}$$

4. (a) No, because $X(t) \geq 0$; (b) no, since $E[X^2(t)] (= t)$ is not a constant.

6. 4.

8. Brownian motion with $\sigma^2 = 2$.

10. (a) No, because $Y(t) \geq 0$; (b) 1; (c) $2s/t$; (d) no, because $\text{Cov}[Y(s), Y(t)]$ is not a function of $|t - s|$.

12. (a) $\Phi(j + 1 - i) - \Phi(j - i)$, where Φ is the distribution function of the $N(0, 1)$ distribution; (b) $\simeq 0.4659$.

14. (a)

$$\frac{1}{\epsilon^2}(t + \epsilon - \min\{t + \epsilon, t + s\})$$

(b) (i) yes; (ii) yes; (iii) no, because $C_X(t, t + s) \neq \sigma^2 t$; (iv) yes.

16. (a) $N(0, 2(1 + e^{-1}))$; (b) (i) $E[U(t)] \equiv 0$ and $V[U(t)] = 2(t + e^{-t} - 1)$; (ii) yes.

18. (b) No, because we have that $E[Y(t)] = \exp\{(\mu + \frac{1}{2}\sigma^2)t\}$, which depends on t (if $\mu \neq -\frac{1}{2}\sigma^2$); if $\mu = -\frac{1}{2}\sigma^2$, then $E[Y(t)] \equiv 1$, but $E[Y(t+s)Y(t)]$ $(= e^{\sigma^2 t})$ is not a function of s.

20. (a) 0; (b) yes, because X and Y are Gaussian random variables and their covariance is equal to zero.

22. $E[Y(t)] \equiv 0$ and

$$C_Y(s,t) = \frac{\sigma^2}{2c}\left(e^{c(s+t)} - e^{c|t-s|}\right) \quad \text{for } s,t \geq 0$$

24. (a) $N(0, 11/3)$; (b)

$$f_{T_1}(t) = \frac{3\sqrt{3}}{\sqrt{2\pi}}t^{-5/2}\exp\left\{-\frac{3}{2t^3}\right\} \quad \text{for } t > 0$$

26. (a) $E[Y(t)] \equiv 0$ and

$$\text{Cov}[Y(t), Y(t+s)] = \frac{t^3}{3} + \frac{t^2}{2}\left(s - \frac{(t+s)^2}{2}\right)$$

(b) (i) Yes; (ii) no, because $\text{Cov}[Y(t), Y(t+s)]$ depends on t; (iii) no, since $\text{Cov}[Y(t), Y(t+s)] \neq t[1 - (t+s)]$; (c) $\simeq 2P[N(0,1) > \sqrt{12}d]$.

28. We find that

$$f_{T_0(d)}(t) = -\frac{2d}{\sqrt{2\pi}}\exp\left\{-\frac{d^2 e^{-t}}{2(1 - e^{-t})}\right\}\frac{d}{dt}\left(\frac{e^{-t/2}}{\sqrt{1 - e^{-t}}}\right) \quad \text{for } t > 0$$

30. (a) 0; (b) $1 + 2t$; (c) no, because $\text{Cov}[Z(t), Z(t+s)]$ depends on t; (d)

$$f_{T_d(z)}(t) = \frac{1}{\sqrt{2\pi}}\frac{2(d - z)}{(1 + 2t)^{3/2}}\exp\left\{-\frac{(d - z)^2}{2(1 + 2t)}\right\} \quad \text{for } t > 0$$

Chapter 5

2. (a) $e^{-2.5}$ ($\simeq 0.0821$); (b) $\simeq 0.0028$; (c) $1/3$; (d)

$$C_X(t_1, t_2) = \frac{10}{t_1 t_2}\min\{t_1, t_2\}$$

4. (a) (i) e^{-2} ($\simeq 0.1353$); (ii) $\simeq 0.1839$; (iii) $\simeq 0.3233$; (b) no, because $E[Z(t)] = \exp\{t(e^{-1} - 1)\}$, which depends on t.

6. e^{-5}.

8. λs.

10. (a) $(1 - e^{-1})^2$; (b) $\simeq 1.64$.

12. (a) $\simeq 0.3535$; (b) $\simeq 0.0119$.

14. (a) $3\lambda/64$; (b) t^2.

16. (a) (i) Mean $= 15$, variance $= 240$; (ii) mean $= 10$, variance $= 520/12$; (b) $\simeq 14.785$.

18. $1/(2\sqrt{\lambda})$.

20. $P[N^*(\delta) = 1] = o(\delta)$.

22. (a) Let $\gamma := \lambda(e^s - 1)$. We find that

$$M_{N(t+S)}(s) = \begin{cases} e^{\gamma t}\dfrac{(e^{2\gamma} - 1)}{2\gamma} & \text{if } s \neq 0 \\ 1 & \text{if } s = 0 \end{cases}$$

(b) $E[N(t + S)] = \lambda(t + 1)$ and $V[N(t + S)] = \lambda(t + 1) + \frac{1}{3}\lambda^2$.

24. (a) $e^{-1}(b - a)/15$; (b) $1/16$; (c) $1/16$.

26. (a) $\simeq 0.6723$; (b) $1/2$; (c) $1/3$.

28. (a) $U(0, s)$; (b) $5e^{-4}$ ($\simeq 0.0916$); (c) $10/3$.

30. (a) (i) $1/(1 - t)$, for $0 < t \leq 1$; (ii) $1/2$; (b) (ii) $P[N(0, 1) \leq 2]$ ($\simeq 0.9772$).

32. (a) $\simeq 0.5590$; (b) 4.75; (c) $\simeq 0.0409$; (d) $\simeq 0.0328$.

34. $\simeq 0.9817$.

36. $1 - e^{-2\lambda}\left(1 + 2\lambda + \frac{3}{2}\lambda^2\right)$.

38. $\simeq 0.217$.

40. (a) $n!/n^n$; (b) $2(1 - t)$ if $0 < t \leq \frac{1}{2}$, and $2\left(t - \frac{1}{2}\right)$ if $\frac{1}{2} < t < 1$.

42. $\{X(t), t \geq 0\}$ is a Poisson process with rate λ.

44. Yes; we have that $p_{i,j} = 0$ if $j < i$ and

$$p_{i,j} = e^{-\lambda}\frac{\lambda^{\sqrt{j} - \sqrt{i}}}{(\sqrt{j} - \sqrt{i})!}$$

if $j \geq i$, where $i, j \in \{0, 1, 2, 4, 9, \ldots\}$.

46. (a) $1/3$; (b) (i) $2x\,e^{-x^2}$, for $x \geq 0$; (ii) $\simeq 0.45$; (c) $9/5$.

48. (a) $\simeq 0.2565$; (b) $1/5$.

50. (a) $\simeq 0.56$; (b) $3/16$.

52. (a) $1 - 2e^{-1}$ ($\simeq 0.2642$); (b) $0.75\,t + 0.5\,t^2$.

54. (a) $1/3$; (b) 588; (c) 232.

56. (a) $\simeq 0.4562$; (b) 2; (c) 2; (d) $1/2$, if $s \in (0, 1]$ or $(2, 3]$.

58. (a) $\ln(s^2 + 1)/\ln 2$; (b) $\simeq 0.6068$.

60. (a) $12t/(1 - e^{-3t})$; (b) $P[N(0, 1) \leq 0.69]$ ($\simeq 0.7549$).

62. (a) Mean $= 8$ and variance $= 40$; (b) 401; (c)

$$P[S \leq s \mid N(8) - N(6) = 1] = \begin{cases} 0 & \text{if } s < 6 \\ \dfrac{s - 6}{5} & \text{if } 6 \leq s \leq 7 \\ \dfrac{4s}{5} - \dfrac{27}{5} & \text{if } 7 < s \leq 8 \\ 1 & \text{if } s > 8 \end{cases}$$

64. (a) (i) Mean $=$ variance $= 21$; (ii) mean $=$ variance $= 24$; (b) $\simeq 0.1562$.

66. (a) $\lambda_1/(2\lambda_1 + \lambda_2)$; (b) $\lambda_1 t$; (c) $1 - \frac{1}{2}e^{-2}$ ($\simeq 0.9323$).

68. (a) $G(\alpha = 2, \lambda)$; that is,

$$f_{\tau_i}(\tau) = \lambda^2 \tau e^{-\lambda\tau} \quad \text{for } \tau \geq 0 \text{ and } i = 1, 2, \ldots$$

(b) we find that

$$P[N(t) = n] = e^{-\lambda t}\left[\frac{(\lambda t)^{2n}}{(2n)!} + \frac{(\lambda t)^{2n+1}}{(2n+1)!}\right] \quad \text{for } n = 0, 1, \ldots$$

(c) $3/(2\lambda)$.

70. (a) $\mu/[\mu + (2 - \alpha)\lambda]$; (b) $(3/\mu)(2 - \alpha)/(3 - \alpha)$.

72. (a) $G(\alpha = 2, \lambda)$; (b) $\simeq 0.81$; (c) $\simeq 1.17$.

74. (b) $\simeq 1/2$; (c) $1/2$.

76. (a) We find that

$$m_N(t) = \exp\left\{\frac{2t - 1}{4}\right\}\left(t + e^{1/4} - \frac{1}{2}\right) - 1 \quad \text{for } 1/2 \leq t < 1$$

(b) $P[N(0, 1) \leq 0.48]$ ($\simeq 0.68$); (c) $5/8$.

78. (a) We calculate

$$m_N(t) = \frac{1}{e}\left(\exp\left\{\frac{e\,t}{e-1}\right\} - 1\right) \quad \text{for } 0 < t < 1$$

(b) $e - e^{1/2}$ ($\simeq 1.07$).

80. No; since the number of events per time unit increases linearly, the time between two consecutive events is (on the average) shorter and shorter, so that the random variables τ_k are not identically distributed (they are not independent either).

82. $\mu_1 > 0$.

84. (a) $1/(2a^2 + 1)$; (b) 0.

Chapter 6

2. (a) $\bar{N} = 11/15$ and $\bar{N}_Q = 4/15$; $\bar{N} \neq \bar{N}_Q + 1$ because $\pi_0 \neq 0$; (b) $1/15$.

4. (a) $\bar{N} = \lambda/\mu$ and $\bar{N}_Q = e^{-\lambda/\mu} + \frac{\lambda}{\mu} - 1$; (b) $1 + \frac{\mu}{\lambda}(e^{-\lambda/\mu} - 1)$.

6. (a) We have

state j	departure rate from j = arrival rate to j
0	$2\lambda\,\pi_0 = \mu\,\pi_1$
1	$(2\lambda + \mu)\,\pi_1 = \mu\,\pi_2 + 2\lambda\,\pi_0$
2	$(\lambda + \mu)\,\pi_2 = \mu\,\pi_3 + 2\lambda\,\pi_1$
$k\ (\geq 3)$	$(\lambda + \mu)\,\pi_k = \mu\,\pi_{k+1} + \lambda\,\pi_{k-1}$

(b) $1/2$; (c) $3/8$.

8. (a) We have

state j departure rate from j = arrival rate to j

$0'$	$10\,\pi_0' = 5\,\pi_0$
0	$(10+5)\,\pi_0 = 10\,\pi_0' + 5\,(\pi_1 + \pi_2 + \pi_3)$
1	$(10+5)\,\pi_1 = 10\,\pi_0$
2	$(10+5)\,\pi_2 = 10\,\pi_1$
3	$5\,\pi_3 = 10\,\pi_2$

(b) $\pi_{0'} = 1/7$, $\pi_0 = 2/7$, $\pi_1 = 4/21$, $\pi_2 = 8/63$, and $\pi_3 = 16/63$; (c) $5/21$.

10. (a) $\pi_n = (n+1)(1/2)^{n+2}$, for $n = 0, 1, \ldots$; (b) $10/11$; (c) $\simeq 0.4180$.

12. (a) $\pi_n = \left(\frac{\mu}{\lambda + \mu}\right)\left(\frac{\lambda}{\lambda + \mu}\right)^n$, for $n = 0, 1, \ldots$; (b) $\frac{\lambda}{\mu^2}(\lambda + \mu)$.

14. $4/(\mu + 4)$.

16. (a) We have

state j departure rate from j = arrival rate to j

0	$\lambda\,\pi_0 = \mu_1\,\pi_{1_1} + \mu_2\,\pi_{1_2}$
1_1	$(\lambda + \mu_1)\,\pi_{1_1} = p_1 \lambda\,\pi_0 + \mu_2\,\pi_2$
1_2	$(\lambda + \mu_2)\,\pi_{1_2} = (1 - p_1)\lambda\,\pi_0 + \mu_1\,\pi_2$
2	$(\lambda + \mu_1 + \mu_2)\,\pi_2 = \lambda\,(\pi_{1_1} + \pi_{1_2}) + (\mu_1 + \mu_2)\,\pi_3$
$3 \leq n \leq 12$	$(\lambda + \mu_1 + \mu_2)\,\pi_n = \lambda\,\pi_{n-1} + (\mu_1 + \mu_2)\,\pi_{n+1}$
13	$(\mu_1 + \mu_2)\,\pi_{13} = \lambda\,\pi_{12}$

(b) the probability requested is given by

$$\frac{1}{(1 - \pi_{13})}\left[p_1\,\pi_0 + \pi_{1_2} + \left(\frac{\mu_1}{\mu_1 + \mu_2}\right)\sum_{n=2}^{12}\pi_n\right]$$

18. (a) We find that

$$\pi_0 = \frac{2\mu - \lambda}{2\mu + \lambda} \quad \text{and} \quad \pi_n = \left(\frac{\lambda}{\mu}\right)^n \frac{1}{2^{n-1}}\pi_0 \quad \text{for } n = 1, 2, \ldots$$

(b) $(\lambda + \mu)\,\pi_{1_2} = \mu\,\pi_2$; (c) the fraction of time that the second counter is open is given by

$$\frac{\lambda^2(2\mu - \lambda)}{2\mu(\lambda + \mu)(2\mu + \lambda)} + \frac{\lambda^2}{\mu(2\mu + \lambda)}$$

20. (a) We have

state j departure rate from j = arrival rate to j

0	$\lambda\,\pi_0 = \mu\,\pi_1$
1	$(\lambda + \mu)\,\pi_1 = \lambda\,\pi_0 + 2\mu\,\pi_2$
2	$(\lambda + 2\mu)\,\pi_2 = \lambda\,\pi_1 + 2\mu\,\pi_3$
3	$2\mu\,\pi_3 = \lambda\,\pi_2$

We find that if $\lambda = 2\mu$, then $\pi_0 = 1/7$ and $\pi_1 = \pi_2 = \pi_3 = 2/7$; (b) 1.2; (c) 2.

22. (a) We find that

$$\pi_{(n,m)} = \left(\frac{25}{36}\right)^n \left(\frac{11}{36}\right) \left(\frac{35}{48}\right)^m \left(\frac{13}{48}\right) \quad \text{for all } n, m \geq 0$$

(b) $\simeq$5.41; (c) 1/5.

24. (a) We define the states

0 : the system is empty

n : there are $n \geq 1$ ordinary customers in the system

n_s: there are $n \geq 1$ customers in the system and the special customer is being served

The balance equations of the system are the following:

state j departure rate from j = arrival rate to j

0	$(\lambda + \lambda_s) \pi_0 = \mu \pi_1 + \mu_s \pi_{1_s}$
1	$(\lambda + \lambda_s + \mu) \pi_1 = \lambda \pi_0 + 2\mu \pi_2 + \mu_s \pi_{2_s}$
1_s	$(\lambda + \mu_s) \pi_{1_s} = \lambda_s \pi_0 + \mu \pi_{2_s}$
$n \geq 2$	$(\lambda + \lambda_s + 2\mu) \pi_n = \lambda \pi_{n-1} + 2\mu \pi_{n+1} + \mu_s \pi_{(n+1)_s}$
n_s $(n \geq 2)$	$(\lambda + \mu_s + \mu) \pi_{n_s} = \lambda_s \pi_{n-1} + \lambda \pi_{(n-1)_s} + \mu \pi_{(n+1)_s}$

(b) we define the states (m, n) and (m_s, n), where $m \in \{0, 1, 2\}$ (respectively, $n \in \{0, 1\}$) is the number of customers in front of server no. 1 (resp., no. 2) and s designates the special customer; there are therefore eight possible states: $(0, 0)$, $(1, 0)$, $(0, 1)$, $(1_s, 0)$, $(1, 1)$, $(1_s, 1)$, $(2_s, 0)$, and $(2_s, 1)$; (c) we can use the state space of part (b); we find that the probability requested is given by

$$\left(\frac{\lambda_s}{\mu + \lambda_s}\right) \frac{\pi_{(1,0)}}{\pi_{(1,0)} + \pi_{(1,1)}} + \left(\frac{\lambda_s}{\mu + \lambda_s}\right) \left(\frac{\mu}{\mu + \lambda_s}\right) \frac{\pi_{(1,1)}}{\pi_{(1,0)} + \pi_{(1,1)}}$$

26. (a) $\pi_0 = 1 - \frac{\lambda}{2\mu}$ and $\pi_n = \left(\frac{\lambda}{2\mu}\right)^n \pi_0$, for $n = 1, 2, \ldots$ (this model is equivalent to the $M/M/1$ model with service rate equal to 2μ rather than μ); (b) 1/3.

28. $\pi_0 = 1/4$ and $\pi_1 = \pi_2 = 3/8$.

30. (a) We define the states (i, j), for $i, j = 0, 1$, where i (respectively, j) is equal to 1 if a customer is being served by the hairdresser (resp., the assistant), and to 0 otherwise; (b) the balance equations of the system are the following:

state j departure rate from j = arrival rate to j

$(0,0)$	$(\lambda_1 + \lambda_2) \pi_{(0,0)} = \mu_1 \pi_{(1,0)} + \mu_2 \pi_{(0,1)}$
$(1,0)$	$(\lambda_1 + \lambda_2 + \mu_1) \pi_{(1,0)} = \lambda_1 \pi_{(0,0)} + \mu_2 \pi_{(1,1)}$
$(0,1)$	$(\lambda_1 + \mu_2) \pi_{(0,1)} = \lambda_2 \pi_{(0,0)} + \mu_1 \pi_{(1,1)}$
$(1,1)$	$(\mu_1 + \mu_2) \pi_{(1,1)} = \lambda_1 \pi_{(0,1)} + (\lambda_1 + \lambda_2) \pi_{(1,0)}$

(c) (i) $\bar{N} = \pi_{(0,1)} + \pi_{(1,0)} + 2\,\pi_{(1,1)}$; (ii) $\bar{T} = \bar{N}/\lambda_e$, where

$$\lambda_e = \lambda_1\,(1 - \pi_{(1,1)}) + \lambda_2\,(\pi_{(0,0)} + \pi_{(1,0)})$$

32. (a) We define the states

   0: the system is empty
   1_1: there is one customer in the system, being served by server no. 1
   1_2: there is one customer in the system, being served by server no. 2
   2: there are two customers in the system, one per server
   2_2: there are two customers in the system, being served by server no. 2
   3: there are three customers in the system, two of them being served
        by server no. 2

(b) the balance equations of the system are given by

state j	departure rate from j = arrival rate to j
0	$\lambda\,\pi_0 = \mu_1\,\pi_{1_1} + \mu_2\,(\pi_{1_2} + \pi_{2_2})$
1_1	$(\lambda + \mu_1)\,\pi_{1_1} = \lambda\,\pi_0 + \mu_2\,(\pi_2 + \pi_3)$
1_2	$(\lambda + \mu_2)\,\pi_{1_2} = \mu_1\,\pi_2$
2	$(\lambda + \mu_1 + \mu_2)\,\pi_2 = \lambda\,(\pi_{1_1} + \pi_{1_2})$
2_2	$(\lambda + \mu_2)\,\pi_{2_2} = \mu_1\,\pi_3$
3	$(\mu_1 + \mu_2)\,\pi_3 = \lambda\,(\pi_2 + \pi_{2_2})$

(c) the fraction of time requested is $(\pi_{2_2} + \pi_3)/(1 - \pi_0 - \pi_{1_1})$.

References

1. Abramowitz, Milton and Stegun, Irene A., *Handbook of Mathematical Functions with Formulas, Graphs, and Mathematical Tables*, Dover, New York, 1965.
2. Breiman, Leo, *Probability and Stochastic Processes: With a View Toward Applications*, Houghton Mifflin, Boston, 1969.
3. Chung, Kai Lai, *Elementary Probability Theory with Stochastic Processes*, Springer-Verlag, New York, 1975.
4. Cox, John C., Ingersoll, Jonathan E., and Ross, Stephen A., An intertemporal general equilibrium model of asset prices, *Econometrica*, vol. 53, pp. 363–384, 1985.
5. Downton, F., Bivariate exponential distributions in reliability theory, *Journal of the Royal Statistical Society*, Series B, vol. 32, pp. 408–417, 1970.
6. Dubins, Lester E. and Savage, Leonard J., *Inequalities for Stochastic Processes: How to Gamble if You Must*, 2nd edition, Dover, New York, 1976.
7. Feller, William, *An Introduction to Probability Theory and Its Applications*, Volume I, 3rd edition, Wiley, New York, 1968.
8. Feller, William, *An Introduction to Probability Theory and Its Applications*, Volume II, 2nd edition, Wiley, New York, 1971.
9. Grimmett, Geoffrey R. and Stirzaker, David R., *Probability and Random Processes*, 2nd edition, Oxford University Press, Oxford, 1992.
10. Hastings, Kevin J., *Probability and Statistics*, Addison-Wesley, Reading, MA, 1997.
11. Helstrom, Carl W., *Probability and Stochastic Processes for Engineers*, 2nd edition, Macmillan, New York, 1991.
12. Hillier, Frederick S. and Lieberman, Gerald J., *Introduction to Stochastic Models in Operations Research*, McGraw-Hill, New York, 1990.
13. Hines, William W. and Montgomery, Douglas C., *Probability and Statistics in Engineering and Management Science*, 3rd edition, Wiley, New York, 1990.
14. Hoel, Paul G., Port, Sidney C., and Stone, Charles J., *Introduction to Stochastic Processes*, Houghton Mifflin, New York, 1972.
15. Hogg, Robert V. and Craig, Allen T., *Introduction to Mathematical Statistics*, 3rd edition, Macmillan, New York, 1970.
16. Karlin, Samuel and Taylor, Howard M., *A Second Course in Stochastic Processes*, Academic Press, New York, 1981.

17. Kaufmann, A. and Cruon, R., *Les Phénomènes d'Attente. Théorie et Applications*, Dunod, Paris, 1961.

18. Lefebvre, M., *Applied Probability and Statistics*, Springer, New York, 2006.

19. Leon-Garcia, Alberto, *Probability and Random Processes for Electrical Engineering*, 2nd edition, Addison-Wesley, Reading, MA, 1994.

20. Lindgren, Bernard W., *Statistical Theory*, 3rd edition, Macmillan, New York, 1976.

21. Medhi, J. Jyotiprasad, *Stochastic Processes*, Wiley, New York, 1982.

22. Papoulis, Athanasios, *Probability, Random Variables, and Stochastic Processes*, 3rd edition, McGraw-Hill, New York, 1991.

23. Parzen, Emmanuel, *Stochastic Processes*, Holden-Day, San Francisco, 1962.

24. Peebles, Peyton Z., *Probability, Random Variables and Random Signal Principles*, 3rd edition, McGraw-Hill, New York, 1993.

25. Roberts, Richard A., *An Introduction to Applied Probability*, Addison-Wesley, Reading, MA, 1992.

26. Ross, Sheldon M., *Stochastic Processes*, 2nd edition, Wiley, New York, 1996.

27. Ross, Sheldon M., *Introduction to Probability Models*, 8th edition, Academic Press, San Diego, 2003.

28. Ruegg, Alan, *Processus Stochastiques. Avec Applications aux Phénomènes d'Attente et de Fiabilité*, Presses Polytechniques Romandes, Lausanne, 1989.

29. Stark, Henry and Woods, John W., *Probability, Random Processes, and Estimation Theory for Engineers*, 2nd edition, Prentice-Hall, Englewood Cliffs, NJ, 1994.

Index

Universitext

Das, A.: The Special Theory of Relativity: A Mathematical Exposition

Debarre, O.: Higher-Dimensional Algebraic Geometry

Deitmar, A.: A First Course in Harmonic Analysis

Demazure, M.: Bifurcations and Catastrophes

Devlin, K. J.: Fundamentals of Contemporary Set Theory

DiBenedetto, E.: Degenerate Parabolic Equations

Diener, F.; Diener, M. (Eds.): Nonstandard Analysis in Practice

Dimca, A.: Sheaves in Topology

Dimca, A.: Singularities and Topology of Hypersurfaces

DoCarmo, M. P.: Differential Forms and Applications

Duistermaat, J. J.; Kolk, J. A. C.: Lie Groups

Dumortier, F.; Llibre, J.; Artes, J. C.: Qualitative Theory of Planar Differential Systems

Edwards, R. E.: A Formal Background to Higher Mathematics Ia, and Ib

Edwards, R. E.: A Formal Background to Higher Mathematics IIa, and IIb

Emery, M.: Stochastic Calculus in Manifolds

Emmanouil, I.: Idempotent Matrices over Complex Group Algebras

Endler, O.: Valuation Theory

Engel, K.; Nagel, R.: A Short Course on Operator Semigroups

Erez, B.: Galois Modules in Arithmetic

Everest, G.; Ward, T.: Heights of Polynomials and Entropy in Algebraic Dynamics

Farenick, D. R.: Algebras of Linear Transformations

Foulds, L. R.: Graph Theory Applications

Franke, J.; Härdle, W.; Hafner, C. M.: Statistics of Financial Markets: An Introduction

Frauenthal, J. C.: Mathematical Modeling in Epidemiology

Freitag, E.; Busam, R.: Complex Analysis

Friedman, R.: Algebraic Surfaces and Holomorphic Vector Bundles

Fuks, D. B.; Rokhlin, V. A.: Beginner's Course in Topology

Fuhrmann, P. A.: A Polynomial Approach to Linear Algebra

Gallot, S.; Hulin, D.; Lafontaine, J.: Riemannian Geometry

Gardiner, C. F.: A First Course in Group Theory

Gårding, L.; Tambour, T.: Algebra for Computer Science

Gärtner, B.; Matousek, J.: Understanding and Using Linear Programming

Godbillon, C.: Dynamical Systems on Surfaces

Godement, R.: Analysis I, and II

Goldblatt, R.: Orthogonality and Spacetime Geometry

Gouvêa, F. Q.: p-Adic Numbers

Gross, M. et al.: Calabi-Yau Manifolds and Related Geometries

Gustafson, K. E.; Rao, D. K. M.: Numerical Range. The Field of Values of Linear Operators and Matrices

Gustafson, S. J.; Sigal, I. M.: Mathematical Concepts of Quantum Mechanics

Hahn, A. J.: Quadratic Algebras, Clifford Algebras, and Arithmetic Witt Groups

Hájek, P.; Havránek, T.: Mechanizing Hypothesis Formation

Heinonen, J.: Lectures on Analysis on Metric Spaces

Hlawka, E.; Schoißengeier, J.; Taschner, R.: Geometric and Analytic Number Theory

Holmgren, R. A.: A First Course in Discrete Dynamical Systems

Howe, R., Tan, E. Ch.: Non-Abelian Harmonic Analysis

Howes, N. R.: Modern Analysis and Topology

Hsieh, P.-F.; Sibuya, Y. (Eds.): Basic Theory of Ordinary Differential Equations

Humi, M., Miller, W.: Second Course in Ordinary Differential Equations for Scientists and Engineers

Hurwitz, A.; Kritikos, N.: Lectures on Number Theory